Microsystems and Nanosystems

Series editors

Roger T. Howe, Stanford, CA, USA

Antonio J. Ricco, Moffett Field, CA, USA

More information about this series at http://www.springer.com/series/11483

John A. Rogers · Roozbeh Ghaffari
Dae-Hyeong Kim

Editors

Stretchable Bioelectronics for Medical Devices and Systems

 Springer

Editors
John A. Rogers
University of Illinois at Urbana–Champaign
Urbana, IL
USA

Roozbeh Ghaffari
MC10 Inc.
Cambridge, MA
USA

Dae-Hyeong Kim
School of Chemical and Biological
Engineering
Seoul National University
Seoul
Korea, Republic of (South Korea)

ISSN 2198-0063   ISSN 2198-0071   (electronic)
Microsystems and Nanosystems
ISBN 978-3-319-28692-1   ISBN 978-3-319-28694-5   (eBook)
DOI 10.1007/978-3-319-28694-5

Library of Congress Control Number: 2016932507

Printed on acid-free paper

This Springer imprint is published by Springer Nature
The registered company is Springer International Publishing AG Switzerland

Contents

Editors and Contributors

About the Editors

Prof. John A. Rogers obtained B.A. and B.S. degrees in Chemistry and in Physics from the University of Texas, Austin, in 1989. From MIT, he received S.M. degrees in Physics and in Chemistry in 1992 and the Ph.D. degree in Physical Chemistry in 1995. From 1995 to 1997, Rogers was a Junior Fellow in the Harvard University Society of Fellows. He joined Bell Laboratories as a Member of Technical Staff in the Condensed Matter Physics Research Department in 1997, and served as Director of this department from the end of 2000–2002. He is currently Swanlund Chair Professor at University of Illinois at Urbana/Champaign, with a primary appointment in the Department of Materials Science and Engineering, and joint appointments in several other departments, including Bioengineering. He is Director of the Seitz Materials Research Laboratory. Rogers' research includes fundamental and applied aspects of materials for unusual electronic and photonic devices, with an emphasis on bio-integrated devices and bio-inspired designs. He has published more than 550 papers and is inventor on over 80 patents, more than 50 of which are licensed or in active use to various startups and large companies. Rogers is a Fellow of the IEEE, APS, MRS, and the AAAS, and he is a Member of the National Academy of Sciences, the National Academy of Engineering, the National Academy of Inventors, and the American Academy of Arts and Sciences. His research has been recognized with many awards, including a MacArthur Fellowship in 2009, the Lemelson-MIT Prize in 2011, the MRS Mid-Career Researcher Award and the Robert Henry Thurston Award (American Society of Mechanical Engineers) in 2013, the 2013 Smithsonian Award for Ingenuity in the Physical Sciences and the 2014 Eringen Medal of the Society for Engineering Science. He received an Honoris Causa Doctorate from the École Polytechnique Fédérale de Lausanne (EPFL) in 2013.

Dr. Roozbeh Ghaffari is Co-Founder and Chief Technology Officer at MC10 Inc. In this role, Roozbeh has shaped the technology vision of the company, and is responsible for defining and developing emerging products from concept phase through clinical validation. These efforts have led to the development and the launch of the BioStamp® and "tattoo-like" electronics as the foundation of MC10's wearable computing platforms. Prior to MC10 Inc., Roozbeh helped the launch of Diagnostics For All, a non-profit organization developing low-cost health diagnostics based on a patented microfluidics technology. Roozbeh is a Founding Editor of the MIT Entrepreneurship Review and is a research staff member at the MIT Research Laboratory of Electronics. Roozbeh's contributions in soft bioelectronics, micro/nano-scale systems, and auditory neuroscience research have been recognized with the MIT Helen Carr Peake Ph.D. Research Prize, the MIT100K Grand Prize, the Harvard Business School Social Enterprise Grand Prize, and the MIT Technology Review Magazine's Top 35 Innovators Under 35. Roozbeh has published over 40 academic papers and is inventor on over 40 patent applications and awards. Roozbeh holds B.S. and M.Eng. degrees in Electrical Engineering from the Massachusetts Institute of Technology, and a Ph.D. in Biomedical Engineering from the Harvard-MIT Division of Health Sciences and Technology.

Prof. Dae-Hyeong Kim obtained B.S. and M.S. degrees in Chemical Engineering from Seoul National University, Korea, in 2000 and 2002, respectively. He received Ph.D. degree in Materials Science and Engineering from University of Illinois at Urbana Champaign, in 2009. From 2009 to 2011, he was a postdoctoral research associate at University of Illinois. He is currently Associate Professor in the School of Chemical and Biological Engineering of Seoul National University. His research aims to develop technologies for high-performance flexible and stretchable electronic devices using high-quality single crystal inorganic materials and novel biocompatible materials that enable a new generation of implantable biomedical systems with novel capabilities and increased performance. In the area of clinical device research, a close collaboration with practicing cardiologists and neurologists demonstrated superb device performances for cardiac electrophysiology and brain–computer interfacing in vivo. These advanced diagnostic and therapeutic tools improve current surgical capabilities for treating cardiac, neural, and other fatal diseases. These achievements were published in high impact journals and honored through several awards. He has published more than 70 papers (including Science, Nature Materials, Nature Neuroscience, Nature Nanotechnology, Nature Communications, and PNAS), 30 international and domestic patents, and four book chapters. He has been recognized with several awards, including George Smith Award (2009) from the IEEE Electron Device Society, MRS Graduate Student Award (Gold Medal, 2009), Green Photonics Award from SPIE (2011), TR 35 award from MIT's Technology Review magazine (2011), and Hong Jin-ki Creative Award (2015).

Contributors

A.J. Aranyosi MC10 Inc., Lexington, MA, USA

Anthony Banks Department of Materials Science and Engineering, Beckman Institute for Advanced Science and Technology, University of Illinois Urbana-Champaign, Urbana, IL, USA

Jennifer Case Purdue University, West Lafayette, IN, USA

R. Chad Webb 3M Company, Saint Paul, MN, USA

Paolo DePetrillo MC10 Inc., Lexington, MA, USA

Michael D. Dickey NC State University, Raleigh, USA

Ray E. Dorsey Department of Neurology, University of Rochester Medical Center, Rochester, NY, USA

Roozbeh Ghaffari MC10 Inc., Lexington, MA, USA

Yonggang Huang Department of Mechanical Engineering and Department of Civil and Environmental Engineering, Northwestern University, Evanston, IL, USA

Yei Hwan Jung Department of Electrical and Computer Engineering, University of Wisconsin-Madison, Madison, USA

Hongki Kang Department of Bio and Brain Engineering, Korea Advanced Institute of Science and Technology (KAIST), Daejeon, Republic of Korea

Dae-Hyeong Kim Center for Nanoparticle Research, Institute for Basic Science (IBS), Seoul, Republic of Korea; School of Chemical and Biological Engineering and Institute of Chemical Processes, Seoul National University, Seoul, Republic of Korea; Interdisciplinary Program for Bioengineering, Seoul National University, Seoul, Republic of Korea

H.S. Kim KU-KIST Graduate School of Converging Science and Technology, Korea University, Seoul, Republic of Korea

I.Y. Kim Department of Biomedical Engineering, Hanyang University, Seoul, Republic of Korea

J.H. Kim KU-KIST Graduate School of Converging Science and Technology, Korea University, Seoul, Republic of Korea

Tae-il Kim Department of Biomedical Engineering, Sungkyunkwan University (SKKU), Suwon, Korea; School of Chemical Engineering, Sungkyunkwan University (SKKU), Suwon, Korea; Center for Neuroscience Imaging Research (CNIR), Institute of Basic Science (IBS), Suwon, Korea

Lauren Klinker MC10 Inc., Lexington, MA, USA

Ja Hoon Koo Center for Nanoparticle Research, Institute for Basic Science (IBS), Seoul, Republic of Korea; Interdisciplinary Program for Bioengineering, Seoul National University, Seoul, Republic of Korea

Rebecca Kramer Purdue University, West Lafayette, IN, USA

Siddharth Krishnan Department of Materials Science and Engineering, University of Illinois at Urbana-Champaign, Urbana, IL, USA

Stéphanie P. Lacour School of Engineering, Centre for Neuroprosthetics, Bertarelli Chair in Neuroprosthetic Technology, Laboratory for Soft Bioelectronic Interfaces, Ecole Polytechnique Fédérale de Lausanne EPFL, Lausanne, Switzerland

Chi Hwan Lee Department of Materials Science and Engineering, Beckman Institute for Advanced Science and Technology, University of Illinois Urbana-Champaign, Urbana, IL, USA

J.H. Lee KU-KIST Graduate School of Converging Science and Technology, Korea University, Seoul, Republic of Korea

Jongsu Lee Center for Nanoparticle Research, Institute for Basic Science (IBS), Seoul, Republic of Korea; School of Chemical and Biological Engineering and Institute of Chemical Processes, Seoul National University, Seoul, Republic of Korea

S.-H. Lee KU-KIST Graduate School of Converging Science and Technology, Korea University, Seoul, Republic of Korea; Department of Biomedical Engineering, College of Health Science, Korea University, Seoul, Republic of Korea

Darren J. Lipomi Department of NanoEngineering, University of California, San Diego, CA, USA

Clifford Liu MC10 Inc., Lexington, MA, USA

Nanshu Lu Department of Aerospace Engineering and Engineering Mechanics, Center for Mechanics of Solids, Structures, and Materials, University of Texas at Austin, Austin, USA; Department of Biomedical Engineering, University of Texas at Austin, Austin, USA

Yinji Ma Department of Mechanical Engineering and Department of Civil and Environmental Engineering, Northwestern University, Evanston, IL, USA

Zhenqiang Ma Department of Electrical and Computer Engineering, University of Wisconsin-Madison, Madison, USA

Moussa Mansour Massachusetts General Hospital, Harvard Medical School, Boston, MA, USA

Ryan McGinnis MC10 Inc., Lexington, MA, USA

Bryan McGrane MC10 Inc., Lexington, MA, USA

Ivan R. Minev School of Engineering, Centre for Neuroprosthetics, Bertarelli Chair in Neuroprosthetic Technology, Laboratory for Soft Bioelectronic Interfaces, Ecole Polytechnique Fédérale de Lausanne EPFL, Lausanne, Switzerland

Jeffrey B. Model MC10 Inc., Lexington, MA, USA

Mohammed Mohammed Purdue University, West Lafayette, IN, USA

Briana Morey MC10 Inc., Lexington, MA, USA

Brian Murphy MC10 Inc., Lexington, MA, USA

Yoonkey Nam Department of Bio and Brain Engineering, Korea Advanced Institute of Science and Technology (KAIST), Daejeon, Republic of Korea

Manuel Ochoa School of Electrical and Computer Engineering, Purdue University, West Lafayette, IN, USA

Timothy F. O'Connor Department of NanoEngineering, University of California, San Diego, CA, USA

Shyamal Patel MC10 Inc., Lexington, MA, USA

Rahim Rahimi School of Electrical and Computer Engineering, Purdue University, West Lafayette, IN, USA

Milan Raj MC10 Inc., Lexington, MA, USA

John A. Rogers Department of Materials Science and Engineering, Beckman Institute for Advanced Science and Technology, University of Illinois Urbana-Champaign, Urbana, IL, USA

Suchol Savagatrup Department of NanoEngineering, University of California, San Diego, CA, USA

Tsuyoshi Sekitani The Institute of Scientific and Industrial Research, Osaka University, Osaka, Ibaraki, Japan

Ellora Sen-Gupta MC10 Inc., Lexington, MA, USA

Nirav Sheth MC10 Inc., Lexington, MA, USA

Marvin Slepian Sarver Heart Center, University of Arizona, Tucson, AZ, USA

Donghee Son Center for Nanoparticle Research, Institute for Basic Science (IBS), Seoul, Republic of Korea; School of Chemical and Biological Engineering and Institute of Chemical Processes, Seoul National University, Seoul, Republic of Korea

Sung Hyuk Sunwoo Department of Biomedical Engineering, Sungkyunkwan University (SKKU), Suwon, Korea; Center for Neuroscience Imaging Research (CNIR), Institute of Basic Science (IBS), Suwon, Korea

Liu Wang Department of Aerospace Engineering and Engineering Mechanics, Center for Mechanics of Solids, Structures, and Materials, University of Texas at Austin, Austin, USA

John A. Wright MC10 Inc., Lexington, MA, USA

Shixuan Yang Department of Aerospace Engineering and Engineering Mechanics, Center for Mechanics of Solids, Structures, and Materials, University of Texas at Austin, Austin, USA

Michelle Yuen Purdue University, West Lafayette, IN, USA

Huilong Zhang Department of Electrical and Computer Engineering, University of Wisconsin-Madison, Madison, USA

Yihui Zhang Department of Engineering Mechanics, Center for Mechanics and Materials, AML, Tsinghua University, Beijing, People's Republic of China

Babak Ziaie School of Electrical and Computer Engineering, Purdue University, West Lafayette, IN, USA

Part I
Materials, Processes, Mechanics, and Devices for Soft/Stretchable Electronics

Chapter 1
Liquid Metals for Soft and Stretchable Electronics

Michael D. Dickey

Abstract Liquid metals are the softest and most deformable class of electrical conductors. They are intrinsically stretchable and can be embedded in elastomeric or gel matrices without altering the mechanical properties of the resulting composite. These composites can maintain metallic electrical conductivity at extreme strains and can form soft, conformal contacts with surfaces. Gallium and several of its alloys, which are liquid metals at or near room temperature, offer a low toxicity alternative to mercury. These metals have negligible vapor pressure (so they do not evaporate) and low viscosity. The surface of these metals reacts rapidly with air to form a thin surface oxide 'skin' that allows these liquids to be patterned despite their large surface tension. For example, liquid metal can be 3D printed, molded, or injected into microchannels. This chapter summarizes the properties, patterning methods, and applications of these remarkable materials to form devices with extremely soft mechanical properties. Liquid metals may be used, for example, as conductors for hyper-elastic wires, stretchable antennas, optical structures, conformal electrodes, deformable interconnects, self-healing wires, components in microsystems, reconfigurable circuit elements, and soft circuit boards. They can also be integrated as functional components in circuits composed entirely of soft materials such as sensors, capacitors, memory devices, and diodes. Research is just beginning to explore ways to utilize these 'softer than skin' materials for biolectronic applications. This chapter summarizes the properties, patterning methods, and applications of liquid metals and concludes with an outlook and future challenges of these materials within this context.

Keywords Liquid metal · Gallium · EGaIn · Eutectic gallium indium · Galinstan · Soft electronics

M.D. Dickey (✉)
NC State University, Raleigh, USA
e-mail: mddickey@ncsu.edu

© Springer International Publishing Switzerland 2016
J.A. Rogers et al. (eds.), *Stretchable Bioelectronics for Medical Devices and Systems*, Microsystems and Nanosystems,
DOI 10.1007/978-3-319-28694-5_1

1.1 Background

Interfacing electronics with the body may enable new types of wearable devices and sensors. There are a number of challenges with interfacing conventional electronics—which are rigid and planar—with biological materials that are moist, curvilinear, dynamic, and deformable. One major challenge arises from the mechanical mismatch between soft biological materials and rigid conventional electronics (Fig. 1.1a). In addition to being rigid, conventional electronic materials fail at 1–3 % strain [1], whereas human skin undergoes strains of ~ 30 % [2] and ~ 100 % at joints [3]. Strategies that enable electronic functionality in soft, elastic, and low modulus materials that can accommodate large strain deformations help address this challenge. In addition, electronic components and circuits that are soft have several additional sources of inspiration and motivation:

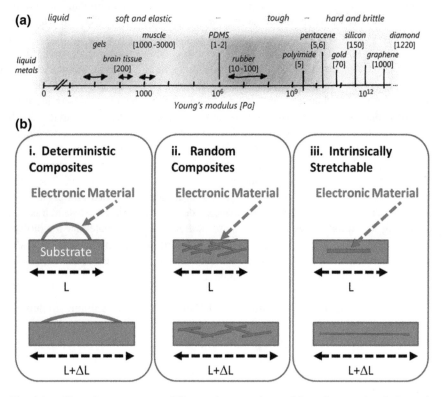

Fig. 1.1 **a** There is an enormous difference between the modulus of conventional electronic materials and the soft materials found in the body. This chapter focuses on liquid metals, which are the softest materials in this plot. Reproduced with permission from MRS Bulletin [13]. **b** There are three general strategies for making electronics that can change from initial length L to an elongated length, L + ΔL. Liquid metals are inherently stretchable and soft, and maintain metallic conductivity to extreme strains

1. Biological 'circuits' are naturally soft. Human skin, for example, provides inspiration by its ability to sense (touch, temperature, humidity), self-heal, and regulate temperature [4, 5].
2. Electronic materials that are intrinsically stretchable have mechanical properties defined by the encasing material (cf. Fig. 1.1b).
3. Materials that are softer than skin can make comfortable, conformal, and 'mechanically transparent' interfaces with the body for 'electronic skin' (e-skin) [6–9] and prosthesis [10].
4. Soft electronics integrate well with soft robotics [11, 12].
5. Liquids are a class of soft materials that can flow to enable 'shape reconfigurable' electronic components and self-healing structures.

The mechanical mismatch between soft and hard materials can be addressed in a number of ways [1, 13–17]. Thin films of otherwise stiff materials [18–24] are flexible and can be rendered stretchable by patterning them into 'deterministic geometries' (e.g., wavy structures [14, 20, 25–28], filamentary and serpentine meshes [17, 29–32], intentionally fractured thin films [33–36], and fractal and kirigami structures [37–42]). Adhering these structures to elastomers (Fig. 1.1b(i)) creates stretchable devices with mechanical properties defined by the elastomer, but deformation limits defined by the stiff electronic materials. It is also possible to create conductive composites by embedding percolated, conductive particles into an elastomeric matrix (Fig. 1.1b(ii)) [43–52]. Relative to solid metal traces, composites have lower conductivity that is maintained until strain values cause loss of percolation. Finally, it is possible to use intrinsically stretchable materials (or 'molecularly stretchable' materials [53]), as shown in Fig. 1.1b(iii). Materials that are intrinsically stretchable are typically soft and easy to deform (e.g., gels, liquids, pastes), although there are exceptions, such as the plasticized organic electronics described by Lipomi in this book.

1.2 Scope

This chapter focuses primarily on liquid metals because (1) liquid metals have superior conductivity to other soft materials, (2) other classes of soft materials, such as gels, have been reviewed elsewhere [54, 55], (3) the author has extensive experience with liquid metals [56, 57], and (4) liquids are extremely soft and intrinsically stretchable (cf. Fig. 1.1). There are impressive examples of soft electronic or ionic devices [54, 55] composed of hydrogels [58] (for ionic diodes [59, 60], soft photovoltaics [61], transparent electrodes [62], capacitors [63], ECG patches), ionic liquids/ionogels [64] (for stretchable electrodes [65] and electrolytes [66]), and pastes/inks (for stretchable batteries [67, 68], conductors [69, 70], rubber-like conductors [3, 43, 45, 52, 71], fluidic conductors [72]). While outside the scope of this chapter, these materials have their own merits, but they do not provide the combination of fluidity and conductivity afforded by liquid metals.

1.3 Motivation of Liquid Metals

According to Wagner and Bauer, the "two most important parameters for stretchable interconnects are high electrical conductance and large critical strain at which conduction is lost [13]." Liquid metals are both intrinsically deformable and highly conductive and have been shown to maintain metallic conductivity up to ~700 % strain [73]. Liquid metals offer new opportunities for fabricating electronic devices with unique mechanical properties, including those that are hyper-elastic, extremely soft, self-healing, and shape reconfigurable. They also provide unique processing and patterning opportunities.

1.4 Liquid Metal Properties

How to choose a liquid metal? There are a very few metals that are liquid at or near room temperature, as shown in Table 1.1.

Hg has been used for a variety of applications including electrodes [76, 77], reconfigurable electromagnetics [78], and microfluidics [79, 80]. However, its toxicity [75] makes Hg inappropriate for bioelectronics and it will not be discussed further. Eliminating from consideration metals that are rare, violently reactive, or radioactive (e.g., Fr, Cs, Rb) leaves only Ga and alloys of Ga.

Ga—discovered in 1875—is considered to have low toxicity [74] and has been utilized in several applications both on and inside the body. Ga has very low solubility in water and therefore most interactions with the body will occur with Ga in an oxidized form. Although Ga has no known physiological function in humans, salts of Ga can be found in the body [81] and have been FDA approved for MRI contrast agents and have proven useful for antibacterial therapy [82], cancer treatment [83], therapeutics [84], and dental amalgams [85]. Alloys of Ga also have been utilized as electrodes to stimulate neurons [86]. Most studies suggest that Ga and salts of Ga

Table 1.1 Metals with low melting points [74]

Metal name	Symbol	Melting point (°C)	Attributes
Mercury	Hg	−38.8	Toxic [75]
Francium	Fr	27.0	Radioactive
Cesium	Cs	28.4	Highly reactive and pyrophoric
Gallium	Ga	29.8	Low toxicity
Rubidium	Rb	39.3	Highly reactive

have low toxicity [87], but at least one organometallic complex has been reported to be toxic [88] and therefore appropriate caution should be taken depending on the context [89]. Ga has no measureable vapor pressure at room temperature and therefore can be handled without concern of it becoming airborne [90].

Although plentiful in the earth, Ga is not mined directly. Instead, it is extracted in trace quantities as a byproduct of aluminum and zinc production [91], making it relatively expensive ($\sim$\$1/g), although it finds many uses in the electronics industry [92]. The applications described in this chapter utilize small volumes of the metal and so cost may not be an issue.

Ga is a liquid over an enormous range of temperatures: its normal melting point is 303 K and normal boiling point is 2676 K [92]. It is technically a solid at room temperature, although it is known to supercool significantly (i.e., it melts at 30 °C, but freezes significantly below 30 °C) [93, 94]. Some elements form binary or ternary eutectic alloys with gallium to achieve a melting point below room temperature. The most common examples of these are eutectic gallium indium (EGaIn, 75.2 % Ga and 24.8 % In, weight percent) [95, 96] and Galinstan (67 % Ga, 20.5 % In and 12.5 % Sn, weight percent) [97]. Table 1.2 summarizes the main physical properties of Ga, EGaIn, and Galinstan and compares them to water. To the first approximation, these metals have similar properties and behavior (and they all form surface oxides, as discussed later). They have desirable properties such as electrical conductivity [98, 99], low viscosity [100, 101], high surface tension [102], and large thermal conductivity [103]. These metals have a viscosity only two times that of water, yet an electrical conductivity three orders of magnitude larger than the most conductive hydrogels (i.e., those doped with graphene [104]), five orders of magnitude greater than salt water [74] and ionic liquids [105], two orders of magnitude greater than rubber-like pastes [52], and six orders of magnitude greater than conductive particles dispersed in oil.

Surface Oxidation Ga-containing liquid metals form a thin oxide 'skin' rapidly [112–114] and spontaneously in the presence of O_2 [115, 116], even at parts per million levels of O_2 [117]. The oxide is 5 Å thick in controlled vacuum conditions [118], although it is likely thicker in ambient conditions [113, 114, 119–123] and is sensitive to moisture [124]. Similar in thickness to the native oxides on common metals (e.g., Al or Cu), the oxide provides an inconsequential barrier to charge

Table 1.2 Physical properties of liquid Ga, EGaIn, Galinstan, and water [97, 99, 103, 106–111]

	Ga	EGaIn	Galinstan	Water
Melting point (°C)	29.8	15.5	10.7	0
Boiling point (°C)	2402	2000	>1300	100
Density (gm/cm^3)	6.91	6.36	6.36	1
Viscosity (10^{-3} kg/m s)	1.969	1.99	2.09	1
Surface tension (mN/m)	750	632	718	72.8
Thermal conductivity (W/m K)	30.5	26.4	25.4	0.6
Electric resistivity (μΩ cm)	27.2	29.4	30.3	20×10^8

transport. For example, the oxide does not influence sensitive electrical measurements that utilize liquid metal electrodes [125].

Although the oxide film is extremely thin, it enables the metal to be patterned into useful shapes. For example, Fig. 1.2a compares EGaIn and Hg injected into microchannels. Mercury retracts spontaneously out of the channel as a result of surface tension, while the oxide on EGaIn keeps it stable within the channels despite having a higher surface tension than mercury [126]. The oxide stabilizes non-spherical shapes like cones (Fig. 1.2b) [95] or stacks of drops (Fig. 1.2c) [127].

Rheological measurements show that the surface oxide dominates the response of the metal to stress at sub-mm length scales [96, 100, 101, 119, 129–132]. The oxide has a surface elastic modulus of 10 N/m and a surface yield stress of ~ 500–600 mN/m [96]. The metal will not flow below this yield stress, while higher stresses rupture the oxide and enable the metal to flow [96]. The presence of water changes the chemical composition, mechanical properties, and physical behavior of the oxide [124, 133].

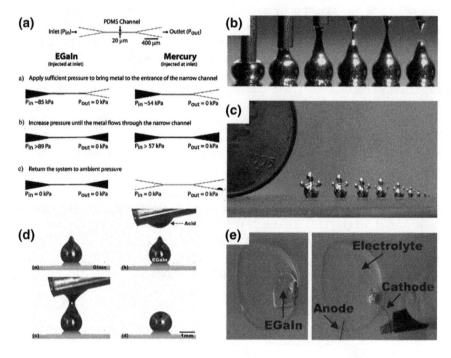

Fig. 1.2 a Comparison between behavior of Ga alloys (EGaIn) and Hg injected in microchannels. The oxide on EGaIn stabilizes it inside the microchannel, while Hg retracts due to surface tension [96], **b** an EGaIn drop bifurcates into a cone shape when stretched [128], and **c** EGaIn "dolls" made by stacking drops of EGaIn [127] demonstrate how the extremely thin oxide skin imparts extraordinary mechanical stability to liquids in shapes otherwise prohibited by surface tension. **d** Strong bases or acids can remove the oxide skin, causing the metal to bead up due to surface tension. **e** An oxide skin maintains the shape of a non-spherical puddle of liquid metal (EGaIn) in electrolyte. Application of a reducing potential to the metal relative to a counter electrode removes the oxide, causing the metal to bead up due to surface tension

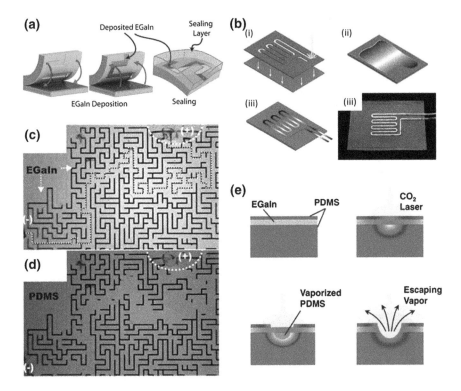

Fig. 1.3 Patterning of liquid metal. **a** Imprint lithography [145]. Copyright Wiley 2014. **b** Stencil printing using a mask to selectively pattern the metal [150]. **c, d** 'Recapillarity' [135] selectively withdraws liquid metals from microfluidic channels by localized electrochemical reduction of the oxide layer. Figures adapted from [135]. Copyright Wiley 2015. **e** Direct laser patterning [151] creates traces of liquid metals **f** and other soft conductors in a rapid, subtractive, and inexpensive fashion

The oxide skin can be etched away in media with pH higher than 10 and lower than 3 (Fig. 1.3d) [134]. The Ga oxide skin can also be removed electrochemically by applying a reducing bias to the liquid metal in electrolyte solution (Fig. 1.3e) [135, 136]. In the absence of the oxide, the liquid metal assumes a spherical shape due to surface tension.

1.5 Methods to Pattern Liquid Metals

A recent review describes state-of-the-art methods for patterning liquid metals [56]. Consequently, the methods will be treated briefly here.

Metals in the liquid state present both challenges and opportunities for patterning. For example, it is possible to inject liquid metals into microchannels, which is not possible with conventional solid metals. However, it is also difficult to use conventional etching and thin film deposition techniques because of the tendency of liquids to flow. Patterning strategies can be organized into four subjective categories. Illustrative examples are as follows:

Injection Alloys of gallium have low viscosity and are readily injected into microchannels by applying sufficient pressure to rupture the oxide skin [96]. Injection (or alternatively, vacuum filling [137]) is a common approach to create soft and stretchable liquid metal structures due to its simplicity, ability to faithfully reproduce the dimensions of the channels [96, 138, 139], and compatibility with microfluidic electronics (cf. Fig. 1.5) [139]. It is possible to inject metal into hollow fibers [73], capillaries with diameters as small as 150 nm [140], and channels with complex geometries [135, 141, 142]. Other fluids such as ionic liquids [143] and solders [144] can also be injected into microchannels.

Lithography-enabled Photolithography can fabricate topographical molds or stencils for patterning liquid metals. For example, elastomeric molds can imprint films of liquid metal [145], as shown in Fig. 1.3a. This technique can produce 2-μm-wide lines stabilized by the oxide skin. It is also possible to pattern lower resolution features (~ 100 μm) by spreading the metal across a stencil mask flush against the target substrate (Fig. 1.3b) [146, 147]. Selective wetting surfaces further enhance the patterning of the metal by varying the composition and roughness of the target substrate [148, 149].

Subtractive Electrochemical reactions can locally remove the oxide layer from the liquid metal within microchannels and induce withdrawal of the metal by capillary action (Fig. 1.3c, d). We call this 'recapillarity' because it utilizes electrochemical reduction to induce capillary withdrawal [135]. Lasers can also ablate the metal embedded in elastomer (Fig. 1.3e) [151].

Additive Direct write and other additive manufacturing techniques deposit materials in predesignated positions, building up an object one pixel at a time. Extruding droplets, cylinders, and other structures of liquid metal directly from a nozzle can produce 2D patterns [152] and 3D structures [127] stabilized by the oxide skin (Fig. 1.4a, b). It is also possible to tape transfer [154], micro-contact print [150], or roller-ball dispense [155] liquid metal. Inkjet printing liquid metals are challenging due to the flow-impediment caused by the oxide. It is, however, possible to inkjet print colloidal particles of liquid metal dispersed in a solvent (Fig. 1.4c) [153]. The resulting structures are not conductive, but can be sintered mechanically at room temperature to render traces conductive [153]. Mechanical sintering can also create conductive pathways in composites with particles dispersed within an elastomeric matrix [156] (Fig. 1.4d–f).

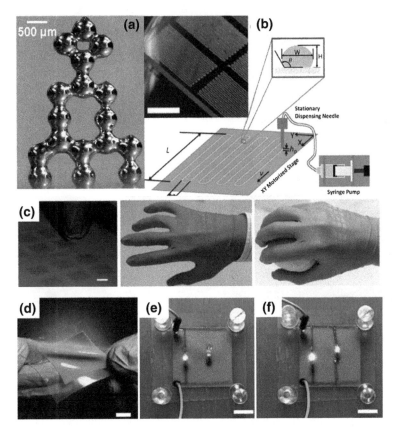

Fig. 1.4 Additive patterning techniques produce free-standing 3D (**a**) [127] and conformal 2D structures (**b**) [152]. Scale bar is 5 mm. Copyright Wiley 2013. **c** Inkjet printing of colloidal suspensions of liquid metal nanoparticles that can be rendered conductive by 'mechanical sintering.' Reproduced from [153]. Copyright Wiley 2015. **d–f** Composites of liquid metal nanoparticles embedded in elastomer can form conductive traces by applying pressure locally to sinter particles. Scale bar is 10 mm. Reproduced from [156]. Copyright Wiley 2015

1.6 Applications of Liquid Metal

There are several hundred papers in the literature that demonstrate applications of liquid metals. This section provides some illustrative examples.

Interconnects Liquid metals form robust and stretchable interconnects between 'islands' of functional circuit components embedded in elastomer without altering the mechanical properties of the overall device. Figure 1.5a illustrates one of the first examples of interconnects formed by injecting liquid metal into microchannels aligned with a LED [157, 158]. It is possible to form interconnects to other electronic components such as sensors [159], integrated circuits (Fig. 1.5b) [160], and

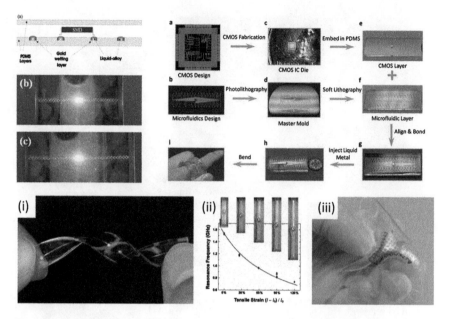

Fig. 1.5 Liquid metal injected into elastomeric microchannels forms interconnects with a (*top, left*) LED and (*top, right*) CMOS chip. Reproduced from [157, 160]. Examples of stretchable liquid metal antennas. (*i*) A dipole antenna [164], (*ii*) radiofrequency antenna [165], (*iii*) loop antenna [166]

super capacitors [161]. Liquid metals can also be utilized for vias for 'reworkable' electronics [162] and to interface with other stretchable conductors [163].

Antennas and Optical Devices Liquid metal injected into elastomeric microchannels forms antennas that are soft, stretchable, and durable. The ability to deform an antenna mechanically offers a new route to change the shape and thereby tune its spectral properties (e.g., frequency) [164], although it is also possible to design antennas to not change frequency during elongation [167]. A wide variety of liquid metal antennas exist including patch antennas [142, 168], coils [169–171], radiofrequency antennas [165, 172], loops [166], spherical caps [173], phase shifting coaxial transmission lines [174], monopoles [175], beam-steering antennas, [176], and reconfigurable antennas [177–186] and filters [187, 188]. Examples of representative antennas are shown in Fig. 1.5. In some geometries these liquid metal antennas perform equivalently to standard copper antennas [164], while in other geometries they perform at a worse, though still acceptable, level [143, 169]. Liquid metals have also been used for mirrors [189], frequency selective surfaces [183], plasmonics [190] (and reconfigurable plasmonics [191]), metamaterials [192] (and reconfigurable or tunable metamaterials [193, 194]), and radiation shielding [195].

Ultra-stretchable Wires In practice, popular elastomers such as polydimethylsiloxane (PDMS) fail mechanically above 40–150 % strain. Nanostructured PDMS

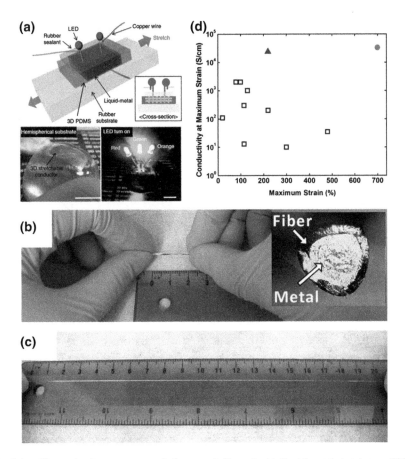

Fig. 1.6 **a** Hyper-elastic nanostructured elastomer infiltrated with liquid metal stretch to ∼220 % strain [196]. **b, c** Ultra-stretchable elastomeric hollow fiber filled with liquid metal maintains metallic conductivity up to ∼700 % strain [73]. **d** A plot adapted from [52] for conductive inks (*squares*) plus the results for liquid metal (*triangle* is from (**a**) and *circle* from (**b**))

infiltrated with liquid metal can stretch to ∼220 % (Fig. 1.6a) [196]. Alternatively, hollow elastomeric fibers composed of a shell of poly(styrene-*b*-(ethylene-co-butylene)-*b*-styrene) injected with metal maintain metallic conductivity up to ∼700 % strain (Fig. 1.6b, c), which is possibly the best combination of strain and conductivity reported to date [73]. These fibers may find applications in electronic textiles or stretchable wiring. These same elastomers can be patterned into microchannels and filled with liquid metal to create stretchable circuits [197].

Typically, there is a trade-off between conductivity and stretchability in composites consisting of conductive materials encased in elastomer. This is most apparent in conductive inks, where increasing the conductive filler content improves the conductivity, but typically at the expense of increased composite stiffness. Liquid metals break this trade-off, as shown in Fig. 1.6d.

100 % Soft Capacitors, Diodes, Memristors Liquid metals have been employed for electronics primarily as conductors, but they may also play an active role in functional circuit elements.

A memristor is a "memory resistor" that stores binary data (i.e., 1's and 0's) based on whether the device is resistive or conductive [198]. Two layers of hydrogel sandwiched between two liquid metal electrodes define a resistive memory device composed entirely of soft, liquid-like materials (Fig. 1.7a) [199].

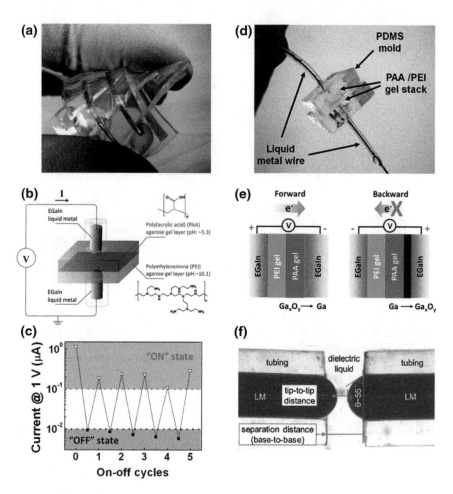

Fig. 1.7 Entirely soft devices utilizing liquid metals with gels and liquids. **a** A resistive memory device composed entirely of soft materials. **b** Schematic of the device. **c** Device turns on and off multiple times [199]. Copyright 2011 John Wiley & Sons, Ltd. **d** A photograph of a soft diode that **e** rectifies ionic current using two layers of hydrogel (PEI is polyethylenimine and PAA is polyacrylic acid) sandwiched by EGaIn liquid metal [200]. Copyright 2012 John Wiley & Sons, Ltd. **f** A soft capacitor featuring two liquid metal electrodes separated by a tunable gap filled with dielectric liquid

The asymmetric pH environment created by the two gels enables one electrode to remain free of surface oxide and therefore conductive at all times. The other electrode can switch between conductive and resistive states via electrochemical surface reactions determined by the pre-applied voltage. Figure 1.7b shows a schematic of a soft memristor, and Fig. 1.7c plots the current in the on and off states through multiple switching cycles. Similar principles can be employed to create soft diodes that only allow current to flow in one direction (Fig. 1.7d, e) [200]. In addition, tunable, soft capacitors can be fashioned from two plugs of the metal separated by a gap filled with dielectric fluid (Fig. 1.7f) [201].

Microfluidics The use of liquid metals for electronic elements in microfluidic systems has been reviewed elsewhere [139, 202]. Liquid metals can also be employed in microfluidics for pumping [80, 203–205], thermal cooling [206], Joule heating [207], energy harvesting [208], electrodes [141], microvalving [209], mixing [210], and stimulating neurons [86].

Soft Electrodes Ga-based liquid metals form excellent, conformal electrodes for electrically contacting and characterizing self-assembled monolayers [211], nanoparticles and quantum dots [212, 213], carbon nanotubes [214], and organic solar cells [215, 216]; this principle can be extended to make electrical contacts for stretchable organic solar cells [217] and e-skin [71]. Liquid metals also serve as deformable electrodes for driving electromechanical instabilities [218, 219].

Bio-electrodes Liquid metal electrodes have been utilized to probe and stimulate neurons [86]. It is possible to inject liquid metal into hydrogel casings inside the body [220] to form electrodes. In principle, this could lead to minimally invasive electrodes. Placing liquid metal directly on the skin creates electrodes [221], although doing so raises several potential issues (the metal smears readily and it has not been tested for biocompatibility with the skin). Ga has also been injected into vasculature for imaging and replication [222, 223].

Reconfigurable Circuits The ability to control the shape or position of liquid metal could enable shape reconfigurable circuits (e.g., switches, tunable antennas, adaptive circuits). Pumping the metal can change its shape [188], but doing so repeatedly requires avoiding adhesion of the oxide to the walls of the channel. This is possible using acid or base to remove the oxide [224–228], utilizing a carrier fluid to create a 'slip layer' around the metal [124, 182], or employing non-wetting surfaces [148, 229, 230]. Voltage can also drive movement of the metal via (1) continuous electrowetting [79, 231, 232] (for micromirrors [233, 234]), (2) electrochemical surface reactions that tune the surface tension over an enormous range [136] (for steering metal through complex channels [235], reconfiguring antennas [181], and switching light valves [236]), (3) recapillary [135], (4) Marangoni effects to propel droplets [237, 238], and (5) electrostatic forces [239]. Although Ga alloys are non-responsive to magnetic fields, it is possible to manipulate droplets magnetically by coating them with magnetic powders [240].

Sensors and E-skin This section will be brief since the chapter by Kramer in this book discusses liquid metals for sensors in soft, electronic skin. Liquid metal-filled microchannels (Fig. 1.8a–d) undergo changes in geometry when deformed. The resulting changes in resistance or capacitance of the traces can create soft sensors responsive to touch, strain, curvature, and other modes of deformation [3, 241–249]. Liquid metals in microchannels can also sense deformation by reflecting light from periodic buckles that form on the walls of the channels due to compressive forces. These buckles diffract light and could be used as sensors or color-changing surfaces (Fig. 1.8e–f) [250].

Soft devices containing liquid metal electrodes can also sense changes in temperature, oxygen concentration, and humidity by monitoring changes in conductance and capacitance through ionic liquids (Fig. 1.8g) [251]. Liquid metals inside carbon nanotubes make extremely small thermometers [252] and junctions between liquid metals and conventional metals can be harnessed to create thermocouples [253].

Self-healing Circuits that heal electrically and mechanically when damaged provide robustness [255]. For example, microdroplets of liquid metal dispersed in polymer release liquid metal when cut. The released liquid metal can span damaged regions of embedded solid metal traces to regain continuity (Fig. 1.9a) [256]. Similarly, injecting liquid metal into microchannels composed of self-healing

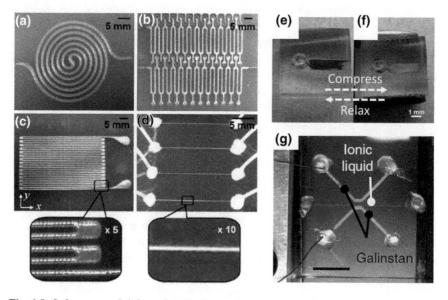

Fig. 1.8 Soft sensors of deformation (strain, touch, curvature, etc.) composed of liquid metal traces embedded in elastomer. **a** Spiral channel, **b** serpentine-shaped channel, **c** strain gauge, **d** small liquid structures [254]. **e, f** Compressing liquid metal in microchannels creates buckles that diffract light [250]. **g** A metal-ionic liquid–metal junction capable of sensing changes in conductivity through the ionic liquid in response to humidity and temperature [251]

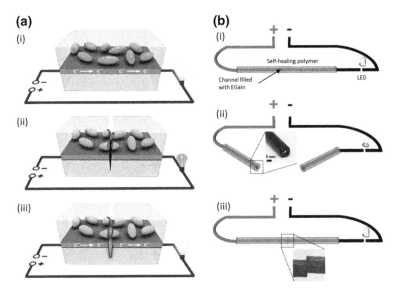

Fig. 1.9 Self-healing circuits utilizing liquid metal. **a** Liquid metal droplets dispersed in polymer help bridge wires damaged during cutting [256]. **b** Liquid metal wires embedded in self-healing polymer can be cut completely apart and put back together [258]

polymer [257] forms self-healing wires [258]. When cut, the wires do not leak liquid metal due to the stabilizing effects of the oxide. When brought back together, the cut wire regains conductivity and the encasing polymer heals the circuit mechanically (Fig. 1.9b). Liquid metals can serve also as self-healing electrodes in Li-ion batteries [259].

Energy Capture/Storage Ga-based liquid metals have been utilized as electrodes for organic solar cells [260] (see section on electrodes), for energy harvesting via reverse electrowetting [208], and as battery electrodes [259].

1.7   Outlook: Opportunities and Challenges

Ga liquid metals have the best combination of conductivity and strechability of any conductors (cf. Fig. 1.6d) and are therefore promising materials for soft, stretchable electronics. They also form surface oxides that allow them to be patterned despite being liquid. This chapter summarizes the properties, patterning methods, and applications of these remarkable materials. These metals can be utilized as antennas, interconnects, and electrodes, but also can be integrated as functional components in circuits composed entirely of soft materials such as sensors, capacitors, memory devices, and diodes. They also have unique capabilities to self-heal, sense, and shape–reconfigure.

Beyond these useful properties, challenges and questions remain:

1. Although Ga appears to have low toxicity (http://www.nature.com/ncomms/2015/151202/ncomms10066/full/ncomms10066.html), more studies are needed to understand the health implications and limitations for bioelectronics. Metals embedded in elastomer have no way to directly contact the body, but what happens if the elastomer fails?
2. Although electronic applications utilize small quantities, Ga is expensive relative to common metals.
3. Ga and its alloys are more resistive than Cu, which is the state of the art metal used for antennas and interconnects.
4. During deformation, liquid metal structures change shape. This can be harnessed for sensors or tunable antennas, but also provides additional considerations for circuit design.
5. Ga liquid metals make excellent electrical contacts [261]. However, the tendency of liquid metals to wet, diffuse into, and alloy with other metals (e.g., Ag [262] or Al [263]) provides additional design considerations for long-term use. The use of graphene to serve as a conductive, physical barrier between metals [264, 265] may be a solution.
6. The oxide layer is critical to many of the applications described in this chapter, yet it changes in the presence of water [124], which provides additional design considerations.
7. The wetting properties of liquid metals to other surfaces are complex due to the presence of the surface oxide, which can break and reform [266]. Recent studies have started to elucidate the wetting behavior [267].
8. There are opportunities to create shape reconfigurable metallic structures, but the tendency of the oxide to adhere to surfaces and disrupt fluidity remains a challenge.

I hope that this book chapter will inspire new applications that take advantage of the unique properties of Ga-based liquid metals and inform new approaches to the challenges and opportunities presented by these unusual materials.

Acknowledgments I am grateful to Chris Trlica for helping to assemble Fig. 1.6 and for editing this chapter. I also thank many students and colleagues whose hard work I have tried to highlight.

References

1. Z. Suo, Mechanics of stretchable electronics and soft machines. MRS Bull. **37**, 218–225 (2012)
2. D.-H. Kim, N. Lu, R. Ma, Y.-S. Kim, R.-H. Kim, S. Wang, J. Wu, S.M. Won, H. Tao, A. Islam, K.J. Yu, T. Kim, R. Chowdhury, M. Ying, L. Xu, M. Li, H.-J. Chung, H. Keum, M. McCormick, P. Liu, Y.-W. Zhang, F.G. Omenetto, Y. Huang, T. Coleman, J.A. Rogers, Epidermal Electronics. Science **333**, 838–843 (2011)

3. Y. Menguc, Y.-L. Park, H. Pei, D. Vogt, P.M. Aubin, E. Winchell, L. Fluke, L. Stirling, R. J. Wood, C.J. Walsh, Wearable soft sensing suit for human gait measurement. Int. J. Robot. Res. **33**, 1748–1764 (2014)
4. R.S. Dahiya, G. Metta, M. Valle, G. Sandini, Tactile sensing-from humans to humanoids. IEEE Trans. Robot. **26**, 1–20 (2010)
5. D.J. Tobin, Biochemistry of human skin—our brain on the outside. Chem. Soc. Rev. **35**, 52–67 (2006)
6. M.L. Hammock, A. Chortos, B.C.-K. Tee, J.B.-H. Tok, Z. Bao, 25th anniversary article: the evolution of electronic skin (e-skin): a brief history, design considerations, and recent progress. Adv. Mater. **25**, 5997–6038 (2013)
7. A. Nathan, A. Ahnood, M.T. Cole, S. Lee, Y. Suzuki, P. Hiralal, F. Bonaccorso, T. Hasan, L. Garcia-Gancedo, A. Dyadyusha, S. Haque, P. Andrew, S. Hofmann, J. Moultrie, D. Chu, A. J. Flewitt, A.C. Ferrari, M.J. Kelly, J. Robertson, G.A.J. Amaratunga, W.I. Milne, Flexible electronics: the next ubiquitous platform. Proc. IEEE **100**, 1486–1517 (2012)
8. V.J. Lumelsky, M.S. Shur, S. Wagner, Sensitive Skin. IEEE Sens. J. **1**, 41–51 (2001)
9. S. Wagner, S.P. Lacour, J. Jones, P.H.I. Hsu, J.C. Sturm, T. Li, Z.G. Suo, Electronic skin: architecture and components. Phys. E-Low-Dimens. Syst. Nanostructures **25**, 326–334 (2004)
10. J. Kim, M. Lee, H.J. Shim, R. Ghaffari, H.R. Cho, D. Son, Y.H. Jung, M. Soh, C. Choi, S. Jung, K. Chu, D. Jeon, S.-T. Lee, J.H. Kim, S.H. Choi, T. Hyeon, D.-H. Kim, Stretchable silicon nanoribbon electronics for skin prosthesis. Nat. Commun. **5**, (2014)
11. C. Majidi, Soft robotics: a perspective—current trends and prospects for the future. Soft Robot. **1**, 5–11 (2013)
12. S. Kim, C. Laschi, B. Trimmer, Soft robotics: a bioinspired evolution in robotics. Trends Biotechnol. **31**, 287–294 (2013)
13. S. Wagner, S. Bauer, Materials for stretchable electronics. MRS Bull. **37**, 207–213 (2012)
14. J.A. Rogers, T. Someya, Y. Huang, Materials and mechanics for stretchable electronics. Science **327**, 1603–1607 (2010)
15. D.-H. Kim, N. Lu, Y. Huang, J.A. Rogers, Materials for stretchable electronics in bioinspired and biointegrated devices. MRS Bull. **37**, 226–235 (2012)
16. T. Sekitani, T. Someya, Stretchable organic integrated circuits for large-area electronic skin surfaces. MRS Bull. **37**, 236–245 (2012)
17. J. Vanfleteren, M. Gonzalez, F. Bossuyt, Y.-Y. Hsu, T. Vervust, I. De Wolf, M. Jablonski, Printed circuit board technology inspired stretchable circuits. MRS Bull. **37**, 254–260 (2012)
18. R.L. Crabb, F.C. Treble, Thin silicon solar cells for large flexible arrays. Nature **213**, 1223–1224 (1967)
19. T.P. Brody, The birth and early childhood of active matrix—a personal memoir. J. Soc. Inf. Disp. **4**, 113 (1996)
20. D.-H. Kim, J.A. Rogers, Stretchable electronics: materials strategies and devices. Adv. Mater. **20**, 4887–4892 (2008)
21. S.D. Theiss, S. Wagner, Amorphous silicon thin-film transistors on steel foil substrates. IEEE Electron Device Lett. **17**, 578–580 (1996)
22. D. Tobjörk, R. Österbacka, Paper electronics. Adv. Mater. **23**, 1935–1961 (2011)
23. G.H. Gelinck, H.E.A. Huitema, E. van Veenendaal, E. Cantatore, L. Schrijnemakers, J.B.P. H. van der Putten, T.C.T. Geuns, M. Beenhakkers, J.B. Giesbers, B.-H. Huisman, E. J. Meijer, E.M. Benito, F.J. Touwslager, A.W. Marsman, B.J.E. van Rens, D.M. de Leeuw, Flexible active-matrix displays and shift registers based on solution-processed organic transistors. Nat. Mater. **3**, 106–110 (2004)
24. J.A. Rogers, Z. Bao, K. Baldwin, A. Dodabalapur, B. Crone, V.R. Raju, V. Kuck, H. Katz, K. Amundson, J. Ewing, P. Drzaic, Paper-like electronic displays: large-area rubber-stamped plastic sheets of electronics and microencapsulated electrophoretic inks. Proc. Natl. Acad. Sci. U. S. A. **98**, 4835–4840 (2001)
25. M. Watanabe, H. Shirai, T. Hirai, Wrinkled polypyrrole electrode for electroactive polymer actuators. J. Appl. Phys. **92**, 4631–4637 (2002)

26. D.S. Gray, J. Tien, C.S. Chen, High-conductivity elastomeric electronics. Adv. Mater. **16**, 393–397 (2004)
27. Y. Sun, W.M. Choi, H. Jiang, Y.Y. Huang, J.A. Rogers, Controlled buckling of semiconductor nanoribbons for stretchable electronics. Nat. Nanotechnol. **1**, 201–207 (2006)
28. D.-H. Kim, J.-H. Ahn, W.M. Choi, H.-S. Kim, T.-H. Kim, J. Song, Y.Y. Huang, Z. Liu, C. Lu, J.A. Rogers, Stretchable and foldable silicon integrated circuits. Science **320**, 507–511 (2008)
29. D. Brosteaux, F. Axisa, M. Gonzalez, J. Vanfleteren, Design and fabrication of elastic interconnections for stretchable electronic circuits. IEEE Electron Device Lett. **28**, 552–554 (2007)
30. D.-H. Kim, J. Song, W.M. Choi, H.-S. Kim, R.-H. Kim, Z. Liu, Y.Y. Huang, K.-C. Hwang, Y. Zhang, J.A. Rogers, Materials and noncoplanar mesh designs for integrated circuits with linear elastic responses to extreme mechanical deformations. Proc. Natl. Acad. Sci. **105**, 18675–18680 (2008)
31. C.F. Guo, T. Sun, Q. Liu, Z. Suo, Z. Ren, Highly stretchable and transparent nanomesh electrodes made by grain boundary lithography. Nat. Commun. **5**, (2014)
32. Y. Zhang, S. Wang, X. Li, J.A. Fan, S. Xu, Y.M. Song, K.-J. Choi, W.-H. Yeo, W. Lee, S.N. Nazaar, B. Lu, L. Yin, K.-C. Hwang, J.A. Rogers, Y. Huang, Experimental and theoretical studies of serpentine microstructures bonded to prestrained elastomers for stretchable electronics. Adv. Funct. Mater. **24**, 2028–2037 (2014)
33. S.P. Lacour, C. Tsay, S. Wagner, An elastically stretchable TFT circuit. IEEE Electron Device Lett. **25**, 792–794 (2004)
34. S.P. Lacour, S. Wagner, Z.Y. Huang, Z. Suo, Stretchable gold conductors on elastomeric substrates. Appl. Phys. Lett. **82**, 2404–2406 (2003)
35. S.P. Lacour, J. Jones, S. Wagner, T. Li, Z. Suo, Stretchable interconnects for elastic electronic surfaces. Proc. IEEE **93**, 1459–1467 (2005)
36. I.R. Minev, P. Musienko, A. Hirsch, Q. Barraud, N. Wenger, E.M. Moraud, J. Gandar, M. Capogrosso, T. Milekovic, L. Asboth, R.F. Torres, N. Vachicouras, Q. Liu, N. Pavlova, S. Duis, A. Larmagnac, J. Vörös, S. Micera, Z. Suo, G. Courtine, S.P. Lacour, Electronic dura mater for long-term multimodal neural interfaces. Science **347**, 159–163 (2015)
37. S. Xu, Z. Yan, K.-I. Jang, W. Huang, H. Fu, J. Kim, F. Wei, M. Flavin, J. McCracken, R. Wang, A. Badea, Y. Liu, D. Xiao, G. Zhou, J. Lee, H.U. Chung, H. Cheng, W. Ren, A. Banks, X. Li, U. Paik, R.G. Nuzzo, Y. Huang, Y. Zhang, J.A. Rogers, Assembly of micro/nanomaterials into complex, three-dimensional architectures by compressive buckling. Science 347, 154–159 (2015)
38. T. Castle, Y. Cho, X. Gong, E. Jung, D.M. Sussman, S. Yang, R.D. Kamien, Making the cut: lattice kirigami rules. Phys. Rev. Lett. **113**, 245502 (2014)
39. M.K. Blees, A.W. Barnard, P.A. Rose, S.P. Roberts, K.L. McGill, P.Y. Huang, A.R. Ruyack, J.W. Kevek, B. Kobrin, D.A. Muller, P.L. McEuen, Graphene kirigami. Nature **524**, 204–207 (2015)
40. D.M. Sussman, Y. Cho, T. Castle, X. Gong, E. Jung, S. Yang, R.D. Kamien, Algorithmic lattice kirigami: a route to pluripotent materials. Proc. Natl. Acad. Sci. U. S. A. **112**, 7449–7453 (2015)
41. T.C. Shyu, P.F. Damasceno, P.M. Dodd, A. Lamoureux, L. Xu, M. Shlian, M. Shtein, S.C. Glotzer, N.A. Kotov, A kirigami approach to engineering elasticity in nanocomposites through patterned defects. Nat. Mater. **14**, 785–789 (2015)
42. J.A. Fan, W.-H. Yeo, Y. Su, Y. Hattori, W. Lee, S.-Y. Jung, Y. Zhang, Z. Liu, H. Cheng, L. Falgout, M. Bajema, T. Coleman, D. Gregoire, R.J. Larsen, Y. Huang, J. A. Rogers, Fractal design concepts for stretchable electronics. Nat. Commun. **5**, (2014)
43. T. Sekitani, Y. Noguchi, K. Hata, T. Fukushima, T. Aida, T. Someya, A rubberlike stretchable active matrix using elastic conductors. Science **321**, 1468–1472 (2008)
44. S. Rosset, M. Niklaus, P. Dubois, H.R. Shea, Metal Ion Implantation for the fabrication of stretchable electrodes on elastomers. Adv. Funct. Mater. **19**, 470–478 (2009)

45. T. Sekitani, H. Nakajima, H. Maeda, T. Fukushima, T. Aida, K. Hata, T. Someya, Stretchable active-matrix organic light-emitting diode display using printable elastic conductors. Nat. Mater. **8**, 494–499 (2009)
46. M. Park, J. Im, M. Shin, Y. Min, J. Park, H. Cho, S. Park, M.-B. Shim, S. Jeon, D.-Y. Chung, J. Bae, J. Park, U. Jeong, K. Kim, Highly stretchable electric circuits from a composite material of silver nanoparticles and elastomeric fibres. Nat. Nanotechnol. **7**, 803–809 (2012)
47. P. Lee, J. Lee, H. Lee, J. Yeo, S. Hong, K.H. Nam, D. Lee, S.S. Lee, S.H. Ko, Highly stretchable and highly conductive metal electrode by very long metal nanowire percolation network. Adv. Mater. **24**, 3326–3332 (2012)
48. Y. Zhang, C.J. Sheehan, J. Zhai, G. Zou, H. Luo, J. Xiong, Y.T. Zhu, Q.X. Jia, Polymer-embedded carbon nanotube ribbons for stretchable conductors. Adv. Mater. **22**, 3027–3031 (2010)
49. K. Liu, Y. Sun, P. Liu, X. Lin, S. Fan, K. Jiang, Cross-stacked superaligned carbon nanotube films for transparent and stretchable conductors. Adv. Funct. Mater. **21**, 2721–2728 (2011)
50. H. Stoyanov, M. Kollosche, S. Risse, R. Wache, G. Kofod, Soft conductive elastomer materials for stretchable electronics and voltage controlled artificial muscles. Adv. Mater. **25**, 578–583 (2013)
51. Y. Kim, J. Zhu, B. Yeom, M. Di Prima, X. Su, J.-G. Kim, S.J. Yoo, C. Uher, N.A. Kotov, Stretchable nanoparticle conductors with self-organized conductive pathways. Nature **500**, 59-U77 (2013)
52. N. Matsuhisa, M. Kaltenbrunner, T. Yokota, H. Jinno, K. Kuribara, T. Sekitani, T. Someya, Printable elastic conductors with a high conductivity for electronic textile applications. Nat. Commun. **6**, 7461 (2015)
53. S. Savagatrup, A.D. Printz, T.F. O'Connor, A.V. Zaretski, D.J. Lipomi, molecularly stretchable electronics. Chem. Mater. **26**, 3028–3041 (2014)
54. H.-J. Koo, O.D. Velev, Ionic current devices—recent progress in the merging of electronic, microfluidic, and biomimetic structures. Biomicrofluidics **7**, 031501 (2013)
55. H. Chun, T.D. Chung, Iontronics. Annu. Rev. Anal. Chem. **88**, 441–462 (2015)
56. I.D. Joshipura, H.R. Ayers, C. Majidi, M.D. Dickey, Methods to pattern liquid metals. J. Mater. Chem. **C3**, 3834–3841 (2015)
57. M.D. Dickey, Emerging applications of liquid metals featuring surface oxides. ACS Appl. Mater. Interfaces **6**, 18369–18379 (2014)
58. E.M. Ahmed, Hydrogel: preparation, characterization, and applications: a review. J. Adv. Res. **6**, 105–121 (2015)
59. O.J. Cayre, S.T. Chang, O.D. Velev, Polyelectrolyte diode: nonlinear current response of a junction between aqueous ionic gels. J. Am. Chem. Soc. **129**, 10801–10806 (2007)
60. J.-H. Han, K.B. Kim, H.C. Kim, T. D. Chung, Ionic circuits based on polyelectrolyte diodes on a microchip. Angew. Chem.Int. Ed. **48**, 3830–3833 (2009)
61. H.-J. Koo, S.T. Chang, J.M. Slocik, R.R. Naik, O.D. Velev, Aqueous soft matter based photovoltaic devices. J. Mater. Chem. **21**, 72–79 (2011)
62. C. Keplinger, J.-Y. Sun, C.C. Foo, P. Rothemund, G.M. Whitesides, Z. Suo, Stretchable, transparent, ionic conductors. Science **341**, 984–987 (2013)
63. J.-Y. Sun, C. Keplinger, G.M. Whitesides, Z. Suo, Ionic skin. Adv. Mater. **26**, 7608–7614 (2014)
64. J. Le Bideau, L. Viau, A. Vioux, Ionogels, ionic liquid based hybrid materials. Chem. Soc. Rev. **40**, 907–925 (2011)
65. B. Chen, J.J. Lu, C.H. Yang, J.H. Yang, J. Zhou, Y.M. Chen, Z. Suo, Highly stretchable and transparent ionogels as nonvolatile conductors for dielectric elastomer transducers. ACS Appl. Mater. Interfaces **6**, 7840–7845 (2014)
66. S. Saricilar, D. Antiohos, K. Shu, P.G. Whitten, K. Wagner, C. Wang, G.G. Wallace, High strain stretchable solid electrolytes. Electrochem. Commun. **32**, 47–50 (2013)
67. M. Kaltenbrunner, G. Kettlgruber, C. Siket, R. Schwoediauer, S. Bauer, Arrays of ultracompliant electrochemical dry gel cells for stretchable electronics. Adv. Mater. **22**, 2065–2067 (2010)

68. G. Kettlgruber, M. Kaltenbrunner, C.M. Siket, R. Moser, I.M. Graz, R. Schwödiauer, S. Bauer, Intrinsically stretchable and rechargeable batteries for self-powered stretchable electronics. J. Mater. Chem. **A1**, 5505 (2013)
69. K. Suganuma, *Introduction to Printed Electronics*, vol. 74 (Springer, New York, 2014)
70. Applications of Organic and Printed Electronics (Springer, US, 2013)
71. D.J. Lipomi, M. Vosgueritchian, B.C.-K. Tee, S.L. Hellstrom, J.A. Lee, C.H. Fox, Z. Bao, Skin-like pressure and strain sensors based on transparent elastic films of carbon nanotubes. Nat. Nanotechnol. **6**, 788–792 (2011)
72. S. Rosset, H.R. Shea, Flexible and stretchable electrodes for dielectric elastomer actuators. Appl. Phys. Mater. Sci. Process. **110**, 281–307 (2013)
73. S. Zhu, J.-H. So, R. Mays, S. Desai, W.R. Barnes, B. Pourdeyhimi, M.D. Dickey, ultrastretchable fibers with metallic conductivity using a liquid metal alloy core. Adv. Funct. Mater. **23**, 2308–2314 (2013)
74. W.M. Haynes, *CRC Handbook of Chemistry and Physics*, (CRC Press/Taylor and Francis, Boca Raton, 2011)
75. T.W. Clarkson, L. Magos, The toxicology of mercury and its chemical compounds. Crit. Rev. Toxicol. **36**, 609–662 (2006)
76. R.E. Holmlin, R. Haag, M.L. Chabinyc, R.F. Ismagilov, A.E. Cohen, A. Terfort, M.A. Rampi, G.M. Whitesides, Electron transport through thin organic films in metal—insulator—metal junctions based on self-assembled monolayers. J. Am. Chem. Soc. **123**, 5075–5085 (2001)
77. D.C. Grahame, Measurement of the capacity of the electrical double layer at a mercury electrode. J. Am. Chem. Soc. **71**, 2975–2978 (1949)
78. T.S. Kasirga, Y.N. Ertas, M. Bayindir, Microfluidics for reconfigurable electromagnetic metamaterials. Appl. Phys. Lett. **95**, 214102 (2009)
79. H.J. Lee, C.-J. Kim, Surface-tension-driven microactuation based on continuous electrowetting. J. Microelectromechanical Syst. **9**, 171–180 (2000)
80. K.-S. Yun, I.-J. Cho, J.-U. Bu, C.-J. Kim, E. Yoon, A surface-tension driven micropump for low-voltage and low-power operations. J. Microelectromechanical Syst. **11**, 454–461 (2002)
81. T.L. Ziegler, K.K. Divine, P.L. Goering, in *Elements and their compounds in the environment*, ed. by E. Merian, M. Anke, M. Ihnat, M. Stoeppler (Wiley-VCH Verlag GmbH, 2004), pp. 775–786
82. C. Bonchi, F. Imperi, F. Minandri, P. Visca, E. Frangipani, Repurposing of gallium-based drugs for antibacterial therapy. BioFactors **40**, 303–312 (2014)
83. M. Frezza, C.N. Verani, D. Chen, Q.P. Dou, The therapeutic potential of gallium-based complexes in anti-tumor drug design. Lett. Drug Des. Discov. **4**, 311–317 (2007)
84. L.R. Bernstein, Mechanisms of therapeutic activity for gallium. Pharmacol. Rev. **50**, 665–682 (1998)
85. H.J. Caul, D.L. Smith, Alloys of gallium with powdered metals as possible replacement for dental amalgam. J. Am. Dent. Assoc. **193953**, 315–324 (1956)
86. N. Hallfors, A. Khan, M.D. Dickey, A.M. Taylor, Integration of pre-aligned liquid metal electrodes for neural stimulation within a user-friendly microfluidic platform. Lab Chip **13**, 522–526 (2013)
87. J.E. Chandler, H.H. Messer, G. Ellender, Cytotoxicity of gallium and indium ions compared with mercuric ion. J. Dent. Res. **73**, 1554–1559 (1994)
88. C.S. Ivanoff, A.E. Ivanoff, T.L. Hottel, Gallium poisoning: a rare case report. Food Chem. Toxicol. **50**, 212–215 (2012)
89. J.L. Domingo, J. Corbella, A review of the health-hazards from gallium exposure. Trace Elem. Med. **8**, 56–64 (1991)
90. F. Geiger, C.A. Busse, R.I. Loehrke, The vapor pressure of indium, silver, gallium, copper, tin, and gold between 0.1 and 3.0 bar. Int. J. Thermophys. **8**, 425–436 (1987)

91. Gray, F., Kramer, D. A., Bliss, J. D. & Updated by Staff. in *Kirk-Othmer Encyclopedia* of *Chemical Technology* (John Wiley & Sons, Inc., 2000). http://onlinelibrary.wiley.com/doi/10.1002/0471238961.0701121219010215.a01.pub3/abstract
92. R.R. Moskalyk, Gallium: the backbone of the electronics industry. Miner. Eng. **16**, 921–929 (2003)
93. L.J. Briggs, Gallium: thermal conductivity; supercooling; negative pressure. J. Chem. Phys. **26**, 784–786 (1957)
94. A. Burdakin, B. Khlevnoy, M. Samoylov, V. Sapritsky, S. Ogarev, A. Panfilov, G. Bingham, V. Privalsky, J. Tansock, T. Humpherys, Melting points of gallium and of binary eutectics with gallium realized in small cells. Metrologia **45**, 75 (2008)
95. R.C. Chiechi, E.A. Weiss, M.D. Dickey, G. M. Whitesides, Eutectic Gallium–Indium (EGaIn): a moldable liquid metal for electrical characterization of self-assembled monolayers. Angew. Chem. Int. Ed. **47**, 142–144 (2008)
96. M.D. Dickey, R.C. Chiechi, R.J. Larsen, E.A. Weiss, D.A. Weitz, G.M. Whitesides, Eutectic gallium-indium (EGaIn): a liquid metal alloy for the formation of stable structures in microchannels at room temperature. Adv. Funct. Mater. **18**, 1097–1104 (2008)
97. N.B. Morley, J. Burris, L.C. Cadwallader, M.D. Nornberg, GaInSn usage in the research laboratory. Rev. Sci. Instrum. **79**, 056107 (2008)
98. N.F. Mott, The resistance of liquid metals. Proc. R. Soc. Lond. Ser. **A146**, 465–472 (1934)
99. D. Zrnic, D.S. Swatik, Resistivity and surface tension of the eutectic alloy of gallium and indium. J. Common Met. **18**, 67–68 (1969)
100. M.R. Hopkins, T.C. Toye, The determination of the viscosity of molten metals. Proc. Phys. Soc. Sect. **B63**, 773–782 (1950)
101. K.E. Spells, Determination of the viscosity of liquid gallium over an extended range of temperature. Proc. Phys. Soc. Lond. **48**, 299–311 (1936)
102. S.C. Hardy, The surface tension of liquid gallium. J. Cryst. Growth **71**, 602–606 (1985)
103. M.J. Duggin, The thermal conductivity of liquid gallium. Phys. Lett. A **29**, 470–471 (1969)
104. L. Zhang, G. Shi, Preparation of highly conductive graphene hydrogels for fabricating supercapacitors with high rate capability. J. Phys. Chem. C **115**, 17206–17212 (2011)
105. S. Zhang, N. Sun, X. He, X. Lu, X. Zhang, Physical properties of ionic liquids: database and evaluation. J. Phys. Chem. Ref. Data **35**, 1475–1517 (2006)
106. H.E. Sostman, Melting point of gallium as a temperature calibration standard. Rev. Sci. Instrum. **48**, 127–130 (1977)
107. K.E. Spells, The determination of the viscosity of liquid gallium over an extended nrange of temperature. Proc. Phys. Soc. **48**, 299 (1936)
108. C. Dodd, The electrical resistance of liquid gallium in the neighbourhood of its melting point. Proc. Phys. Soc. Sect. **B63**, 662–664 (1950)
109. Y. Plevachuk, V. Sklyarchuk, S. Eckert, G. Gerbeth, R. Novakovic, Thermophysical properties of the Liquid Ga–In–Sn eutectic Alloy. J. Chem. Eng. Data **59**, 757–763 (2014)
110. S. Yu, M. Kaviany, Electrical, thermal, and species transport properties of liquid eutectic Ga-In and Ga-In-Sn from first principles. J. Chem. Phys. **140**, 064303 (2014)
111. W.H. Hoather, The density and coefficient of expansion of liquid gallium over a wide range of temperature. Proc. Phys. Soc. **48**, 699 (1936)
112. Y.L. Wang, S.J. Lin, Spatial and temporal scaling of oxide cluster aggregation on a liquid-gallium surface. Phys. Rev. B: Condens. Matter **53**, 6152–6157 (1996)
113. A. Plech, U. Klemradt, H. Metzger, J. Peisl, In situ x-ray reflectivity study of the oxidation kinetics of liquid gallium and the liquid alloy. J. Phys.: Condens. Matter **10**, 971 (1998)
114. Y.L. Wang, Y.Y. Doong, T.S. Chen, J.S. Haung, Oxidation of liquid gallium surface—nonequilibrium growth-kinetics of in 2+1 dimensions. J. Vac. Sci. Technol. Vac. Surf. Films **12**, 2081–2086 (1994)
115. A.J. Downs, *Chemistry of Aluminium, Gallium, Indium, and Thallium*, 1st edn. (Blackie Academic & Professional, 1993)
116. I.A. Sheka, I.S. Chaus, T. T. Mitiureva, *The Chemistry of Gallium* (Elsevier, 1966)

117. T. Liu, P. Sen, C.-J. Kim, Characterization of nontoxic liquid-metal alloy galinstan for applications in microdevices. J. Microelectromechanical Syst. **21**, 443–450 (2012)
118. E.H. Kawamoto, S. Lee, P.S. Pershan, M. Deutsch, N. Maskil, B.M. Ocko, X-ray reflectivity study of the surface of liquid gallium. Phys. Rev. B Condens. Matter Mater. Phys. **47**, 6847–6850 (1993)
119. F. Scharmann, G. Cherkashinin, V. Breternitz, C. Knedlik, G. Hartung, T. Weber, J.A. Schaefer, Viscosity effect on GaInSn studied by XPS. Surf. Interface Anal. **36**, 981–985 (2004)
120. J.M. Chabala, Oxide-growth kinetics and fractal-like patterning across liquid gallium surfaces. Phys. Rev. B **46**, 11346–11357 (1992)
121. D. Tonova, M. Patrini, P. Tognini, A. Stella, P. Cheyssac, R. Kofman, Ellipsometric study of optical properties of liquid Ga nanoparticles. J. Phys. Condens. Matter **11**, 2211 (1999)
122. M.W. Knight, T. Coenen, Y. Yang, B.J.M. Brenny, M. Losurdo, A.S. Brown, H.O. Everitt, A. Polman, Gallium plasmonics: deep subwavelength spectroscopic imaging of single and interacting gallium nanoparticles. ACS Nano **9**, 2049–2060 (2015)
123. M. Yarema, M. Wörle, M.D. Rossell, R. Erni, R. Caputo, L. Protesescu, K.V. Kravchyk, D. N. Dirin, K. Lienau, F. von Rohr, A. Schilling, M. Nachtegaal, M.V. Kovalenko, Monodisperse colloidal gallium nanoparticles: synthesis, low temperature crystallization, surface plasmon resonance and li-ion storage. J. Am. Chem. Soc. **136**, 12422–12430 (2014)
124. M.R. Khan, C. Trlica, J.-H. So, M. Valeri, M.D. Dickey, Influence of water on the interfacial behavior of gallium liquid metal alloys. ACS Appl. Mater. Interfaces **6**, 22467–22473 (2014)
125. W.F. Reus, M.M. Thuo, N.D. Shapiro, C.A. Nijhuis, G.M. Whitesides, The SAM, not the electrodes, dominates charge transport in metal-monolayer//Ga2O3/gallium-indium eutectic junctions. ACS Nano **6**, 4806–4822 (2012)
126. M.J. Regan, H. Tostmann, P.S. Pershan, O.M. Magnussen, E. DiMasi, B.M. Ocko, M. Deutsch, X-ray study of the oxidation of liquid-gallium surfaces. Phys. Rev. B **55**, 10786–10790 (1997)
127. C. Ladd, J.-H. So, J. Muth, M.D. Dickey, 3D printing of free standing liquid metal microstructures. Adv. Mater. **25**, 5081–5085 (2013)
128. R.C. Chiechi, E.A. Weiss, M.D. Dickey, G.M. Whitesides, Eutectic gallium-indium (EGaIn): a moldable liquid metal for electrical characterization of self-assembled monolayers. Angew. Chem. **120**, 148–150 (2008)
129. R.J. Larsen, M.D. Dickey, G.M. Whitesides, D.A. Weitz, Viscoelastic properties of oxide-coated liquid metals. J. Rheol. **53**, 1305–1326 (2009)
130. S.H. Elahi, H. Abdi, H.R. Shahverdi, A new method for investigating oxidation behavior of liquid metals. Rev. Sci. Instrum. **85**, 015115 (2014)
131. M. Jeyakumar, M. Hamed, S. Shankar, Rheology of liquid metals and alloys. J. Non-Newton. Fluid Mech. **166**, 831–838 (2011)
132. Q. Xu, N. Oudalov, Q. Guo, H.M. Jaeger, E. Brown, Effect of oxidation on the mechanical properties of liquid gallium and eutectic gallium-indium. *Phys. Fluids 1994-Present* **24**, 063101 (2012)
133. N. Horasawa, H. Nakajima, S. Takahashi, T. Okabe, Behavior of pure gallium in water and various saline solutions. Dent. Mater. **J16**, 200–208 (1997)
134. M. Pourbaix, *Atlas of Electrochemical Equilibria in Aqueous Solutions*, vol. 16.1 (Natl Assn of Corrosion, 1974)
135. M.R. Khan, C. Trlica, M.D. Dickey, Recapillarity: electrochemically controlled capillary withdrawal of a liquid metal alloy from microchannels. Adv. Funct. Mater. **25**, 671–678 (2015)
136. M.R. Khan, C.B. Eaker, E.F. Bowden, M.D. Dickey, Giant and switchable surface activity of liquid metal via surface oxidation. Proc. Natl. Acad. Sci. **111**, 14047–14051 (2014)
137. A. Fassler, C. Majidi, 3D Structures of liquid-phase GaIn alloy embedded in PDMS with freeze casting. Lab Chip **13**, 4442–4450 (2013)
138. H.-J. Kim, C. Son, B. Ziaie, A multiaxial stretchable interconnect using liquid-alloy-filled elastomeric microchannels. *Appl. Phys. Lett.* **92**, 011904–011904-3 (2008)

139. S. Cheng, Z. Wu, Microfluidic electronics. Lab. Chip **12**, 2782 (2012)
140. W. Zhao, J.L. Bischof, J. Hutasoit, X. Liu, T.C. Fitzgibbons, J.R. Hayes, P.J.A. Sazio, C. Liu, J.K. Jain, J.V. Badding, M.H.W. Chan, Single-fluxon controlled resistance switching in centimeter-long superconducting gallium-indium eutectic nanowires. Nano Lett. **15**, 153–158 (2015)
141. J.-H. So, M.D. Dickey, Inherently aligned microfluidic electrodes composed of liquid metal. Lab Chip **11**, 905–911 (2011)
142. G.J. Hayes, J.-H. So, A. Qusba, M.D. Dickey, G. Lazzi, Flexible liquid metal alloy (EGaIn) microstrip patch antenna. IEEE Trans. Antennas Propag. **60**, 2151–2156 (2012)
143. H. Ota, K. Chen, Y. Lin, D. Kiriya, H. Shiraki, Z. Yu, T.-J. Ha, A. Javey, Highly deformable liquid-state heterojunction sensors. Nat. Commun. **5**, (2014)
144. A.C. Siegel, D.A. Bruzewicz, D.B. Weibel, G.M. Whitesides, Microsolidics: fabrication of three-dimensional metallic microstructures in poly(dimethylsiloxane). Adv. Mater. **19**, 727–733 (2007)
145. B.A. Gozen, A. Tabatabai, O.B. Ozdoganlar, C. Majidi, High-density soft-matter electronics with micron-scale line width. Adv. Mater. **26**, 5211–5216 (2014)
146. R.K. Kramer, C. Majidi, R.J. Wood, Masked deposition of gallium-indium alloys for liquid-embedded elastomer conductors. Adv. Funct. Mater. **23**, 5292–5296 (2013)
147. S.H. Jeong, A. Hagman, K. Hjort, M. Jobs, J. Sundqvist, Z. Wu, Liquid alloy printing of microfluidic stretchable electronics. Lab Chip **12**, 4657 (2012)
148. R.K. Kramer, J.W. Boley, H.A. Stone, J.C. Weaver, R.J. Wood, Effect of microtextured surface topography on the wetting behavior of eutectic gallium-indium alloys. Langmuir **30**, 533–539 (2014)
149. G. Li, X. Wu, D.-W. Lee, Selectively plated stretchable liquid metal wires for transparent electronics. Sens. Actuators B Chem. **221**, 1114–1119 (2015)
150. A. Tabatabai, A. Fassler, C. Usiak, C. Majidi, Liquid-phase gallium-indium alloy electronics with microcontact printing. Langmuir **29**, 6194–6200 (2013)
151. T. Lu, L. Finkenauer, J. Wissman, C. Majidi, Rapid prototyping for soft-matter electronics. Adv. Funct. Mater. **24**, 3351–3356 (2014)
152. J.W. Boley, E.L. White, G.T.-C. Chiu, R.K. Kramer, Direct writing of gallium-indium alloy for stretchable electronics. Adv. Funct. Mater. **24**, 3501–3507 (2014)
153. J.W. Boley, E.L. White, R.K. Kramer, Mechanically sintered gallium-indium nanoparticles. Adv. Mater. **27**, 2355–2360 (2015)
154. S.H. Jeong, K. Hjort, Z. Wu, Tape transfer printing of a liquid metal alloy for stretchable RF electronics. Sensors **14**, 16311–16321 (2014)
155. Y. Zheng, Q. Zhang, J. Liu, Pervasive liquid metal based direct writing electronics with roller-ball pen. AIP Adv. **3**, 112117 (2013)
156. A. Fassler, C. Majidi, Liquid-phase metal inclusions for a conductive polymer composite. Adv. Mater. **27**, 1928–1932 (2015)
157. H.-J. Kim, C. Son, B. Ziaie, A multiaxial stretchable interconnect using liquid-alloy-filled elastomeric microchannels. Appl. Phys. Lett. **92**, 011904 (2008)
158. H.-J. Kim, T. Maleki, P. Wei, B. Ziaie, A biaxial stretchable interconnect with liquid-alloy-covered joints on elastomeric substrate. J. Microelectromechanical Syst. **18**, 138–146 (2009)
159. H. Hu, K. Shaikh, C. Liu, Super flexible sensor skin using liquid metal as interconnect. in *2007 IEEE Sens.* 815–817 (2007)
160. B. Zhang, Q. Dong, C.E. Korman, Z. Li, M.E. Zaghloul, Flexible packaging of solid-state integrated circuit chips with elastomeric microfluidics. Sci. Rep. **3**, (2013)
161. B.Y. Lim, J. Yoon, J. Yun, D. Kim, S.Y. Hong, S.-J. Lee, G. Zi, J.S. Ha, Biaxially stretchable, integrated array of high performance microsupercapacitors. ACS Nano. **8**(11), 11639–11650 (2014)
162. G.A. Hernandez, D. Martinez, C. Ellis, M. Palmer, M.C. Hamilton, Through Si vias using liquid metal conductors for re-workable 3D electronics, in *2013 IEEE 63rd Electronic Components* and *Technology Conference (ECTC)*, (2013), pp. 1401–1406

163. J. Jang, B. Kim, I. You, J. Park, S. Shin, G. Jeon, J.K. Kim, U. Jeong, Interfacing liquid metals with stretchable metal conductors. ACS Appl. Mater. Interfaces **7**(15), 7920–7926 (2015)
164. J. So, J. Thelen, A. Qusba, G.J. Hayes, G. Lazzi, M.D. Dickey, Reversibly deformable and mechanically tunable fluidic antennas. Adv. Funct. Mater. **19**, 3632–3637 (2009)
165. M. Kubo, X. Li, C. Kim, M. Hashimoto, B.J. Wiley, D. Ham, G.M. Whitesides, Stretchable microfluidic radiofrequency antennas. Adv. Mater. **22**, 2749–2752 (2010)
166. S. Cheng, A. Rydberg, K. Hjort, Z. Wu, Liquid metal stretchable unbalanced loop antenna. Appl. Phys. Lett. **94**, 144103–144103-3 (2009)
167. Y. Huang, Y. Wang, L. Xiao, H. Liu, W. Dong, Z. Yin, Microfluidic serpentine antennas with designed mechanical tunability. Lab Chip **14**, 4205–4212 (2014)
168. B. Aissa, M. Nedil, M.A. Habib, E. Haddad, W. Jamroz, D. Therriault, Y. Coulibaly, F. Rosei, Fluidic patch antenna based on liquid metal alloy/single-wall carbon-nanotubes operating at the S-band frequency. Appl. Phys. Lett. **103**, 063101 (2013)
169. A. Qusba, A.K. RamRakhyani, J.-H. So, G.J. Hayes, M.D. Dickey, G. Lazzi, On the design of microfluidic implant coil for flexible telemetry system. IEEE Sens. J. **14**, 1074–1080 (2014)
170. N. Lazarus, C.D. Meyer, S.S. Bedair, H. Nochetto, I.M. Kierzewski, Multilayer liquid metal stretchable inductors. Smart Mater. Struct. **23**, 085036 (2014)
171. A. Fassler, C. Majidi, Soft-matter capacitors and inductors for hyperelastic strain sensing and stretchable electronics. Smart Mater. Struct. **22**, 055023 (2013)
172. S. Cheng, Z. Wu, Microfluidic stretchable RF electronics. Lab. Chip **10**, 3227–3234 (2010)
173. M. Jobs, K. Hjort, A. Rydberg, Z. Wu, A tunable spherical cap microfluidic electrically small antenna. Small **9**, 3230–3234 (2013)
174. G.J. Hayes, S.C. Desai, Y. Liu, P. Annamaa, G. Lazzi, M.D. Dickey, Microfluidic coaxial transmission line and phase shifter. Microw. Opt. Technol. Lett. **56**, 1459–1462 (2014)
175. A.M. Morishita, C.K.Y. Kitamura, A.T. Ohta, W.A. Shiroma, Two-octave tunable liquid-metal monopole antenna. Electron. Lett. **50**, 19–20 (2014)
176. D. Rodrigo, L. Jofre, B.A. Cetiner, Circular beam-steering reconfigurable antenna with liquid metal parasitics. IEEE Trans. Antennas Propag. **60**, 1796–1802 (2012)
177. R.A. Liyakath, A. Takshi, G. Mumcu, Multilayer stretchable conductors on polymer substrates for conformal and reconfigurable antennas. IEEE Antennas Wirel. Propag. Lett. **12**, 603–606 (2013)
178. A. Pourghorban Saghati, J. Singh Batra, J. Kameoka, K. Entesari, A microfluidically-reconfigurable dual-band slot antenna with a frequency coverage ratio of 3:1. IEEE Antennas Wirel. Propag. Lett. 1–1 (2015)
179. Y. Damgaci, B.A. Cetiner, A frequency reconfigurable antenna based on digital microfluidics. Lab Chip **13**, 2883–2887 (2013)
180. A.J. King, J.F. Patrick, N.R. Sottos, S.R. White, G.H. Huff, J.T. Bernhard, Microfluidically switched frequency-reconfigurable slot antennas. IEEE Antennas Wirel. Propag. Lett. **12**, 828–831 (2013)
181. M. Wang, C. Trlica, M.R. Khan, M.D. Dickey, J.J. Adams, A reconfigurable liquid metal antenna driven by electrochemically controlled capillarity. J. Appl. Phys. **117**, 194901 (2015)
182. C. Koo, B.E. LeBlanc, M. Kelley, H.E. Fitzgerald, G.H. Huff, A. Han, Manipulating liquid metal droplets in microfluidic channels with minimized skin residues toward tunable RF applications. J. Microelectromechanical Syst. 1–1 (2015)
183. M. Li, B. Yu, N. Behdad, Liquid-tunable frequency selective surfaces. IEEE Microw. Wirel. Compon. Lett. **20**, 423–425 (2010)
184. T. Bhattacharjee, H. Jiang, N.A. Behdad, Fluidically-tunable, dual-band patch antenna with closely-spaced bands of operation. IEEE Antennas Wirel. Propag. Lett. 1–1 (2015)
185. A. Pourghorban Saghati, J.S. Batra, J. Kameoka, K. Entesari, A miniaturized microfluidically reconfigurable coplanar waveguide bandpass filter with maximum power handling of 10 watts. IEEE Trans. Microw. Theory Tech. 1–11 (2015)

186. B. Wu, M. Okoniewski, C. Hayden, A pneumatically controlled reconfigurable antenna with three states of polarization. IEEE Trans. Antennas Propag. **62**, 5474–5484 (2014)
187. M.R. Khan, G.J. Hayes, S. Zhang, M.D. Dickey, G. Lazzi, A Pressure responsive fluidic microstrip open stub resonator using a liquid metal alloy. IEEE Microw. Wirel. Compon. Lett. **22**, 577–579 (2012)
188. M.R. Khan, G.J. Hayes, J.-H. So, G. Lazzi, M.D. Dickey, A frequency shifting liquid metal antenna with pressure responsiveness. Appl. Phys. Lett. **99**, 013501–013503 (2011)
189. E.F. Borra, G. Tremblay, Y. Huot, J. Gauvin, Gallium liquid mirrors: basic technology, optical-shop tests, and observations. PASP **109**, 319–325 (1997)
190. J. Wang, S. Liu, Z.V. Vardeny, A. Nahata, Liquid metal-based plasmonics. Opt. Express **20**, 2346–2353 (2012)
191. J. Wang, S. Liu, A. Nahata, Reconfigurable plasmonic devices using liquid metals. Opt. Express **20**, 12119–12126 (2012)
192. K. Ling, K. Kim, S. Lim, Flexible liquid metal-filled metamaterial absorber on polydimethylsiloxane (PDMS). Opt. Express **23**, 21375 (2015)
193. T.S. Kasirga, Y.N. Ertas, M. Bayindir, Microfluidics for reconfigurable electromagnetic metamaterials. Appl. Phys. Lett. **95**, 214102–214102-3 (2009)
194. P. Liu, S. Yang, A. Jain, Q. Wang, H. Jiang, J. Song, T. Koschny, C.M. Soukoulis, L. Dong, Tunable meta-atom using liquid metal embedded in stretchable polymer. J. Appl. Phys. **118**, 014504 (2015)
195. Y. Deng, J. Liu, Liquid metal based stretchable radiation-shielding film. J. Med. Devices-Trans. ASME **9**, 014502 (2015)
196. J. Park, S. Wang, M. Li, C. Ahn, J.K. Hyun, D.S. Kim, D.K. Kim, J.A. Rogers, Y. Huang, S. Jeon, Three-dimensional nanonetworks for giant stretchability in dielectrics and conductors. Nat. Commun. **3**, 916 (2012)
197. K.P. Mineart, Y. Lin, S.C. Desai, A.S. Krishnan, R.J. Spontak, M.D. Dickey, Ultrastretchable, cyclable and recyclable 1- and 2-dimensional conductors based on physically cross-linked thermoplastic elastomer gels. Soft Matter **9**, 7695–7700 (2013)
198. D.B. Strukov, G.S. Snider, D.R. Stewart, R.S. Williams, The missing memristor found. Nature **453**, 80–83 (2008)
199. H.-J. Koo, J.-H. So, M.D. Dickey, O.D. Velev, Towards all-soft matter circuits: prototypes of quasi-liquid devices with memristor characteristics. Adv. Mater. **23**, 3559–3564 (2011)
200. J.-H. So, H.-J. Koo, M.D. Dickey, O.D. Velev, Ionic current rectification in soft-matter diodes with liquid-metal electrodes. Adv. Funct. Mater. **22**, 625–631 (2012)
201. S. Liu, X. Sun, O.J. Hildreth, K. Rykaczewski, Design and characterization of a single channel two-liquid capacitor and its application to hyperelastic strain sensing. Lab Chip **15**, 1376–1384 (2015)
202. A.C. Siegel, S.K.Y. Tang, C.A. Nijhuis, M. Hashimoto, S.T. Phillips, M.D. Dickey, G.M. Whitesides, Cofabrication: a strategy for building multicomponent microsystems. Acc. Chem. Res. **43**, 518–528 (2010)
203. S.-Y. Tang, K. Khoshmanesh, V. Sivan, P. Petersen, A.P. O'Mullane, D. Abbott, A. Mitchell, K. Kalantar-zadeh, Liquid metal enabled pump. Proc. Natl. Acad. Sci. **111**, 3304–3309 (2014)
204. M. Gao, L. Gui, A handy liquid metal based electroosmotic flow pump. Lab Chip **14**, 1866–1872 (2014)
205. M. Knoblauch, J.M. Hibberd, J.C. Gray, A.J. van Bel, A galinstan expansion femtosyringe for microinjection of eukaryotic organelles and prokaryotes. Nat. Biotechnol. **17**, 906–909 (1999)
206. M. Hodes, R. Zhang, L.S. Lam, R. Wilcoxon, N. Lower, On the potential of galinstan-based minichannel and minigap cooling. IEEE Trans. Compon. Packag. Manuf. Technol. **4**, 46–56 (2014)
207. J. Je, J. Lee, Design, fabrication, and characterization of liquid metal microheaters. J. Microelectromechanical Syst. **23**, 1156–1163 (2014)

208. T. Krupenkin, J.A. Taylor, Reverse electrowetting as a new approach to high-power energy harvesting. Nat. Commun. **2**, 448 (2011)
209. N. Pekas, Q. Zhang, D. Juncker, Electrostatic actuator with liquid metal-elastomer compliant electrodes used for on-chip microvalving. J. Micromechanics Microengineering **22**, 097001 (2012)
210. S.-Y. Tang, V. Sivan, P. Petersen, W. Zhang, P.D. Morrison, K. Kalantar-zadeh, A. Mitchell, K. Khoshmanesh, Liquid metal actuator for inducing chaotic advection. Adv. Funct. Mater. **24**, 5851–5858 (2014)
211. Y. Zhang, Z. Zhao, D. Fracasso, R.C. Chiechi, Bottom-up molecular tunneling junctions formed by self-assembly. Isr. J. Chem. **54**, 513–533 (2014)
212. K. Du, E. Glogowski, M.T. Tuominen, T. Emrick, T.P. Russell, A.D. Dinsmore, Self-assembly of gold nanoparticles on gallium droplets: controlling charge transport through microscopic devices. Langmuir **29**, 13640–13646 (2013)
213. E.A. Weiss, R.C. Chiechi, S.M. Geyer, V.J. Porter, D.C. Bell, M.G. Bawendi, G.M. Whitesides, Size-dependent charge collection in junctions containing single-size and multi-size arrays of colloidal CdSe quantum dots. J. Am. Chem. Soc. **130**, 74–82 (2008)
214. M.M. Yazdanpanah, S. Chakraborty, S.A. Harfenist, R.W. Cohn, B.W. Alphenaar, Formation of highly transmissive liquid metal contacts to carbon nanotubes. Appl. Phys. Lett. **85**, 3564–3566 (2004)
215. ADu Pasquier, S. Miller, M. Chhowalla, On the use of Ga-In eutectic and halogen light source for testing P3HT-PCBM organic solar cells. Sol. Energy Mater. Sol. Cells **90**, 1828–1839 (2006)
216. F. Ongul, S.A. Yuksel, S. Bozar, G. Cakmak, H.Y. Guney, D.A.M. Egbe, S. Gunes, Vacuum-free processed bulk heterojunction solar cells with E-GaIn cathode as an alternative to Al electrode. J. Phys. Appl. Phys. **48**, 175102 (2015)
217. D.J. Lipomi, B.C.-K. Tee, M. Vosgueritchian, Z. Bao, Stretchable organic solar cells. Adv. Mater. **23**, 1771–1775 (2011)
218. Q. Wang, X. Niu, Q. Pei, M.D. Dickey, X. Zhao, Electromechanical instabilities of thermoplastics: theory and in situ observation. Appl. Phys. Lett. **101**, 141911 (2012)
219. Y. Liu., M. Gao, S. Mei, Y. Han, J. Liu, Ultra-compliant liquid metal electrodes with in-plane self-healing capability for dielectric elastomer actuators. Appl. Phys. Lett. **103**, 064101–064101-4 (2013)
220. C. Jin, J. Zhang, X. Li, X. Yang, J. Li, J. Liu, Injectable 3-D fabrication of medical electronics at the target biological tissues. Sci. Rep. **3**, (2013)
221. Y. Yu, J. Zhang, J. Liu, Biomedical Implementation of liquid metal ink as drawable ECG electrode and skin circuit. PLoS ONE **8**, e58771 (2013)
222. H.J. Meiselman, G.R. Cokelet, Fabrication of hollow vascular replicas using a gallium injection technique. Microvasc. Res. **9**, 182–189 (1975)
223. M. Bradley, A.H. Sacks, A technique for casting and fabricating hollow slide models of the microcirculation. Microvasc. Res. **22**, 210–218 (1981)
224. D. Kim, P. Thissen, G. Viner, D.-W. Lee, W. Choi, Y.J. Chabal, J.-B. Lee, Recovery of nonwetting characteristics by surface modification of gallium-based liquid metal droplets using hydrochloric acid vapor. ACS Appl. Mater. Interfaces **5**, 179–185 (2013)
225. D. Kim, R.G. Pierce, R. Henderson, S.J. Doo, K. Yoo, J.-B. Lee, Liquid metal actuation-based reversible frequency tunable monopole antenna. Appl. Phys. Lett. **105**, 234104 (2014)
226. G. Li, M. Parmar, D. Kim, J.-B. Lee, D.-W. Lee, PDMS based coplanar microfluidic channels for the surface reduction of oxidized Galinstan. Lab. Chip **14**, 200 (2014)
227. B. Cumby, J. Heikenfeld, D. Mast, C. Tabor, M. Dickey, Robust pressure-actuated liquid metal devices showing reconfigurable electromagnetic effects at GHz frequencies, in *2014 IEEE Antennas and Propagation Society International Symposium APSURSI* 553–554 (2014)

228. B.L. Cumby, G.J. Hayes, M.D. Dickey, R.S. Justice, C.E. Tabor, J.C. Heikenfeld, Reconfigurable liquid metal circuits by Laplace pressure shaping. Appl. Phys. Lett. **101**, 174102 (2012)
229. G. Li, M. Parmar, D.-W. Lee, An oxidized liquid metal-based microfluidic platform for tunable electronic device applications. Lab Chip **15**, 766–775 (2015)
230. D. Kim, D. Jung, J.H. Yoo, Y. Lee, W. Choi, G.S. Lee, K. Yoo, J.-B. Lee, Stretchable and bendable carbon nanotube on PDMS super-lyophobic sheet for liquid metal manipulation. J. Micromechanics Microengineering **24**, 055018 (2014)
231. G. Beni, S. Hackwood, J.L. Jackel, Continuous electrowetting effect. Appl. Phys. Lett. **40**, 912–914 (1982)
232. R.C. Gough, A.M. Morishita, J.H. Dang, W. Hu, W.A. Shiroma, A.T. Ohta, Continuous electrowetting of non-toxic liquid metal for RF applications. IEEE Access **2**, 874–882 (2014)
233. H.J. Zeng, A.D. Feinerman, Z.L. Wan, P.R. Patel, Piston-motion micromirror based on electrowetting of liquid metals. J. Microelectromechanical Syst. **14**, 285–294 (2005)
234. Z. Wan, H. Zeng, A. Feinerman, Area-tunable micromirror based on electrowetting actuation of liquid-metal droplets. Appl. Phys. Lett. **89**, 201107–201107-3 (2006)
235. S.-Y. Tang, Y. Lin, I.D. Joshipura, K. Khoshmanesh, M.D. Dickey, Steering liquid metal flow in microchannels using low voltages. Lab Chip **15**, 3905–3911 (2015)
236. J.T.H. Tsai, C.-M. Ho, F.-C. Wang, C.-T. Liang, Ultrahigh contrast light valve driven by electrocapillarity of liquid gallium. Appl. Phys. Lett. **95**, 251110 (2009)
237. S.-Y. Tang, V. Sivan, K. Khoshmanesh, A.P. O'Mullane, X. Tang, B. Gol, N. Eshtiaghi, F. Lieder, P. Petersen, A. Mitchell, K. Kalantar-zadeh, Electrochemically induced actuation of liquid metal marbles. Nanoscale **5**, 5949–5957 (2013)
238. X. Tang, S.-Y. Tang, V. Sivan, W. Zhang, A. Mitchell, K. Kalantar-zadeh, K. Khoshmanesh, Photochemically induced motion of liquid metal marbles. Appl. Phys. Lett. **103**, 174104 (2013)
239. L. Latorre, J. Kim, J. Lee, P.P. de Guzman, H.J. Lee, P. Nouet, C.J. Kim, Electrostatic actuation of microscale liquid-metal droplets. J. Microelectromechanical Syst. **11**, 302–308 (2002)
240. D. Kim, J.-B. Lee, Magnetic-field-induced liquid metal droplet manipulation. J. Korean Phys. Soc. **66**, 282–286 (2015)
241. R.K. Kramer, C. Majidi, R.J. Wood, Wearable tactile keypad with stretchable artificial skin. in *2011 IEEE International Conference on Robotics and Automation ICRA*, (2011), pp. 1103–1107
242. F.L. Hammond, R.K. Kramer, Q. Wan, R.D. Howe, R.J. Wood, Soft tactile sensor arrays for force feedback in micromanipulation. IEEE Sens. J. **14**, 1443–1452 (2014)
243. R.K. Kramer, C. Majidi, R. Sahai, R.J. Wood, Soft curvature sensors for joint angle proprioception. in *2011 IEEERSJ International Conference on Intelligent Robots and Systems (IROS)* (2011), pp. 1919–1926
244. Y.-L. Park, B.-R. Chen, R.J. Wood, Design and fabrication of soft artificial skin using embedded microchannels and liquid conductors. IEEE Sens. J. **12**, 2711–2718 (2012)
245. R. Matsuzaki, K. Tabayashi, Highly stretchable, global, and distributed local strain sensing line using gainsn electrodes for wearable electronics. Adv. Funct. Mater. **25**, 3806–3813 (2015)
246. K. Noda, E. Iwase, K. Matsumoto, I. Shimoyama, Stretchable liquid tactile sensor for robot-joints, in *2010 IEEE International Conference on Robotics and Automation (ICRA)* (2010), pp. 4212–4217
247. J.T.B. Overvelde, Y. Mengüç, P. Polygerinos, Y. Wang, Z. Wang, C.J. Walsh, R.J. Wood, K. Bertoldi, Mechanical and electrical numerical analysis of soft liquid-embedded deformation sensors analysis. Extreme Mech. Lett. **1**, 42–46 (2014)
248. J. Park, I. You, S. Shin, U. Jeong, Material approaches to stretchable strain sensors. ChemPhysChem **16**, 1155–1163 (2015)
249. R.D.P. Wong, J.D. Posner, V.J. Santos, Flexible microfluidic normal force sensor skin for tactile feedback. Sens. Actuators Phys. **179**, 62–69 (2012)

250. M.G. Mohammed, M.D. Dickey, Strain-controlled diffraction of light from stretchable liquid metal micro-components. Sens. Actuators Phys. **193**, 246–250 (2013)
251. H. Ota, K. Chen, Y. Lin, D. Kiriya, H. Shiraki, Z. Yu, T.-J. Ha, A. Javey, Highly deformable liquid-state heterojunction sensors. Nat. Commun. **5**, 5032 (2014)
252. Y. Gao, Y. Bando, Nanotechnology: carbon nanothermometer containing gallium. Nature **415**, 599 (2002)
253. H. Li, Y. Yang, J. Liu, Printable tiny thermocouple by liquid metal gallium and its matching metal. Appl. Phys. Lett. **101**, 073511 (2012)
254. Y.-L. Park, C. Majidi, R. Kramer, P. Bérard, R.J. Wood, Hyperelastic pressure sensing with a liquid-embedded elastomer. J. Micromechanics Microengineering **20**, 125029 (2010)
255. S.J. Benight, C. Wang, J.B.H. Tok, Z. Bao, Stretchable and self-healing polymers and devices for electronic skin. Prog. Polym. Sci. **38**, 1961–1977 (2013)
256. B.J. Blaiszik, S.L.B. Kramer, M.E. Grady, D.A. McIlroy, J.S. Moore, N.R. Sottos, S.R. White, Autonomic restoration of electrical conductivity. Adv. Mater. **24**, 398–401 (2012)
257. P. Cordier, F. Tournilhac, C. Soulie-Ziakovic, L. Leibler, Self-healing and thermoreversible rubber from supramolecular assembly. Nature **451**, 977–980 (2008)
258. E. Palleau, S. Reece, S.C. Desai, M.E. Smith, M.D. Dickey, Self-healing stretchable wires for reconfigurable circuit wiring and 3d microfluidics. Adv. Mater. **25**, 1589–1592 (2013)
259. R.D. Deshpande, J. Li, Y.-T. Cheng, M.W. Verbrugge, Liquid metal alloys as self-healing negative electrodes for lithium ion batteries. J. Electrochem. Soc. **158**, A845–A849 (2011)
260. D.J. Lipomi, Z. Bao, Stretchable, elastic materials and devices for solar energy conversion. Energy Environ. Sci. **4**, 3314–3328 (2011)
261. V.J. King, Liquid alloy for making contacts to metallic and nonmetallic surfaces. Rev. Sci. Instrum. **32**, 1407 (1961)
262. E. Glickman, M. Levenshtein, L. Budic, N. Eliaz, Interaction of liquid and solid gallium with thin silver films: synchronized spreading and penetration. Acta Mater. **59**, 914–926 (2011)
263. E. Pereiro-Lopez, W. Ludwig, D. Bellet, Discontinuous penetration of liquid Ga into grain boundaries of Al polycrystals. Acta Mater. **52**, 321–332 (2004)
264. D. Prasai, J.C. Tuberquia, R.R. Harl, G.K. Jennings, K.I. Bolotin, Graphene: corrosion-inhibiting coating. ACS Nano **6**, 1102–1108 (2012)
265. P. Ahlberg, S.H. Jeong, M. Jiao, Z. Wu, U. Jansson, S.-L. Zhang, Z.-B. Zhang, Graphene as a diffusion barrier in galinstan-solid metal contacts. IEEE Trans. Electron Devices **61**, 2996–3000 (2014)
266. J.V. Naidich, J.N. Chuvashov, Wettability and contact interaction of gallium-containing melts with non-metallic solids. J. Mater. Sci. **18**, 2071–2080 (1983)
267. K. Doudrick, S. Liu, E.M. Mutunga, K.L. Klein, V. Damle, K.K. Varanasi, K. Rykaczewski, Different shades of oxide: from nanoscale wetting mechanisms to contact printing of gallium-based liquid metals. Langmuir **30**, 6867–6877 (2014)

Chapter 2
Stretchability, Conformability, and Low-Cost Manufacture of Epidermal Sensors

Nanshu Lu, Shixuan Yang and Liu Wang

Abstract Epidermal sensors and electronics represent a class of artificial devices whose thickness, mass density, and mechanical stiffness are well-matched with human epidermis. They can be applied as temporary transfer tattoos on the surface of any part of human body for physiological measurements, electrical or thermal stimulation, as well as wireless communications. Except for comfort and wearability, epidermal sensors can offer unprecedented signal quality even under severe skin deformation. This chapter tries to address two fundamental mechanics challenges for epidermal sensors: first, how to predict and improve the stretchability and compliance when epidermal devices are made out of intrinsically brittle and rigid inorganic electronic materials; and second, when laminating on human skin, how to predict and improve the conformability between epidermal devices and the microscopically rough skin surfaces. Since the ideal use of epidermal devices would be one-time, disposable patches, a low cost, high throughput manufacture process called the "cut-and-paste" method is introduced at the end of this chapter.

Keywords Epidermal sensors · Stretchable · Serpentine · Conformability · Low-cost manufacture

2.1 Introduction

Development of high-performance wearable vital sign monitors is the central focus for the emerging field of mobile health (mHealth), one of the highest anticipated markets based on the Internet of Things (IoT) concept. High-fidelity sensing holds

N. Lu (✉) · S. Yang · L. Wang
Department of Aerospace Engineering and Engineering Mechanics,
Center for Mechanics of Solids, Structures, and Materials,
University of Texas at Austin, Austin, USA
e-mail: nanshulu@utexas.edu

N. Lu
Department of Biomedical Engineering, University of Texas at Austin, Austin, USA

© Springer International Publishing Switzerland 2016
J.A. Rogers et al. (eds.), *Stretchable Bioelectronics for Medical Devices and Systems*, Microsystems and Nanosystems,
DOI 10.1007/978-3-319-28694-5_2

31

the key to the success of wearable health monitors, which essentially demands both high signal-to-noise ratio and low susceptibility to motion artifacts. While wafer based sensors, electronics, optics, and conventional power sources exhibit high performance and superior chemical stability, their hard, planar, and rigid form factors are not compatible with the soft, curvilinear, and dynamic human body [1–3]. As a result, state-of-the-art wearable devices such as the wrist or arm bands and chest straps are not capable of intimate-skin integration and hence cannot provide medical-grade data.

Epidermal sensors and electronics represent a class of artificial devices that is fundamentally different from current wearable devices in terms of the form factor, wearability, and signal quality [4]. Through mechanics optimization [5–7] and innovation in materials processing [2, 8], stiff and brittle inorganic electronic materials can now be made soft and stretchable, which enabled tattoo-like patches with the thickness, mass density, and mechanical stiffness well matched with human epidermis, as demonstrated in Fig. 2.1a [4]. The key is to pattern intrinsi- cally stiff and blanket high-performance nanomembranes into structurally soft and stretchable filamentary serpentines, and transfer print them from rigid wafers to ultrathin (e.g., 30 μm thick), skin-like elastomeric substrates, as shown in Fig. 2.1b left frame [4]. Although the device is made of intrinsically stiff gold (Au) and silicon (Si) nanomembranes with modulus in the 100 GPa regime, the overall mechanical stiffness of this epidermal sensor is measured to be 150 kPa, which is literally as soft as epidermis (160 kPa) [4]. Since serpentine is the enabling geometry for skin-soft sensors and electronics, the fundamental mechanics that predicts and optimizes the stretchability and compliance of serpentine is going to be the first topic of this chapter.

Due to the ultimate thinness and softness of epidermal sensors, laminating them on microscopically rough skin surface leads to fully conformal contact, as evi- denced in Fig. 2.1b right frame [9]. Such conformal contact enhances contact area, which results in reduced skin-electrode interface impedance and improved sensor-skin adhesion. As a result, when skin deforms, even by van der Waals adhesion alone, the epidermal sensor is able to follow all kinds of skin distortion without any delamination, sliding, or imposing any mechanical constraint to the skin, essentially behaving like a secondary skin. Hence the second topic of this chapter will reveal the key parameters that govern the conformability between the device patch and the corrugated skin surface.

The suppressed interfacial sliding leads to the most distinctive advantage of epidermal sensors—the immunity from motion artifacts. Motion-induced noise and signal degradation is a ubiquitous problem in conventional health monitors, such as the long term, dry, capacitive electrocardiogram (ECG) electrodes [10]. Figure 2.1c [11] shows a direct comparison of ECGs simultaneously measured by conventional capacitive flat electrodes (upper frame) and epidermal electrodes (lower frame) from the same human subject. When the subject moved his upper body, motion artifacts were clearly present in the upper frame (as indicated by the red arrow), whereas the

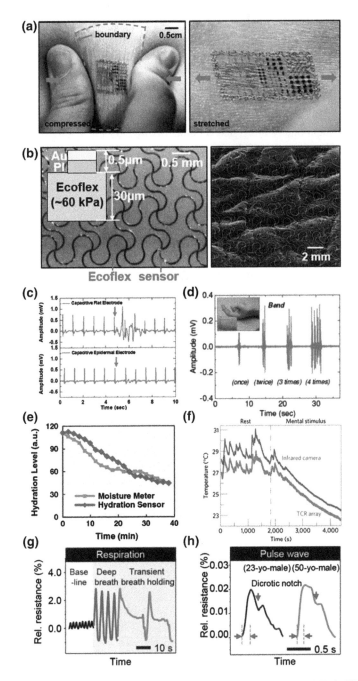

Fig. 2.1 Examples of epidermal sensors. **a** The first epidermal electronics [4]. **b** Filamentary serpentine design of gold electrodes showing extraordinary conformability on rough skin replica [9]. **c** Conventional dry ECG electrode is susceptible to motion artifacts (upper frame), whereas epidermal ECG sensor (bottom frame) is not affected at all [11]. **d** Epidermal EMG sensor measuring finger bend [12]. **e** Epidermal hydration sensor [13]. **f** Epidermal temperature detector [16]. **g** Epidermal strain gauge sensing respiration on the chest [17]. **h** Epidermal pulse wave sensor on the wrist [17]

signal acquired by the epidermal electrodes was not affected at all. For similar reasons, the electromyogram (EMG) measured from severe finger bend demonstrates extraordinary signal-to-noise ratio (Fig. 2.1d [12]). Not only electrophysiological signals, but also skin hydration measured by impedance sensors [13–15] (e.g., Fig. 2.1e [13]) and skin temperature by resistance temperature detectors (e.g., Fig. 2.1f [16]) are well compared to gold standards. However, not all motions should be excluded from epidermal sensors. For example, deformation sensitive soft strain gauges are intentionally integrated on epidermal patches to accurately detect respiration from the chest (Fig. 2.1g [17]), pulse wave (Fig. 2.1f [17]) and tremor [18] from the wrist, as well as hand gesture [19] because they are able to closely follow skin deformation. Other than examples displayed in Fig. 2.1, epidermal electronics have also inspired epidermal biochemical sensors [20, 21], skin stiffness sensors [22], wearable antenna for wireless near field communication (NFC) [15, 23, 24], and so on. In summary, epidermal sensors and electronics have been demonstrated as a platform technology with unlimited potential applications.

With so many exciting applications in mind, it is important not to forget that the thinness and softness of epidermal patches would lead to collapsing and crumpling after they are peeled off from human skin, making their ideal use to be disposable devices. As a result, the success of epidermal devices hinges on the realization of low cost, high throughput manufacture, which is going to be the last topic of this chapter.

2.2 Stretchability of Serpentines

Currently, the most popular strategy to build in-plane stretchable circuits and sensors out of intrinsically stiff materials is to pattern the stiff materials into meandering serpentine ribbons [25–28]. When thin serpentines ribbons are stretched, the ribbons can rotate in plane as well as buckle out of plane to accommodate the applied deformation, resulting in greatly reduced strains as well as much lower effective stiffness [26]. In addition to metal [25, 27–32] or graphene interconnects [33], electrophysiological or thermal sensing electrodes [4, 9, 12, 16], microheaters [18, 34], silicon-based solar cells and amplifiers [4], as well as zinc oxide-based nanogenerators [35] can all be patterned into serpentine shapes using conventional photolithography and etching methods [4, 28, 33]. Except for polymer-bonded serpentines, freestanding stretchable serpentine network has been applied in deployable sensor networks [36, 37] and expandable coronary stents [38]. In these two cases, the serpentine thickness is much larger than the ribbon width, and the expandability just comes from in-plane rigid body rotation of the serpentine arms. Although serpentines are so popular, their stretchability vary over a wide range and their designs lack rational guidance. According to the existing studies, the applied strain-to-rupture of metallic serpentine ribbons varies from 54 to 1600 % [25, 27–

29, 32, 36], depending on the geometric parameters such as ribbon width, arc radius, arm length substrate support, and so on. To obtain a fundamental understanding of their effects, and to optimize their layout under practical constraints, we have performed systematic theoretical, numerical, and experimental studies, as discussed in the follows.

2.2.1 Freestanding, Non-buckling Serpentines

It is possible to obtain analytical solutions to freestanding, non-buckling serpentine ribbons through curved beam (CB) theory and elasticity theory, as shown in Fig. 2.2 [39, 40]. Figure 2.2a depicts a unit cell cut out of a one-directional periodic serpentine ribbon whose geometry can be completely defined by four parameters: the ribbon width w, the arc radius R, the arc angle α, and the arm length l. The end-to-end distance of a unit cell is denoted by S, which is a function of the four basic parameters. When this unit cell is subjected to a tensile displacement u_0 at each end, the effective applied strain ε_{app} is defined as

$$\varepsilon_{app} = \frac{2u_0}{S}. \qquad (2.1)$$

Therefore a straight ribbon (i.e. $\alpha = -90°$) of length S should experience a uniform strain of ε_{app} everywhere inside the ribbon if the end effects are neglected. Taking advantage of symmetry, a unit cell can ultimately be represented by a quarter cell with fixed boundary at the axis of symmetry and a displacement of $u_0/2$ at the end, as shown in Fig. 2.2b. The reaction force is named P in Fig. 2.2b. Assuming linear elastic material and small deformation, through CB theory, the normalized stiffness and maximum strain in the serpentine can be obtained analytically as [39]:

$$\frac{P}{P'} = \frac{\frac{w}{R}\left(\cos\alpha - \frac{l}{2R}\sin\alpha\right)}{2\left[\begin{array}{c}\cos^2\alpha\left(\frac{l^3}{2R^3} + 3\left(\frac{\pi}{2}+\alpha\right)\frac{l^2}{R^2} + 12\frac{l}{R} - 12\left(\frac{\pi}{2}+\alpha\right)\right) \\ + \sin 2\alpha\left(6\left(\frac{\pi}{2}+\alpha\right)\frac{l}{R}+9\right) \\ + \frac{w^2}{R}\left[\left(\frac{\pi}{2}+\alpha\right)\left(\frac{l}{2R}\cos\alpha + \sin\alpha\right)^2 + \frac{l}{2R}\left(\sin\alpha + \frac{3\overline{E}}{2G}\cos\alpha\right)\right] + 18\left(\frac{\pi}{2}+\alpha\right)\end{array}\right]}, \qquad (2.2)$$

where $P' = 2\overline{E}wu_0/S$ represents the reaction force needed for the linear counterpart of the serpentine to elongate by $2u_0$, and

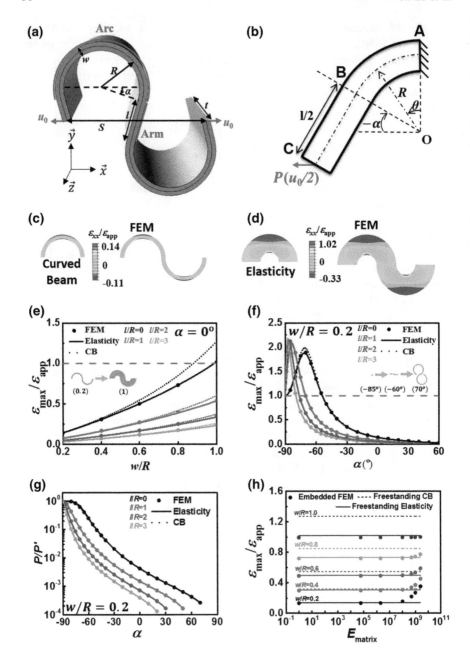

◀ **Fig. 2.2** 2D plane strain model of non-buckling freestanding and polymer-embedded serpentines [39, 40]. **a** Geometric parameters and boundary conditions of the unit cell of a freestanding serpentine ribbon. **b** Simplified model due to symmetry. **c** Comparison between curved beam (CB) theory and FEM results of strain distribution in a narrow serpentine. **d** Comparison between elasticity theory and FEM results of strain distribution in a wide serpentine. **e** Normalized maximum strain in the serpentine is plotted as a function of normalized ribbon width for serpentines with various arm lengths, which suggests smaller width and longer arms yield lower strains. **f** Normalized maximum strain in the serpentine is plotted as a function of arc angle for various arm lengths, which suggests nonmonotonic arc angle effect and that some serpentines can develop higher strains than straight ribbons. **g** Serpentine stiffness normalized by its linear counterpart is plotted as a function of arc angle for various arm lengths, which indicates that there can be orders of magnitude reduction when patterning straight ribbons into serpentine shapes. **h** Normalized maximum strain in an embedded serpentine as a function of matrix modulus, which suggests that analytical solutions to freestanding serpentines is valid for serpentines embedded in soft polymer matrix (e.g., $E_{\text{matrix}} < 100$ MPa)

$$\frac{\varepsilon_{\text{max}}}{\varepsilon_{\text{app}}} = \frac{\frac{w}{R}\left[\frac{12}{2-\frac{w}{R}} + \left(\frac{12}{2-\frac{w}{R}} - \frac{w}{R}\right)\left(\sin\alpha + \frac{l}{2R}\cos\alpha\right)\right]\left(\cos\alpha - \frac{l}{2R}\sin\alpha\right)}{\begin{bmatrix}\cos^2\alpha\left(\frac{l^3}{2R^3} + 3\left(\frac{\pi}{2}+\alpha\right)\frac{l^2}{R^2} + 12\frac{l}{R} - 12\left(\frac{\pi}{2}+\alpha\right)\right) \\ + \sin 2\alpha\left(6\left(\frac{\pi}{2}+\alpha\right)\frac{l}{R} + 9\right) \\ + \frac{w^2}{R}\left[\left(\frac{\pi}{2}+\alpha\right)\left(\frac{l}{2R}\cos\alpha + \sin\alpha\right)^2 + \frac{l}{2R}\left(\sin\alpha + \frac{3\overline{E}}{2G}\cos\alpha\right)\right] + 18\left(\frac{\pi}{2}+\alpha\right)\end{bmatrix}}.$$

$$(2.3)$$

Such closed-form analytical results have found excellent agreement with finite element modeling (FEM) results as shown in Figs. 2.3c, e–g, as long as the serpentine width is small enough compared to its arc radius, e.g., $w/R < 0.5$. According to the contour plots of strain distribution in Fig. 2.3c, it is obvious that the maximum strain always occurs at the inner crest of the arc, which is exactly where fracture occurred in our experiments [39]. It is evident in Fig. 2.3e that serpentines with smaller width and longer arms exhibit lower maximum strains and hence better stretchability. Figure 2.3e also suggests that the CB solutions (dashed curves) match the FEM results (solid markers) well when $w/R < 0.5$. To obtain accurate solutions for wide serpentines, we recently developed the elasticity solution [40] whose contour plot is compared with FEM in Fig. 2.3d and strain and stiffness results are plotted as solid curves in Figs. 2.3e–h. Through the comparison, it is evident that elasticity solution can fully capture the FEM results for all ribbon widths. Unfortunately, unlike the CB solution, the elasticity solutions cannot be expressed as short equations. The effect of the arc angle α is illustrated by Fig. 2.3f. Two observations can be readily made: first, while the effects of w/R and l/R are monotonic, the effect of α is not always monotonic; second, when α is close to $-90°$, i.e., when the serpentine approaches a straight ribbon, the maximum strain in the serpentine may exceed the applied strain, meaning the serpentine is less stretchable compared with their straight counterpart. As a result, it is important for us to realize that not all serpentines can help reduce strains and the design of the serpentine needs to be rationalized. Fig. 2.3g plots the effective stiffness of the

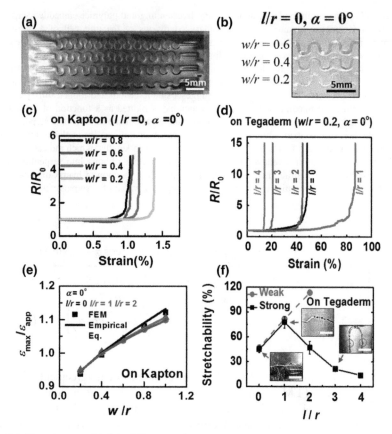

Fig. 2.3 Experiments of brittle ITO serpentines supported by stiff [50] and soft [51] polymer substrates. **a** Top view of ITO serpentines directly sputtered on a Kapton substrate through a stencil. **b** Top view of transparent ITO-PET serpentines transfer printed on a Tegaderm substrate. **c** Resistance of ITO serpentines of different widths measured when Kapton is pulled by uniaxial tensile strain, which indicates that ITO serpentines on stiff polymer like Kapton are not very stretchable. **d** Resistance of ITO serpentines of different arm lengths measured when Tegaderm is pulled by uniaxial tensile strain, which indicates much enhanced stretchability. **e** Normalized FEM maximum strain in Kapton-supported serpentine as a function of normalized serpentine width. **f** Measured stretchability of Tegaderm-supported serpentine as a function of serpentine arm length

serpentines normalized by that of a straight ribbon. It is impressive that the effective stiffness can be reduced by orders of magnitude by simply changing a straight ribbon into serpentine shapes—this is why metal or silicon-based serpentines can be made as soft as skins [4] and tissues [41].

When thick, non-buckling serpentines of stiff metal are embedded in a polymer matrix [25], we have performed FEM to reveal their ε_{max} when ε_{app} is applied to pull the polymer matrix. Figure 2.3h plots $\varepsilon_{max}/\varepsilon_{app}$ as a function of the matrix modulus E_{matrix} against the CB (dashed line) and elasticity (solid line) solutions for freestanding serpentines, which are flat because they are independent of E_{matrix}.

When E_{matrix} < 100 MPa, the FEM results fall right on the analytical solutions for freestanding serpentines, which indicates that soft polymer matrix has almost no effect on the deformation of non-buckling serpentine. Hence the strain in metal serpentines embedded in polydimethylsiloxane (PDMS) (e.g., 10:1 Sylgard 184 PDMS has a Young's modulus of 3 MPa [42]) can be well captured by the analytical solutions for freestanding serpentines when metal deformation is within the elastic regime.

When serpentine ribbons' thickness is much smaller compared to their width, as in most epidermal sensors, freestanding serpentines will buckle out of plane to avoid in-plane bending by developing out-of-plane bending and twisting, which stores much lower strain energy. Buckling and postbuckling theories and FEM have been developed to address this problem [26, 43, 44]. To enhance the areal coverage of functional serpentines without compromising the multidirectional stretchability, a concept of self-similar or fractal serpentines have been proposed, which has also greatly enhanced the topologies of serpentine designs [24, 32, 45, 46]. So far, the optimization theories for freestanding and interlaced serpentine networks are still lacking.

2.2.2 Substrate-Supported Serpentines

The mechanical behaviors of polymer-bonded serpentines are expected to be very different from the freestanding ones. A few experiments and FEM have been conducted to provide insights into the shape-dependent mechanical behaviors of polymer-supported metal-based serpentines [27–30, 36, 47–49]. Other than metallic serpentines, ceramic serpentines start to gain popularity as stretchable solar cells [4], amplifiers [4], and nanogenerators [35]. But so far there is little experimental mechanics investigation to reveal the stretchability of polymer-bonded brittle serpentine thin films due to the difficulty to fabricate and handle brittle serpentines on soft polymer substrates. We have used indium tin oxide (ITO) as a model brittle material to study the mechanics of polymer-bonded brittle serpentines [50, 51].

Thin ITO films have been a popular electrode material in flat panel displays [52] and solar cells [53] attributing to their combined high electrical conductivity and optical transparency. However, ITO is not mechanically favorable in flexible/stretchable electronics due to its brittle nature. Cracks were observed at applied tensile strains around 1 % in polymer-supported blanket thin ITO films [38, 54]. Resistance versus applied strain curves have been widely adopted to indicate the stretchability of conductive thin films such as metal [55–57] and ITO [38]. We have sputtered ITO serpentines on Kapton® (DuPont) substrates through stencils as shown in Fig. 2.3a [50] and "cut-and-pasted" ITO-PET (polyethylene terephthalate) bilayer serpentines on soft stretchable Tegaderm® (3M medical tape) substrates as displayed in Fig. 2.3b [51]. Through the measurement of electrical resistance as a function of tensile strain applied to the substrate (Fig. 2.3c for ITO on Kapton and Fig. 2.3d for ITO-PET on Tegaderm), the strain-to-rupture (a.k.a. stretchability) can be determined as the strain at which the resistance starts to blow up.

A series of FEMs are performed for systematically varied serpentine shapes when the substrate is Kapton (E_{Kapton} = 2.5 GPa). An empirical equation is fitted based on the FEM results as discussed in [50]. Figure 2.3e plots $\varepsilon_{max}/\varepsilon_{app}$ as a function of w/r with different l/r using both the FEM results and the empirical equation. It is evident that w/r always has a monotonic effect on $\varepsilon_{max}/\varepsilon_{app}$—smaller strains in narrower ribbons. Another important finding is that when w/r is beyond about 0.4, $\varepsilon_{max}/\varepsilon_{app}$ will go beyond 1, which means the stretchability of the serpentine will actually be lower than its straight counterpart, indicating a strain augmentation instead of strain reduction effect. Compared with the effect of w/r, the effects of arm length l/r and α (not shown) are not as significant, especially when w/r is small. The fundamental reason is that the in-plane rotation and out-of-plane buckling/twisting of the serpentine ribbons are fully constrained by the stiff polymer substrates. Hence we conclude that stiff serpentines directly bonded to stiff polymer substrates like Kapton or PET are in general not much more stretchable than their linear counterpart and many shapes actually can be even worse.

Brittle serpentines are expected to be more stretchable when bonded to a very soft substrate like the Tegaderm ($E_{Tegaderm}$ = 7 MPa [58]) because the constraint from the substrate is much reduced. However, directly sputtering thin ITO films on elastomeric substrates resulted in surface wrinkles [59] and from our own experience, ITO films easily cracked with even very gentle handling. The idea we came up with is to first sputter ITO on transparent PET sheet and then transfer ITO-coated PET serpentines on Tegaderm [51]. Depending on serpentine shape and adhesion to the Tegaderm substrate, ITO ribbons can remain conductive when stretched beyond 100 %, as shown in Fig. 2.3f. Generally speaking, weakly-bonded serpentines debond from the substrate before rupture, and are always more stretchable with narrower ribbon width, longer arms, larger arc radius, and larger arc angle. The stretchability of well-bonded serpentines, however, show non-monotonic dependence on those geometric parameters due to stronger substrate constraints. It is found that serpentines with long arms rupture by transverse buckling due to Poisson's contraction in Tegaderm, which degrades stretchability when the transverse dimension is large.

2.3 Conformability of Epidermal Sensors on Rough Skin Surfaces

Noninvasive but intimate contact with epidermis is required for superior signal-to-noise ratio and suppressed motion artifacts in epidermal electrophysiological (EP) sensors, hydration sensors, and mechanical sensors [4, 9, 12, 13, 22]. Moreover, conformability-enabled efficient heat and mass transfer between device and skin has led to very precise skin temperature detectors [16], low power wearable heaters [60, 61], effective sweat monitors [20, 62], and on demand drug delivery patches [18]. A comprehensive mechanistic understanding on the conformability of thin device sheets on rough epidermis can offer important insights into the design of the mechanical properties of epidermal devices.

Conformability between epidermal electronics and skin has been analytically studied through the energy minimization method [63, 64], which can successfully predict non-conformed and fully conformed scenarios. Jeong et al. [12] carried out a series of experiments to study the conformability of elastomer membranes (Ecoflex, Smooth-On, USA) on an Ecoflex skin replica. Membrane-substrate conformability is clearly revealed by the cross-sectional scanning electron microscopy (SEM) images as displayed in Fig. 2.4a [12]: 5 μm thick patch achieved full conformability to the substrate, 36 μm thick one only partially conformed to the substrate, whereas patches with thickness of 100 μm and 500 μm remained non-conformed at all. For illustration purpose, 2D schematics for fully conformed (FC), partially conformed (PC), and non-conformed (NC) scenarios are offered in Fig. 2.4b–d, respectively. We therefore developed an extended theory that can

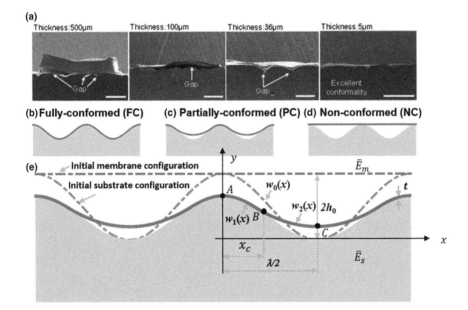

Fig. 2.4 Illustration of epidermal device conforming to rough skin surface. **a** Cross-sectional SEM pictures showing 5 μm Ecoflex is fully conformed to Ecoflex-based skin replica, 36 μm one achieved partial conformability, whereas 100 μm and 500 μm ones did not conform at all [12]. **b–d** offer schematics of fully conformed (FC), partially conformed (PC), and non-conformed (NC) scenarios. **e** Schematic of a generic partially conformed membrane with geometric parameters and characteristic points labeled: the initial amplitude and wavelength of the substrate is $2h_0$ and λ, respectively; after membrane lamination, the substrate surface within the contact zone deforms to a new sinusoidal shape with amplitude $2h_1$ (not labeled in the figure) and unchanged wavelength; x_c is the horizontal projection of the contact zone; Point B denotes the delaminating point

predict all three scenarios by successfully finding the substrate elastic energy under partially conformed contact [65].

A 2D schematic for a generic partially conformed configuration is given in Fig. 2.4e. For simplicity, the membrane is modeled as a uniform linear elastic membrane with plane strain modulus $\overline{E}_m$ and thickness t. The skin is assumed to be a pre-corrugated linear elastic half space with plane strain modulus $\overline{E}_s$. Within the Cartesian coordinate system xy defined in Fig. 2.4e, the surface profile of the undeformed skin as outlined by the dashed peachpuff curve is simply characterized by a sinusoidal function $w_0(x)$ with amplitude $2h_0$ and wavelength λ. When the elastic membrane is laminated on the skin substrate and starts to conform to the substrate due to interface adhesion, a contact zone with horizontal projection denoted as x_c is labeled in Fig. 2.4e. Therefore $x_c = \lambda/2$ represents FC scenario (Fig. 2.4b), $0 < x_c < \lambda/2$ PC scenario (Fig. 2.4c), and $x_c = 0$ NC scenario (Fig. 2.4d). Due to membrane-substrate interaction, the skin substrate deforms. Here, we simply postulate that the surface profile of the skin within the contact zone deforms from the initial sinusoidal shape $w_0(x)$ to a new sinusoidal shape $w_1(x)$ (within $0 \leq x \leq x_c$) with the same wavelength but a different amplitude, $2h_1$. This assumption holds all the way till $x_c = \lambda/2$, which means the overall skin surface deforms from one sinusoidal profile to another with the same wavelength but different amplitude. The profile of a partially conformed membrane, $w_2(x)$ as illustrated by the magenta curve in Fig. 2.4e is sectional: from A to B, i.e., when $0 \leq x \leq x_c$, the membrane fully conforms to the substrate and thus $w_2(x) = w_1(x)$; from B to C, i.e., when $x_c \leq x \leq \lambda/2$, the membrane is suspended and $w_2(x)$ is assumed to be a hyperbolic shape which will decay to a parabolic shape when normal strain in the membrane is small, i.e., a pure bending condition is assumed [66, 67].

To solve for x_c and h_1, energy minimization method is adopted. The total energy per period of the system, U_{total}, consists of the following four energies:

$$U_{\text{total}} = U_{\text{bending}} + U_{\text{membrane}} + U_{\text{adhesion}} + U_{\text{substrate}}, \qquad (2.4)$$

where U_{bending} is the bending energy of the membrane, U_{membrane} is the membrane energy associated with tensile strain in the membrane, U_{adhesion} is the interface adhesion energy between the membrane and the substrate which is proportional to the interface work of adhesion, γ, and $U_{\text{substrate}}$ is the elastic energy stored in the substrate, which equals to the work done to the substrate by the membrane.

Detailed derivations of the four energies yielded four dimensionless governing parameters for this problem [65]: $\beta = 2\pi h_0/\lambda, \eta = t/\lambda, \alpha = \overline{E}_m/\overline{E}_s$, and $\mu = \gamma/(\overline{E}_s \lambda)$, which are physically interpreted as normalized roughness of the corrugated skin surface (β), normalized membrane thickness (η), membrane-to-substrate modulus ratio (α), and normalized membrane-substrate interface adhesion (μ), respectively. In addition, there are two unknown dimensionless parameters: $\hat{x}_c = 2x_c/\lambda$ and $\xi = h_1/h_0$, which once solved can yield the contact zone size and the amplitude of the deformed substrate. By fixing β, α, μ, and η, minimization of

Eq. (2.4) with respect to $\widehat{x}_c$ and ξ within the domain confined by $0 \le \widehat{x}_c \le 1$ and $0 \le \xi \le 1$ will give us the equilibrium solution, which can be visualized as the global minimum of a 3D plot of the normalized total energy landscape as a function of $\widehat{x}_c$ and ξ.

Our analytical solutions can be directly compared with Jeong's conformability experiment [12]. Basic parameters that can be extracted from the experiment are: substrate roughness $h_0 = 50\,\mu m$, $\lambda = 250\,\mu m$, plane strain moduli of Ecoflex membrane and substrate $\overline{E}_s = \overline{E}_m = 92\,KPa$ [4]. The interface work of adhesion is taken as $\gamma = 50\,mJ/m^2$ according to our recent experimental measurements of elastomer adhesion [42]. Based on those given parameters, the four dimensionless parameters are computed as: $\beta = 1.2$, $\alpha = 1$, $\mu = 0.003$, and $\eta = 0.02, 0.144, 0.04$, and 2, which corresponds to the four different experimental thicknesses of the membrane $t = 5, 36, 100, 500\,\mu m$, respectively. By fixing the substrate morphology $\beta = 1.2$ (i.e., $h_0 = 50\,\mu m$, $\lambda = 250\,\mu m$), Fig. 2.5 predicts the conformability as a function of the other three parameters α, μ and η. The 3D plot in Fig. 2.5a shows two critical surfaces dividing FC/PC and PC/NC. The conclusion is consistent with

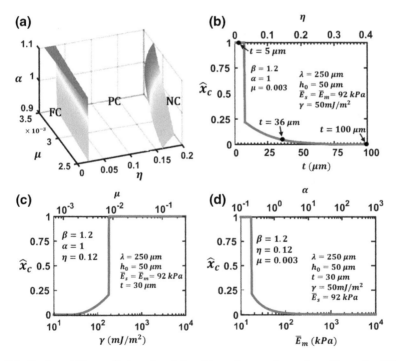

Fig. 2.5 Conformability prediction when fixing skin surface roughness: $h_0 = 50\,\mu m$, $\lambda = 250\,\mu m$ [65]. **a** Surfaces dividing FC/PC and PC/NC. **b** Contact area $\widehat{x}_c$ versus t (or η) when $\beta = 1.2, \alpha = 1, \mu = 0.003$, which agrees well with Jeong's experiments [12] (labeled as black dots). **c** Contact area $\widehat{x}_c$ versus γ (or μ) when $\beta = 1.2, \alpha = 1, \eta = 0.12$. **d** Contact area $\widehat{x}_c$ versus $\overline{E}_m$ (or α) when $\beta = 1.2, \mu = 0.003, \eta = 0.12$

our intuition that the FC condition can be achieved at small η, i.e., thin membrane, small α, i.e., compliant membrane, and large μ, i.e., strong membrane-substrate adhesion. On the contrary, NC condition occurs at large α, large η, and small μ.

To better illustrate the effect of individual variables, we chose to fix three variables and only change one at a time. For example, in Fig. 2.5b, $\hat{x}_c$ is plotted as a function of t in the bottom axis and η in the top axis with $\beta = 1.2, \alpha = 1, \mu = 0.003$ fixed. It is evident that as the film thickness grows from 0, the conformability goes from FC to PC and finally NC. While the transition from PC to NC is smooth, the transition from FC to PC is abrupt, which suggests a significant drop (>77 %) of contact area from FC to PC. Similar jumps have been observed for few layer graphene (FLG) conforming to silicon substrate [68] and elastic membrane laminated on rigid, corrugated substrate [66]. More analysis on how different substrate morphologies affect snap-through transition of conformability can be found in [67]. Quantitatively, full conformability requires $\eta < 0.03$, i.e., $t < 7.5\,\mu m$. When $\eta > 0.28$, i.e., $t > 70\,\mu m$, there is no conformability at all. When $0.03 < \eta < 0.28$, i.e., when $7.5\,\mu m < t < 70\,\mu m$, the contact area of the PC scenario can be determined. The three black dots in Fig. 2.5b indicate the three different membrane thicknesses tested in the Jeong's experiment [12], which are fully consistent with our prediction.

Since the original epidermal electronics was fabricated on 30 μm thick Ecoflex [4], the conformability of a 30 μm thick Ecoflex on an Ecoflex skin replica has been predicted. In order to show the effect of adhesion energy and membrane modulus over wide ranges, $\hat{x}_c$ versus γ (or μ) and $\hat{x}_c$ versus $\overline{E}_m$ (or α) are plotted with $\log x$ scale in Figs. 2.5c, d, respectively, with the rest three variables fixed. In Fig. 2.5c, it is evident that when $\mu > 0.008$, i.e., $\gamma > 138\,mJ/m^2$, FC mode can be achieved but when $\mu < 0.0016$, i.e., $\gamma < 30\,mJ/m^2$, the membrane would not conform to the substrate at all. Figure 2.5d indicates that when $\alpha < 0.2$, i.e., $\overline{E}_m < 10\,kPa$, FC is guaranteed but when $\alpha > 2.5$, i.e., $\overline{E}_m > 125\,kPa$, there is no conformability. It is also noted that the abrupt transition from FC to PC is also present in Fig. 2.5c, d, with the same maximum contact area (23 % of total surface area) under PC. In summary, Fig. 2.5 offers a quantitative guideline toward conformable skin-mounted electronics in a four-dimensional design space.

2.4 "Cut-and-Paste" Manufacture of Epidermal Sensors

Disposable and widely accessible epidermal sensors rely on the realization of low cost, high throughput manufacture. Current manufacture leverages standard microelectronics fabrication processes including vacuum deposition of films, spin coating, photolithography, wet and dry etching, as well as transfer-printing [4, 23, 69]. Although it has been proved effective, it requires access to cleanroom facilities and the time and labor costs are burdensome. Plus the overall size of the device is limited to the scale of the rigid handling wafer.

Our newly invented "cut-and-paste" method offers a very simple and immediate solution to these limitations [58]. Instead of vacuum deposition, thin metal-on-polymer laminates of various thicknesses can be directly purchased from industrial manufacturers. Instead of using photolithography patterning, a benchtop electronic cutter plotter is used to mechanically carve out the designed patterns, with excess being removed, which is a freeform, subtractive manufacturing process, inverse to the popular freeform, additive manufacturing technology [70]. The cutter plotter can pattern on thin sheet metals and polymers up to 12 inches wide and several feet long, largely exceeding lab-scale wafer sizes. Since the patterns can be carved with the support of thermal release tapes (TRT), the patterned films can be directly printed onto a variety of tattoo adhesives and medical tapes with almost 100 % yield. The whole process can be completed on an ordinary bench without any wet process within 10 min, which allows rapid prototyping. Equipment used in this process only includes a desktop cutter plotter and a hot plate for TRT heating, which enables portable manufacture. Since no rigid handle wafer is needed throughout the process, the "cut-and-paste" method is intrinsically compatible with roll-to-roll manufacture. The only drawback of this method is the limited patterning resolution of state-of-the-art cutter plotters, which is universally around 200 μm.

The detailed "cut-and-paste" process is illustrated in Fig. 2.6 [58]. Since thin metal films supported by stiff polymers such as Kapton and PET are more stretchable than freestanding metal foils [55], we always use metal-PET laminates

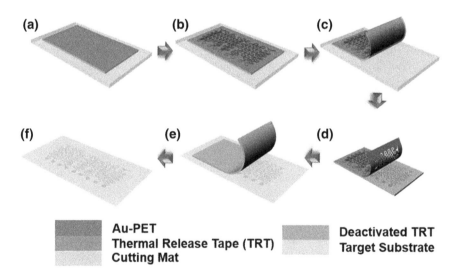

Fig. 2.6 Schematics for the cost and time effective "cut-and-paste" process [58]. **a** Au-PET-TRT (APT) laminated on the cutting mat with PET being the topmost layer. **b** Carving designed seams in the Au-PET layer by an automated cutter plotter. **c** Peeling APT off from the cutting mat. **d** Removing excessive Au-PET flakes after deactivating the TRT on hot plate. **e** Printing patterned Au-PET onto target substrate. **f** Resulted epidermal sensor systems with Au being the topmost layer. Steps (**a**)–(**e**) can be repeated for other materials

as the starting materials, which can be readily purchased from industrial suppliers such as Sheldahl (Northfield, MN) and Neptco (Pawtucket, RI). To manufacture gold (Au)-based stretchable electrophysiological (EP) electrodes, resistance temperature detectors (RTD), and impedance sensors, 100 nm-Au-on-13 μm-PET sheet was uniformly bonded to a flexible, single-sided TRT (Semiconductor Equipment Corp., USA) with Au side touching the adhesive of the TRT. The other side of the TRT was then adhered to a tacky flexible cutting mat, as shown in Fig. 2.6a. The cutting mat was fed into an electronic cutter plotter (Silhouette Cameo, USA) with the PET side facing the cutting blade. By importing our AutoCAD design into the Silhouette Studio software, the cutter plotter can automatically carve the Au-PET sheet with designed seams within minutes (Fig. 2.6b). Once seams were formed, the TRT was gently peeled off from the cutting mat (Fig. 2.6c) and slightly baked on a 115 °C hotplate for 1–2 min. Heat deactivated the adhesives on the TRT such that the excessive flakes can be easily peeled off by tweezers (Fig. 2.6d), leaving only the desired device patterns loosely resting on the TRT. The patterned devices were finally printed on a target substrate with native adhesives, which could be a temporary tattoo paper (Silhouette) or a medical tape, such as the 3M Tegaderm transparent dressing or the 3M kind removal silicone tape (KRST) (Fig. 2.6e), yielding a Au-based epidermal sensing patch (Fig. 2.6f). Steps illustrated by Figs. 2.6a–e can be repeated for other thin sheets of metals and polymers, which can be printed on the same target substrate with alignment markers, rendering a multimaterial, multiparametric epidermal sensing patch ready for skin mounting.

To demonstrate the "cut-and-paste" method, multimaterial epidermal sensor systems are fabricated and applied to measure EP signals such as ECG (Fig. 2.7a), EMG (Fig. 2.7b), and electroencephalogram (EEG) (Fig. 2.7c), skin temperature (Fig. 2.7d), skin hydration (Fig. 2.7e), and respiratory rate (Fig. 2.7f). Those measurements are compared favorably to state-of-the-art gold standards. Except that the soft strain gauges for respiratory rate sensing are made out of electrically conductive rubber (ECR)(Elastosil LR 3162, Wacker Silicones), all other sensors are Au-based passive sensors. In addition to sensors, a planar stretchable coil of 9 μm-Al-13 μm-PET ribbons exploiting the double-stranded serpentine design is also integrated on the same patch as a low frequency, wireless strain gauge, which can also serve as NFC antenna in the future.

Note that the above mentioned "cut-and-paste" method has proved effective in patterning metal-on-polymer laminates and elastomeric sheets, but it is not directly applicable to ceramic-coated polymer as indentation of the cutting blade would easily shatter intrinsically brittle ceramic film. However, the stretchable ITO-PET serpentines on Tegaderm shown in Fig. 2.3b were fabricated through a variation of the "cut-and-paste" method [51]. Instead of cutting blanket ITO-PET sheets, we first cut on a bare PET sheet, deposited thin ITO film on the cut PET (ITO deposited over the seam trenches are discontinuous), and then removed the excessive. In this way, no cutting is performed on ITO and the pattern in the ITO-PET bilayer is formed by a process very similar to the "lift-off" process in microfabrication. ITO is just one example of brittle materials. Such a cost and time effective process can also be applied to pattern other brittle materials.

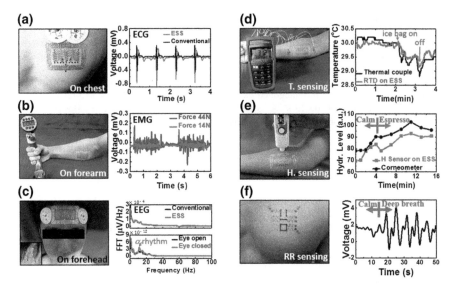

Fig. 2.7 ECG, EMG, EEG, skin temperature, skin hydration, and respiratory rate measurements by low-cost epidermal sensors [58]. **a** ECG simultaneously measured by ESS (*red*) and Ag/AgCl gel electrodes (*black*). Stronger ECG signals were obtained by the epidermal electrodes. **b** Epidermal electrodes attached on human forearm for EMG measurement when the subject is gripping a commercial dynamometer with different forces. Higher EMG amplitude corresponded to higher gripping force. **c** EEG measured on human forehead by both epidermal electrodes and Ag/AgCl gel electrodes. Two frequency spectrums of the EEG are well overlapped. 10 Hz alpha rhythm measured by epidermal electrodes is clearly visible when eyes are closed. **d** Skin temperature changes measured by both epidermal resistance temperature detector (RTD) and thermocouple found good correlation. **e** Real-time skin hydration before and after Espresso intake measured by both commercial corneometer and epidermal impedance sensor. **f** Voltage outputs from the electrically conductive rubber (ECG)-based Wheatstone bridge during normal and deep breath

2.5 Conclusions

This chapter presents our recent studies on the stretchability, conformability, and low-cost manufacture of epidermal sensors, which is a compelling example of bio-integrated electronics. After a decade of development, the intrinsic stiffness and brittleness of electronic materials is no longer a road block to stretchable electronics. We want to emphasize that in addition to stretchability of the sensors, conformability to soft but rough tissue surface is another key requirement for bio-integrated electronics. While examples in this chapter are based on metal and ceramic materials, the fundamental mechanics and ubiquitous "cut-and-paste" manufacturing process are also applicable to other classes of materials. Except epidermal applications, soft sensor and electronic sheets have also found many other exciting applications *in vivo* as well as in tissue engineering, and there will be many more to come.

References

1. D.H. Kim, R. Ghaffari, N.S. Lu, J.A. Rogers, Flexible and stretchable electronics for bio-integrated devices. Annu. Rev. Biomed. Eng. **14**, 113–128 (2012)
2. D.H. Kim, N.S. Lu, R. Ghaffari, J.A. Rogers, Inorganic semiconductor nanomaterials for flexible and stretchable bio-integrated electronics. NPG Asia Materials **4**, e15 (2012)
3. J. van den Brand, M. de Kok, A. Sridhar, M. Cauwe, R. Verplancke, F. Bossuyt, et al., Flexible and stretchable electronics for wearable healthcare, in *Proceedings of the 2014 44th European Solid-State Device Research Conference* (Essderc, 2014), pp. 206–209
4. D.H. Kim, N.S. Lu, R. Ma, Y.S. Kim, R.H. Kim, S.D. Wang et al., Epidermal electronics. Science **333**, 838–843 (2011)
5. Z.G. Suo, Mechanics of stretchable electronics and soft machines. MRS Bull. **37**, 218–225 (2012)
6. N. Lu, S. Yang, Mechanics for stretchable sensors. Curr. Opin. Solid State Mater. Sci. in press (2015)
7. J. Song, Mechanics of stretchable electronics. Curr. Opin. Solid State Mater. Sci. **19**, 160–170 (2015)
8. D.H. Kim, N.S. Lu, Y.G. Huang, J.A. Rogers, Materials for stretchable electronics in bioinspired and biointegrated devices. MRS Bull. **37**, 226–235 (2012)
9. W.-H. Yeo, Y.-S. Kim, J. Lee, A. Ameen, L. Shi, M. Li et al., Multifunctional electronics: multifunctional epidermal electronics printed directly onto the skin. Adv. Mater. **25**, 2772 (2013)
10. J.S. Lee, J. Heo, W.K. Lee, Y.G. Lim, Y.H. Kim, K.S. Park, Flexible capacitive electrodes for minimizing motion artifacts in ambulatory electrocardiograms. Sensors **14**, 14732–14743 (2014)
11. J.W. Jeong, M.K. Kim, H.Y. Cheng, W.H. Yeo, X. Huang, Y.H. Liu et al., Capacitive epidermal electronics for electrically safe, long-term electrophysiological measurements. Adv. Healthc. Mater. **3**, 642–648 (2014)
12. J.W. Jeong, W.H. Yeo, A. Akhtar, J.J.S. Norton, Y.J. Kwack, S. Li et al., Materials and optimized designs for human-machine interfaces via epidermal electronics. Adv. Mater. **25**, 6839–6846 (2013)
13. X. Huang, W.H. Yeo, Y. Liu, J.A. Rogers, Epidermal differential impedance sensor for conformal skin hydration monitoring. Biointerphases **7**, 52 (2012)
14. X. Huang, H. Cheng, K. Chen, Y. Zhang, Y. Zhang, Y. Liu et al., Epidermal impedance sensing sheets for precision hydration assessment and spatial mapping. IEEE Trans. Biomed. Eng. **60**, 2848–2857 (2013)
15. X. Huang, Y.H. Liu, H.Y. Cheng, W.J. Shin, J.A. Fan, Z.J. Liu et al., Materials and designs for wireless epidermal sensors of hydration and strain. Adv. Funct. Mater. **24**, 3846–3854 (2014)
16. R.C. Webb, A.P. Bonifas, A. Behnaz, Y.H. Zhang, K.J. Yu, H.Y. Cheng et al., Ultrathin conformal devices for precise and continuous thermal characterization of human skin. Nat. Mater. **12**, 938–944 (2013)
17. M.K. Choi, O.K. Park, C. Choi, S. Qiao, R. Ghaffari, J. Kim, et al., Cephalopod-inspired miniaturized suction cups for smart medical skin. Adv. Healthc. Mater. (2015)
18. D. Son, J. Lee, S. Qiao, R. Ghaffari, J. Kim, J.E. Lee et al., Multifunctional wearable devices for diagnosis and therapy of movement disorders. Nat. Nanotechnol. **9**, 397–404 (2014)
19. J. Kim, M. Lee, H.J. Shim, R. Ghaffari, H.R. Cho, D. Son, et al., Stretchable silicon nanoribbon electronics for skin prosthesis. Nat. Commun. **5** (2014)
20. A.J. Bandodkar, D. Molinnus, O. Mirza, T. Guinovart, J.R. Windmiller, G. Valdes-Ramirez et al., Epidermal tattoo potentiometric sodium sensors with wireless signal transduction for continuous non-invasive sweat monitoring. Biosens. Bioelectron. **54**, 603–609 (2014)
21. W. Jia, A.J. Bandodkar, G. Valdes-Ramirez, J.R. Windmiller, Z. Yang, J. Ramirez et al., Electrochemical tattoo biosensors for real-time noninvasive lactate monitoring in human perspiration. Anal. Chem. **85**, 6553–6560 (2013)

22. C. Dagdeviren, Y. Shi, P. Joe, R. Ghaffari, G. Balooch, K. Usgaonkar, et al., Conformal piezoelectric systems for clinical and experimental characterization of soft tissue biomechanics. Nat. Mater. **14** (2015)
23. J. Kim, A. Banks, H. Cheng, Z. Xie, S. Xu, K.-I. Jang, et al., Epidermal electronics with advanced capabilities in near-field communication. Small (2014)
24. J.A. Fan, W.H. Yeo, Y.W. Su, Y. Hattori, W. Lee, S.Y. Jung, et al., Fractal design concepts for stretchable electronics. Nat. Commun. **5** (2014)
25. D.S. Gray, J. Tien, C.S. Chen, High-conductivity elastomeric electronics. Adv. Mater. **16**, 393 (2004)
26. T. Li, Z.G. Suo, S.P. Lacour, S. Wagner, Compliant thin film patterns of stiff materials as platforms for stretchable electronics. J. Mater. Res. **20**, 3274–3277 (2005)
27. D. Brosteaux, F. Axisa, M. Gonzalez, J. Vanfleteren, Design and fabrication of elastic interconnections for stretchable electronic circuits. IEEE Electron Device Lett. **28**, 552–554 (2007)
28. D.H. Kim, J.Z. Song, W.M. Choi, H.S. Kim, R.H. Kim, Z.J. Liu et al., Materials and noncoplanar mesh designs for integrated circuits with linear elastic responses to extreme mechanical deformations. Proc. Natl. Acad. Sci. U.S.A. **105**, 18675–18680 (2008)
29. Y.Y. Hsu, M. Gonzalez, F. Bossuyt, F. Axisa, J. Vanfleteren, I. De Wolf, In situ observations on deformation behavior and stretching-induced failure of fine pitch stretchable interconnect. J. Mater. Res. **24**, 3573–3582 (2009)
30. Y.Y. Hsu, M. Gonzalez, F. Bossuyt, J. Vanfleteren, I. De Wolf, Polyimide-enhanced stretchable interconnects: design, fabrication, and characterization. IEEE Trans. Electron Devices **58**, 2680–2688 (2011)
31. D.H. Kim, N.S. Lu, R. Ghaffari, Y.S. Kim, S.P. Lee, L.Z. Xu et al., Materials for multifunctional balloon catheters with capabilities in cardiac electrophysiological mapping and ablation therapy. Nat. Mater. **10**, 316–323 (2011)
32. S. Xu, Y.H. Zhang, J. Cho, J. Lee, X. Huang, L. Jia, et al., Stretchable batteries with self-similar serpentine interconnects and integrated wireless recharging systems. Nat. Commun. **4** (2013)
33. R.H. Kim, M.H. Bae, D.G. Kim, H.Y. Cheng, B.H. Kim, D.H. Kim et al., Stretchable, transparent graphene interconnects for arrays of microscale inorganic light emitting diodes on rubber substrates. Nano Lett. **11**, 3881–3886 (2011)
34. C.J. Yu, Z. Duan, P.X. Yuan, Y.H. Li, Y.W. Su, X. Zhang et al., Electronically programmable, reversible shape change in two- and three-dimensional hydrogel structures. Adv. Mater. **25**, 1541–1546 (2013)
35. T. Ma, Y. Wang, R. Tang, H. Yu, H. Jiang, Pre-patterned ZnO nanoribbons on soft substrates for stretchable energy harvesting applications. J. Appl. Phys. **113** (2013)
36. G. Lanzara, N. Salowitz, Z.Q. Guo, F.K. Chang, A spider-web-like highly expandable sensor network for multifunctional materials. Adv. Mater. **22**, 4643–4648 (2010)
37. M. Gonzalez, F. Axisa, F. Bossuyt, Y.Y. Hsu, B. Vandevelde, J. Vanfleteren, Design and performance of metal conductors for stretchable electronic circuits. Circuit World **35**, 22–29 (2009)
38. G. Mani, M.D. Feldman, D. Patel, C.M. Agrawal, Coronary stents: a materials perspective. Biomaterials **28**, 1689–1710 (2007)
39. T. Widlund, S. Yang, Y.-Y. Hsu, N. Lu, Stretchability and compliance of freestanding serpentine-shaped ribbons. Int. J. Solids Struct. **51**, 4026–4037 (2014)
40. S. Yang, S. Qiao, N. Lu, Elasticity solutions to freestanding, non-buckling serpentine ribbons. To be submitted (2016)
41. D.H. Kim, R. Ghaffari, N.S. Lu, S.D. Wang, S.P. Lee, H. Keum et al., Electronic sensor and actuator webs for large-area complex geometry cardiac mapping and therapy. Proc. Natl. Acad. Sci. U.S.A. **109**, 19910–19915 (2012)
42. Y.L. Yu, D. Sanchez, N.S. Lu, Work of adhesion/separation between soft elastomers of different mixing ratios. J. Mater. Res. **30**, 2702–2712 (2015)

43. Y.W. Su, J. Wu, Z.C. Fan, K.C. Hwang, J.Z. Song, Y.G. Huang et al., Postbuckling analysis and its application to stretchable electronics. J. Mech. Phys. Solids 60, 487–508 (2012)
44. Y.H. Zhang, S. Xu, H.R. Fu, J. Lee, J. Su, K.C. Hwang et al., Buckling in serpentine microstructures and applications in elastomer-supported ultra-stretchable electronics with high areal coverage. Soft Matter 9, 8062–8070 (2013)
45. Y. Zhang, H. Fu, Y. Su, S. Xu, H. Cheng, J.A. Fan, et al., Mechanics of ultra-stretchable self-similar serpentine interconnects. Acta Mater. 61, 7816–7827, (2013)
46. Y. Zhang, H. Fu, S. Xu, J.A. Fan, K.-C. Hwang, J. Jiang, et al., A hierarchical computational model for stretchable interconnects with fractal-inspired designs. J. Mech. Phys. Solids, 72, 115–130 (2014)
47. Y.Y. Hsu, M. Gonzalez, F. Bossuyt, F. Axisa, J. Vanfleteren, I. DeWolf, The effect of pitch on deformation behavior and the stretching-induced failure of a polymer-encapsulated stretchable circuit. J. Micromech. Microeng. 20 (2010)
48. Y.Y. Hsu, M. Gonzalez, F. Bossuyt, F. Axisa, J. Vanfleteren, B. Vandevelde et al., Design and analysis of a novel fine pitch and highly stretchable interconnect. Microelectron. Int. 27, 33–38 (2010)
49. M. Gonzalez, B. Vandevelde, W. Christiaens, Y.Y. Hsu, F. Iker, F. Bossuyt et al., Design and implementation of flexible and stretchable systems. Microelectron. Reliab. 51, 1069–1076 (2011)
50. S. Yang, B. Su, G. Bitar, N. Lu, Stretchability of indium tin oxide (ITO) serpentine thin films supported by Kapton substrates. Int. J. Fract. 190, 99–110, (2014)
51. S. Yang, E. Ng, N. Lu, Indium tin oxide (ito) serpentine ribbons on soft substrates stretched beyond 100%. Extreme Mech. Lett. 2, 37–45 (2015)
52. U. Betz, M.K. Olsson, J. Marthy, M.F. Escola, F. Atamny, Thin films engineering of indium tin oxide: Large area flat panel displays application. Surf. Coat. Technol. 200, 5751–5759 (2006)
53. H. Schmidt, H. Flugge, T. Winkler, T. Bulow, T. Riedl, W. Kowalsky, Efficient semitransparent inverted organic solar cells with indium tin oxide top electrode. Appl. Phys. Lett. 94 (2009)
54. Y. Leterrier, L. Medico, F. Demarco, J.A.E. Manson, U. Betz, M.F. Escola et al., Mechanical integrity of transparent conductive oxide films for flexible polymer-based displays. Thin Solid Films 460, 156–166 (2004)
55. N.S. Lu, X. Wang, Z. Suo, J. Vlassak, Metal films on polymer substrates stretched beyond 50%. Appl. Phys. Lett. 91, 221909 (2007)
56. R.M. Niu, G. Liu, C. Wang, G. Zhang, X.D. Ding, J. Sun, Thickness dependent critical strain in submicron Cu films adherent to polymer substrate. Appl. Phys. Lett. 90 (2007)
57. N.S. Lu, Z.G. Suo, J.J. Vlassak, The effect of film thickness on the failure strain of polymer-supported metal films. Acta Mater. 58, 1679–1687 (2010)
58. S. Yang, Y.C. Chen, L. Nicolini, P. Pasupathy, J. Sacks, J. Becky et al., "Cut-and-paste" manufacture of multiparametric epidermal sensor systems. Adv. Mater. (2015). doi:10.1002/adma.201502386
59. M.D. Casper, A.Ö. Gözen, M.D. Dickey, J. Genzer, J.-P. Maria, Surface wrinkling by chemical modification of poly(dimethylsiloxane)-based networks during sputtering. Soft Matter 9, 7797 (2013)
60. S. Choi, J. Park, W. Hyun, J. Kim, J. Kim, Y.B. Lee et al., Stretchable Heater Using Ligand-Exchanged Silver Nanowire Nanocomposite for Wearable Articular Thermotherapy. ACS Nano 9, 6626–6633 (2015)
61. S. Hong, H. Lee, J. Lee, J. Kwon, S. Han, Y.D. Suh et al., Highly stretchable and transparent metal nanowire heater for wearable electronics applications. Adv. Mater. 27, 4744–4751 (2015)
62. X. Huang, Y.H. Liu, K.L. Chen, W.J. Shin, C.J. Lu, G.W. Kong et al., Stretchable, wireless sensors and functional substrates for epidermal characterization of sweat. Small 10, 3083–3090 (2014)

63. Z.Y. Huang, W. Hong, Z. Suo, Nonlinear analyses of wrinkles in a film bonded to a compliant substrate. J. Mech. Phys. Solids **53**, 2101–2118 (2005)
64. J. Xiao, A. Carlson, Z.J. Liu, Y. Huang, J.A. Rogers, Analytical and experimental studies of the mechanics of deformation in a solid with a wavy surface profile. J. Appl. Mech. **77**, 011003 (2009)
65. L. Wang, N. Lu, Conformability of a thin elastic membrane laminated on a soft substrate with slightly wavy surface. J. Appl. Mech. under review (2016)
66. S. Qiao, J.-B. Gratadour, L. Wang, N. Lu, Conformability of a thin elastic membrane laminated on a rigid substrate with corrugated surface
67. T.J. Wagner, D. Vella, The sensitivity of graphene "snap-through" to substrate geometry. Appl. Phys. Lett. **100**, 233111 (2012)
68. S. Scharfenberg, N. Mansukhani, C. Chialvo, R.L. Weaver, N. Mason, Observation of a snap-through instability in graphene. Appl. Phys. Lett. **100**, 021910 (2012)
69. N. Lu, D.H. Kim, Flexible and stretchable electronics paving the way for soft robotics. Soft Robotics **1**, 53–62 (2013)
70. M.A. Pacheco, C.L. Marshall, Review of dimethyl carbonate (DMC) manufacture and its characteristics as a fuel additive. Energy Fuels **11**, 2–29 (1997)

Chapter 3
Mechanics and Designs of Stretchable Bioelectronics

Yihui Zhang

Abstract This chapter reviews mechanics-guided designs that enable highly deformable forms of bioelectronics, through soft, conformal integration of hard functional components with soft elastomeric substrates. Three representative strategies, including wavy, wrinkled design, island-bridge design, and Origami/Kirigami-inspired design, are summarized, highlighting the key design concepts, unique mechanical behaviors, and analytical/computational mechanics models that guide the design optimization. Finally, some perspectives are provided on the remaining challenges and opportunities.

Keywords Stretchable electronics · Mechanics · Wrinkling · Postbuckling · Serpentine interconnects · Fractal design · Origami and Kirigami

3.1 Introduction

During the last decade, fast developments and substantial achievements have been made on various aspects of stretchable bioelectronics, which has been reshaping this new field. This class of electronics, while maintaining the functionalities of established technologies, could offer superior mechanical attributes that are inaccessible to traditional electronics, e.g., stretched like a rubber band, twisted like a rope, and bent around a pencil tip, without mechanical fatigue or any significant change in operating characteristics [1, 2]. Those superior mechanical characteristics pave the way to a range of bio-inspired and bio-integrated applications [3–11] that could not be addressed with any other approach.

Y. Zhang (✉)
Department of Engineering Mechanics, Center for Mechanics and Materials,
AML, Tsinghua University, Beijing 100084, People's Republic of China
e-mail: yihuizhang@tsinghua.edu.cn; yihui.zhang2011@gmail.com

© Springer International Publishing Switzerland 2016
J.A. Rogers et al. (eds.), *Stretchable Bioelectronics for Medical Devices and Systems*, Microsystems and Nanosystems,
DOI 10.1007/978-3-319-28694-5_3

There are two general routes to stretchable forms of bioelectronics [1, 12, 13]: (1) developing novel materials that are intrinsically stretchable to serve as functional components of devices; (2) devising novel structural designs for heterogeneous integration of hard (e.g., moduli approaching 1 TPa for SWNTs and graphene [14, 15]) inorganic, functional components and soft (e.g., modulus ~ 3 kPa for certain cellular forms of elastomers [16], and ~ 60 kPa for ecoflex [17]) elastomeric platforms to result in device systems that are stretchable. This chapter will focus on the latter route, aiming to provide a review on the recent advances of the mechanics-guided designs and models. In this route, the quantitative mechanics design has been playing crucial roles, at a level of importance that is comparable to circuit design in conventional electronics [12].

3.2 Wavy, Wrinkled Design

The wavy, wrinkled design represents the first type of structural design introduced in stretchable inorganic electronics. This design was initially proposed to achieve a stretchable form of single-crystal silicon for high-performance electronics on a rubber substrate [18]. The key of this design strategy is to introduce an initial strain in a soft substrate (e.g., silicone), either by thermally induced expansion [19] or mechanical pre-stretch [18, 20], and then utilize this prestrain to drive the formation of wavy, wrinkled patterns in hard thin films (e.g., metal or semiconductor). Figure 3.1a presents a schematic illustration of the fabrication procedure for wavy silicon ribbons on an elastomeric substrate (e.g., PDMS) [18]. Figure 3.1b presents scanning electron micrographs (SEM) of highly periodic, wavy, single-crystal Si ribbons generated with this strategy.

Many mechanics models have been developed to predict the configurations of the wavy ribbon structures, as well as their stretchabilities. Several excellent review papers on these models can be found in literature [21–23], and therefore, only some key results will be introduced herein. In the regime of small prestrain (e.g., $0.5 \% < \varepsilon_{pre} < 5 \%$), an energy approach based on the small deformation theory was developed to determine the buckled configurations [24, 25]. In this approach, a sinusoidal profile of out-of-plane displacement is assumed for the bucked ribbon whose wavelength (λ_0) and amplitude (A_0) can be determined, through the minimization of total strain energy, as

$$\lambda_0 = \frac{\pi h_f}{\sqrt{\varepsilon_c}} = 2\pi h_f \left(\frac{\bar{E}_f}{3\bar{E}_s}\right)^{1/3}, \ A_0 = h_f \sqrt{\frac{\varepsilon_{pre}}{\varepsilon_c} - 1}, \tag{3.1}$$

where the subscripts 'f' and 's' denote the ribbon and substrate, respectively; $\bar{E} = E/(1 - v^2)$ is the plane-strain modulus, with v denoting the Poisson ratio; h_f is the thickness of the ribbon; and $\varepsilon_c = \left(3\bar{E}_s/\bar{E}_f\right)^{2/3}/4$ is the critical strain to trigger the buckling, which is $\sim 0.034 \%$ for the Si ribbon on a PDMS substrate [18].

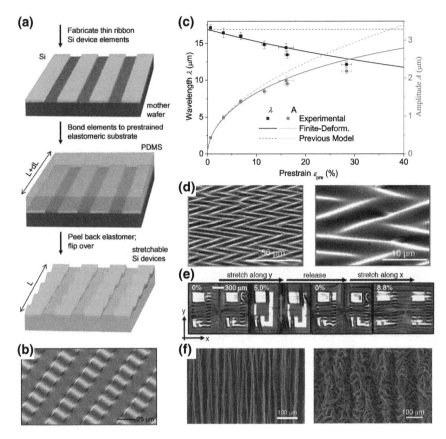

Fig. 3.1 a Schematic illustration of the fabricating process for wavy, wrinkled single-crystal Si ribbons on an elastomeric substrate. **b** SEM images of wavy, wrinkled, single-crystal Si ribbons fabricated with the process in **a**. **c** Dependences of wavelength and amplitude on the prestrain for wrinkled structures of Si (100 nm thickness) on PDMS. **d** SEM images of a 2-D wavy Si nanomembrane fully bonded on PDMS. **e** Optical images of wavy Si complementary metal–oxide–semiconductor (CMOS) inverters under tensile strain along the *x* and *y* directions. **f** Parallel ridges (*left panel*) and crumpled patterns (*right panel*) of graphene papers after the biaxially pre-stretched (∼400 %) elastomeric substrate is uniaxiallly and biaxially relaxed, respectively. **a** and **b** are adapted with permission from Ref. [18], Copyright 2006, American Association for the Advancement of Science. **c** is adapted with permission from Ref. [26], Copyright 2007, National Academy of Sciences. **d** is adapted with permission from Ref. [30], Copyright 2007, American Chemical Society. **e** is adapted with permission from [31], Copyright 2008, American Association for the Advancement of Science. **f** is adapted with permission from Ref. [34], Copyright 2014, Nature Publishing Group

Equation (3.1) indicates that the wavelength is independent on the prestrain (referred to as 'previous model' in Fig. 3.1c), which deviates evidently from the experimental measurements at a large prestrain (e.g., >5 %). In this regime (e.g., 5 % < ε_{pre} < 30 %), the experimental observations showed a reduced wavelength with the increase of prestrain [26], which is mainly attributed to the finite deformation in the elastomeric substrate. To account for such effect during post-buckling, Jiang et al. [26] and Song et al. [27] developed analytic models that yielded corrected solutions to the buckling wavelength and amplitude, i.e.,

$$\lambda = \frac{\lambda_0}{(1 + \varepsilon_{pre})(1 + \xi)^{1/3}}, \quad A = \frac{A_0}{\sqrt{1 + \varepsilon_{pre}}(1 + \xi)^{1/3}}, \quad (3.2)$$

where λ_0 and A_0 represent the wavelength and amplitude in the small deformation condition and $\xi = 5\varepsilon_{pre}(1 + \varepsilon_{pre})/32$. Figure 3.1c shows that the new solutions [Eq. (3.2)] could well capture the phenomenon of strain-dependent wavelength, and offer a more accurate prediction of amplitude as compared to small deformation solutions. The peak strain in this regime can be approximated by $\varepsilon_{peak} \approx 2\sqrt{\varepsilon_{pre}\varepsilon_c}(1 + \xi)^{1/3}/\sqrt{1 + \varepsilon_{pre}}$, from which the maximum prestrain to avoid fracture in the ribbons can be determined. For example, it gives a maximum prestrain of ~ 29 % for Si with a fracture strain of ~ 1.8 %, and therefore indicates the stretchability of the wavy Si ribbons to be as large as ~ 29 %. Using a similar approach, Cheng et al. [28] investigated the stretchable ribbons formed with a heterogeneous bilayer substrate, in which the top, soft layer reduces the peak strain of ribbons to enable a high stretchability, whereas the relatively stiff layer at the bottom offers robustness and high strength to the system. Jiang et al. [29] extended the above energy approach to a three-dimensional (3-D) model, and established an analytic model to study the effect of finite ribbon width on the amplitude and wavelength during postbuckling.

The above mechanical strategy was further exploited by Choi et al. [30] and Kim et al. [31] to generate two-dimensional (2-D), wavy, electronic devices (as shown in Figs. 3.1d, e), which offer excellent stretchability along various directions. In the case of biaxial prestrain, the determination of buckling configurations is quite complicated, as shown by the computational and theoretical studies in Refs. [24, 25, 32, 33]. Introducing this design concept in graphene films on elastomeric substrate, Zang et al. [15, 34] demonstrated that ridged or crumpled patterns of grapheme films (as shown in Fig. 3.1f) can be formed by using extremely large prestrains (e.g., up to 400 %).

3.3 Island-Bridge Design

The island-bridge design represents another general strategy which is widely used in stretchable bioelectronics. The coplanar island-bridge design was initially proposed by Lacour et al. [35], using stretchable wavy, metal electrodes to interconnect active devices, while the noncoplanar island-bridge design was first introduced by Kim et al. [36] to significantly increase the stretchability. In this design, the functional components usually reside on the island, while the electrical interconnects form the bridge. Under stretching, the rigid islands (which have high effective stiffness) keep almost undeformed (e.g., with <1 % strain) to ensure the mechanical integrity of functional materials, while the interconnects (which have low effective stiffness) deform to provide the stretchability. Therefore, the design of bridge structure is the key to the effective mechanical properties (e.g., modulus, stretchability, etc.) of the system. Depending on the geometric nature of the bridge, the island-bridge design can be further classified into three groups, which are elaborated in the following subsections.

3.3.1 Island-Bridge Design with Straight Interconnects

In this type of island-bridge design, the straight interconnects can be either strongly bonded (via chemical covalent bonding) or weakly bonded (via van der Waals interaction) to a pre-stretched elastomeric substrate, resulting in coplanar wavy interconnects [35] or noncoplanar arc-shaped interconnects [36], upon release of the prestrain. Since the wavy configurations have been introduced in detail in Sect. 3.2, this subsection focuses on the noncoplanar arc-shaped interconnects, as shown in Figs. 3.2a, b. Such design was adopted in many stretchable electronic systems, e.g, the eyeball-like digital cameras [6, 37] and photovoltaic devices [38, 39].

Song et al. [40] developed an analytic model to investigate the buckling physics of island-bridge structure with straight interconnects. A sinusoidal expression was assumed to characterize the arc-shaped profile during postbuckling, in which the unknown amplitude was determined from minimization of the total strain energy. By applying the reaction forces and bending moments at the ends of interconnect to the membrane-shaped island, an approximation solution to the peak strain of island can be also obtained. This model [40] applies mainly to the case in which the maximum deflection (i.e., the out-of-plane displacement) of the interconnect is relatively small compared to the bridge length, since the assumption of sinusoidal postbuckling configuration will not be accurate enough in the regime of large interconnect deflection. To overcome this issue, Li et al. [41] developed a finite deformation model that discards the assumption of sinusoidal profile for the bridge deformation and the approximation of dimensional analysis for the island deformation. An accurate solution of the interconnect deflection and peak strains (both in the bridge and in the island) can be obtained, as shown in Figs. 3.2c, d. The results show that a thin and long interconnect is preferred to increase the stretchability.

The island-bridge structure sometimes undergoes stretching along a direction deviated from the axial directions of the interconnects. In this case, the lateral buckling might be triggered in the straight interconnects, which involves not only bending deformation, but also twisting deformation. Su et al. [42] established a theoretical framework for the postbuckling of beams that may involve complex buckling modes such as lateral buckling, and adopted the perturbation method to obtain the amplitude of buckled interconnects. Chen et al. [43] derived analytical solutions for describing the critical lateral bucking and postbuckling configuration of interconnects under shear.

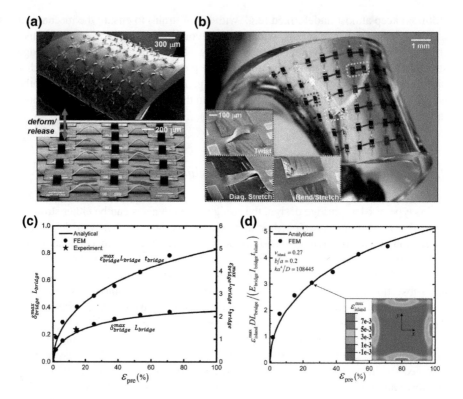

Fig. 3.2 **a** SEM images of an array of CMOS inverters that adopt an island-bridge configuration, in an undeformed state (*lower panel*, formed with ∼20 % prestrain) and in a deformed state associated with a complex twisting motion (*upper panel*). **b** Optical image of a stretchable array of CMOS inverters under combined bending and twisting deformations, with the insets (colorized) highlighting three different types of deformation mode: diagonal stretching, twisting, and bending. **c** Normalized maximum deflection and maximum strain of the bridge versus the prestrain of the substrate, where t_{bridge} and L_{bridge} denote the thickness and length of the bridge, respectively. **d** Normalized maximum strain of the island versus the prestrain of the substrate, where D and t_{island} are the flexural rigidity and thickness of the island, respectively, and $E_{bridge}I_{bridge}$ is the bending stiffness of the bridge. **a** and **b** are adapted with permission from Ref. [36] (Copyright 2008, National Academy of Sciences). **c** and **d** are adapted with permission from Ref. [41] (Copyright 2013, Royal Society of Chemistry)

3.3.2 Island-Bridge Design with Serpentine Interconnects

Interconnects with a filamentary, serpentine configuration are more flexible than straight interconnects, thereby with a potential to offer a higher stretchability, even without any use of pre-stretch in the elastomeric substrate. Based on the different requirements of target applications, the serpentine interconnect can be fully bonded, partially bonded, or completely nonbonded to the substrate by means of different fabrication processes. Under different bonding conditions, the mechanics behaviors of serpentine interconnect could be qualitatively different, as detailed below.

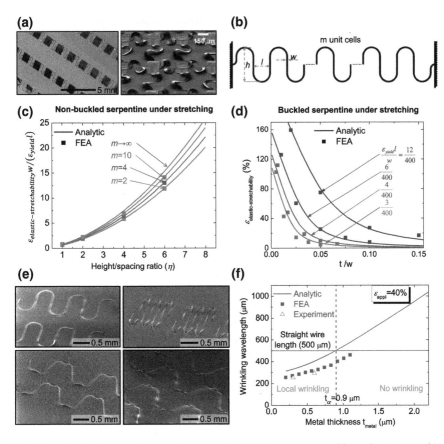

Fig. 3.3 **a** Optical image (*left panel*) of an island-bridge structure with coplanar serpentine interconnects, and SEM image (*right panel*) of an island-bridge structure with noncoplanar serpentine interconnect enabled by a strategy of substrate pre-stretch. **b** Schematic illustration of geometric parameters for a representative serpentine interconnect with *m* unit cells. **c** Normalized elastic stretchability of non-buckled, thick, serpentine interconnect as a function of height/spacing aspect ratio for different number of unit cells, in which ε_{yield} is the yield strain of the interconnect material. **d** Elastic stretchability of buckled, thin, serpentine interconnect as a function of thickness/width aspect ratio for various $\varepsilon_{yield}l/w$, $m = 1$ and $\eta = 4$. **e** SEM images of serpentine interconnects fully bonded onto an elastomeric substrate, in the undeformed state (*upper left panel*), compressed state (*upper right panel*) with the use of substrate pre-stretch, and stretched states [without any wrinkling for a thick (4 μm Cu) interconnect, *lower left panel*; with local wrinkling for a thin (0.3 μm Cu) interconnect, *lower right panel*]. **f** Experiment measurement, FEA simulation and analytic modeling of wrinkling wavelength at different metal thicknesses of the interconnects that are sandwiched by two polyimide layers (1.2 μm), and mounted on a soft (60 kPa) substrate (0.5 mm). **a** is adapted with permission from Refs. [44] (Copyright 2013, Royal Society of Chemistry) and [36] (Copyright 2008, National Academy of Sciences). **b** and **d** are adapted from with permission from Ref. [44], Copyright 2013, Royal Society of Chemistry. **c** is adapted with permission from Ref. [45], Copyright 2013, Elsevier Science Ltd. **e** and **f** are adapted with permission from Ref. [49], Copyright 2014, WILEY-VCH Verlag GmbH & Co. KGaA, Weinheim

The completely nonbonded serpentine interconnects are usually clamped at two ends by relatively rigid islands (in the left panel of Fig. 3.3a) that are fully bonded to the substrates or raised surface relief structures of the substrates [44]. Noncoplanar configurations of the serpentine interconnect can be realized by use of prestrain in the substrate [36], as shown in Fig. 3.3a (right panel). In general, a serpentine interconnect could contain a number (m) of periodically distributed unit cells, as illustrated in Fig. 3.3b, in which the representative unit cell comprises two half circles connected by straight lines, with height h and spacing l. Upon stretching from the two ends, two different deformation modes could occur in the nonbonded serpentine interconnects, depending on the thickness/width ratio (t/w) [44]: (i) non-buckled, in-plane deformations, for large t/w (typically comparable to or larger than 1); (ii) buckled deformations, involving not only in-plane but also out-of-plane deformations, for small t/w (typically smaller than 1/5). The critical strain to trigger buckling in the serpentine interconnect can be solved analytically [44], which shows a proportional dependence on the square of thickness/width ratio (t^2/w^2).

For non-buckled serpentine interconnects under stretching, Zhang et al. [45] obtained an analytic solution of the elastic stretchability based on the beam theory. The normalized stretchability (Fig. 3.3c) increases with increase in the height/spacing ratio η or number of unit cell m. Based on the 2-D elasticity theory, Widlund et al. [46] derived a different solution of stretchability for non-buckled serpentine structures, which could provide a more accurate prediction, in particular for a relatively large width/spacing ratio (w/l). For buckled serpentine interconnects under stretching, Zhang et al. [44] proposed a theoretical model that gives a semi-analytic solution of elastic stretchability. This solution agrees well with the results of finite element analyses (FEA) (Fig. 3.3d), indicating that the elastic stretchability increases with decreasing serpentine thickness (via t/w) or increasing spacing (via $\varepsilon_{yield}l/w$). The analytic modeling of relevant buckled configurations is quite challenging, which can be complemented by FEA [36, 44, 47].

The fully bonded serpentine interconnects can be integrated with freestanding or pre-stretched substrate, resulting in two different configurations (upper left and right panels of Fig. 3.3e). The deformation mode of fully bonded serpentine interconnect is more complicated than the nonbonded counterpart, because of the constraints from the substrate. For this reason, the existing studies on the mechanical performances mainly relied on computational approaches, e.g., FEA [3, 48–54]. Zhang et al. [49] studied the effects of key geometric and materials parameters on the elastic stretchability and the enhancements enabled by the prestrain strategy. A drastic decrease in the elastic stretchability with increasing metal thickness was reported, which can be mainly attributed to the changes in the buckling mode, as illustrated in Fig. 3.3e (lower left and right panels), f.

3.3.3 Island-Bridge Design with Fractal-Inspired Interconnects

The fractal-inspired interconnect design was initially introduced by Xu et al. [17] to realize an ultra-stretchable lithium-ion battery, and was later explored in several other stretchable bio-electronic devices [5, 55–57]. This design could make full use of a limited space by increasing the fractal order, which offers a great advantage in enhancing, simultaneously, the areal coverage of functional components and the system-level stretchability. In this design, it is important to devise an appropriate fractal layouts of the interconnects that could fit well with the other components of the device. Figure 3.4a provides an example of the fractal layout that starts from the serpentine configuration as the first order structure.

Similar to the serpentine interconnects, the fractal-inspired interconnects (simply referred to as fractal interconnects) can be also fabricated in a manner that the interconnects remain fully bonded or completely nonbonded to the substrate. Upon stretching, the nonbonded fractal interconnects also undergo non-buckled, in-plane deformations for a large t/w (typically comparable to or larger than 1), and buckled deformations for a small t/w (typically smaller than 1/5). For non-buckled fractal rectangular and serpentine interconnects, Zhang et al. [45] developed analytical models of flexibility and elastic stretchability (Fig. 3.4b, c), through establishing the recursive formulae at different fractal orders. Su et al. [58] proposed an analytic approach to determine the tensile stiffness for the fractal interconnects of an arbitrary shape, e.g., zigzag, sinusoidal shapes. For buckled fractal serpentine interconnects under stretching, an interesting deformation mechanism of ordered unraveling (shown in Fig. 3.4d) was observed in mechanics modeling and experiment measurement [17]. Based on this unique mechanism, Zhang et al. [59] developed a hierarchical computational model (HCM) for postbuckling analysis of the fractal interconnects, which could substantially reduce the computational efforts and costs as compared to conventional FEA. Figure 3.4e shows a huge increase of elastic stretchability from ~ 10.7 % for the first order, to ~ 2140 % for the fourth order, for the buckled fractal serpentine interconnects.

For fully bonded fractal interconnects, Fan et al. [60] studied the deformations of various fractal layouts (e.g., Peano, Greek cross, Hilbert, etc.), as shown in Fig. 3.5, through combined FEA and experiment. They also introduced a high precision approach to measure the elastic-plastic transition (or the elastic stretchability) for the fractal interconnects which shows reasonable accordance with FEA calculations [60]. Due to the additional loading from the substrate, the elastic stretchability of fully bonded fractal interconnects is usually much lower than that of the completely nonbonded counterpart.

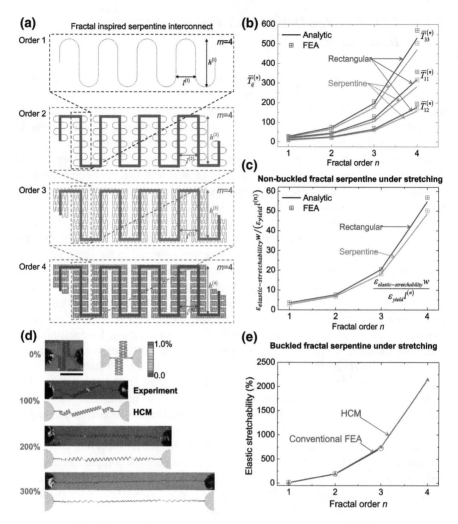

Fig. 3.4 a Geometric construction of fractal-inspired serpentine interconnects from order 1 to 4. **b** Three normalized flexibility components, and **c** normalized elastic stretchability versus the fractal order, for non-buckled, thick, fractal serpentine interconnects with $w/l^{(1)} = 0.4$, where $l^{(n)}$ denotes the spacing of the highest order, as illustrated in **a**. **d** Optical images and corresponding predictions of buckled configurations based on HCM for a second order fractal interconnect under various stages of stretching. **e** Elastic stretchability versus the fractal order for buckled, thin, fractal serpentine interconnect from $n = 1$ to 4, with $(m, \eta) = (4, 8/\sqrt{11})$, the thickness/width aspect ratio ($t/w = 0.03$), and the width to spacing ratio ($w/l^{(1)} = 0.4$), for structures of different fractal orders. The color in HCM results shown in **d** represents the magnitude of maximum principal strain. **a–c** are adapted with permission from Ref. [45], Copyright 2013, Elsevier Science Ltd. **d** and **e** are adapted from permission from Refs. [17] (Copyright 2013, Nature Publishing Group) and [59] (Copyright 2014, Elsevier Science Ltd)

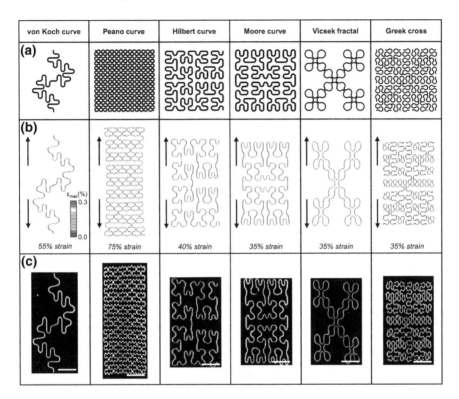

von Koch curve	Peano curve	Hilbert curve	Moore curve	Vicsek fractal	Greek cross

Fig. 3.5 a Fractal-inspired interconnects with six different topologies, which are fully bonded onto elastomeric substrates. **b** FEA calculations and **c** corresponding experimental MicroXCT images (*scale bars* 2 mm) of each structure under elastic tensile strain. The interconnects consist of a gold layer (300 nm) sandwiched by two polyimide layers (1.2 μm), and are mounted on a soft (50 kPa) substrate (0.5 mm). Adapted from permission from Ref. [60], Copyright 2014, Nature Publishing Group

3.4 Origami/Kirigami-Inspired Designs

Origami and Kirigami represent an ancient art of paper folding and cutting in which strategically designed creases and cuts could transform an initially flat paper into desired 3-D configuration with different topologies. The origami design concept was recently introduced in stretchable batteries by Cheng et al. [61] and Song et al. [62], to achieve high areal energy density and unprecedented deformability. A representative example of Miura folding [62] is shown in Fig. 3.6a, in which many identical parallelogram faces are connected by 'mountain' and 'valley' creases. Under external loadings, the parallelogram faces themselves usually remain undeformed while the creases undergo folding and/or unfolding leading to a strain concentration at the crease region. Thereby, a critical task is to reduce the strain level at the crease region, so as to avoid material fracture or plastic yielding. Because of the complex 3-D geometry of origami structures, only some effective

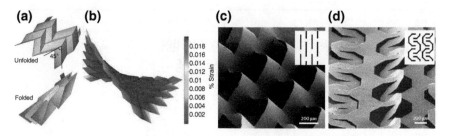

Fig. 3.6 **a** Schematic illustration of a representative origami patterns using Miura folding. **b** FEA results of a 45° Miura pattern under twisting (with ∼90° twisting angle per unit cell). **c, d** Two examples of microscale kirigami patterns in graphene oxide/PVA nanocomposites after photolithography. **a** and **b** are adapted with permission from Ref. [62], Copyright 2014, Nature Publishing Group. **c** and **d** are adapted with permission from Ref. [66], Copyright 2015, Nature Publishing Group

mechanical properties (e.g., bending rigidity and Poisson ratio) were studied through analytic modeling [63, 64], while the detailed deformation and strain distribution were examined mainly through FEA [65], with an example shown in Fig. 3.6b.

The Kirigami-inspired design was introduced in stretchable electronics recently [65–67]. Cho et al. [65] proposed a set of fractal-inspired cut for a relatively thick silicone sheet with deposition of a conductive film of multiwall carbon nanotubes, which could create non-buckled, rotating units under stretching. This approach enables a stretchable electrode that could expand to >800 % of the original area. Shyu et al. [66] showed that a network of cuts made in thin, rigid nanocomposite sheets induces lateral buckling (Fig. 3.6c, d) and prevents local failure, thereby increasing the ultimate strain of the sheets from 4 to 370 %. Song et al. [67] combined the use of folding and cutting in producing lithium-ion batteries that can be stretched >150 %. In all of the above cases, FEA serves as an important tool to guide the design of layout and geometry for the cuts.

3.5 Concluding Remarks

This chapter reviews the mechanics-guided designs and models developed for stretchable electronics, involving three general strategies. While many analytical and computational models have been developed for each type of strategy, most of them were built with certain idealizations/assumptions, and may not be applicable for some extreme conditions. As such, there are still many open challenges and opportunities for future research. For example, development of theoretical models that account for the interfacial delamination is desirable for serpentine and fractal interconnects bonded onto or encapsulated in a soft elastomer. For the emerging design strategy inspired by Origami/Kirigami concepts, the topology optimization of creases and cuts to identify a best layout for prescribed loading conditions remains as a virgin area of research. In addition, most of the current designs mainly rely on silicone rubbers to serve as the substrate, which, although soft, does not match well the nonlinear mechanical properties of biological tissues in many bio-electronic applications [68]. Thereby, the development of advanced forms of soft assembly platform in the substrate and/or encapsulation for high quality biointegration represents another important direction to explore.

References

1. J.A. Rogers, T. Someya, Y.G. Huang, Mater. Mech. Stretchable Electron. Sci. **327**, 1603–1607 (2010)
2. T. Sekitani, T. Someya, Stretchable large-area organic electronics. Adv. Mater. **22**, 2228–2246 (2010)
3. D.H. Kim, N.S. Lu, R. Ma, Y.S. Kim, R.H. Kim, S.D. Wang et al., Epidermal electronics. Science **333**, 838–843 (2011)
4. M. Kaltenbrunner, T. Sekitani, J. Reeder, T. Yokota, K. Kuribara, T. Tokuhara et al., An ultra-lightweight design for imperceptible plastic electronics. Nature **499**, 458–463 (2013)
5. S. Xu, Y.H. Zhang, L. Jia, K.E. Mathewson, K.I. Jang, J. Kim et al., Soft microfluidic assemblies of sensors, circuits, and radios for the skin. Science **344**, 70–74 (2014)
6. H.C. Ko, M.P. Stoykovich, J.Z. Song, V. Malyarchuk, W.M. Choi, C.J. Yu et al., A hemispherical electronic eye camera based on compressible silicon optoelectronics. Nature **454**, 748–753 (2008)
7. Y.M. Song, Y.Z. Xie, V. Malyarchuk, J.L. Xiao, I. Jung, K.J. Choi et al., Digital cameras with designs inspired by the arthropod eye. Nature **497**, 95–99 (2013)
8. S.P. Lacour, J. Jones, Z. Suo, S. Wagner, Design and performance of thin metal film interconnects for skin-like electronic circuits. IEEE Electron Device Lett. **25**, 179–181 (2004)
9. T. Someya, T. Sekitani, S. Iba, Y. Kato, H. Kawaguchi, T. Sakurai, A large-area, flexible pressure sensor matrix with organic field-effect transistors for artificial skin applications. PNAS **101**, 9966–9970 (2004)
10. S.J. Benight, C. Wang, J.B.H. Tok, Z.A. Bao, Stretchable and self-healing polymers and devices for electronic skin. Prog. Polym. Sci. **38**, 1961–1977 (2013)
11. D.H. Kim, J. Viventi, J.J. Amsden, J.L. Xiao, L. Vigeland, Y.S. Kim et al., Dissolvable films of silk fibroin for ultrathin conformal bio-integrated electronics. Nat. Mater. **9**, 511–517 (2010)
12. D.H. Kim, N.S. Lu, Y.G. Huang, J.A. Rogers, Materials for stretchable electronics in bioinspired and biointegrated devices. MRS Bull. **37**, 226–235 (2012)
13. S. Wagner, S. Bauer, Materials for stretchable electronics. MRS Bull. **37**, 207–217 (2012)
14. C.J. Yu, C. Masarapu, J.P. Rong, B.Q. Wei, H.Q. Jiang, Stretchable supercapacitors based on buckled single-walled carbon nanotube macrofilms. Adv. Mater. **21**, 4793–4797 (2009)
15. J.F. Zang, S. Ryu, N. Pugno, Q.M. Wang, Q. Tu, M.J. Buehler et al., Multifunctionality and control of the crumpling and unfolding of large-area graphene. Nat. Mater. **12**, 321–325 (2013)
16. K.I. Jang, S.Y. Han, S. Xu, K.E. Mathewson, Y.H. Zhang, J.W. Jeong et al., Rugged and breathable forms of stretchable electronics with adherent composite substrates for transcutaneous monitoring. Nat. Commun. **5**, 4779 (2014)
17. S. Xu, Y.H. Zhang, J. Cho, J. Lee, X. Huang, L. Jia et al., Stretchable batteries with self-similar serpentine interconnects and integrated wireless recharging systems. Nat. Commun. **4**, 1543 (2013)

18. D.Y. Khang, H.Q. Jiang, Y. Huang, J.A. Rogers, A stretchable form of single-crystal silicon for high-performance electronics on rubber substrates. Science **311**, 208–212 (2006)
19. N. Bowden, S. Brittain, A.G. Evans, J.W. Hutchinson, G.M. Whitesides, Spontaneous formation of ordered structures in thin films of metals supported on an elastomeric polymer. Nature **393**, 146–149 (1998)
20. J. Jones, S.P. Lacour, S. Wagner, Z.G. Suo, Stretchable wavy metal interconnects. J. Vac. Sci. Technol. A **22**, 1723–1725 (2004)
21. D.Y. Khang, J.A. Rogers, H.H. Lee, Mechanical buckling: mechanics, metrology, and stretchable electronics. Adv. Funct. Mater. **19**, 1526–1536 (2009)
22. J. Song, H. Jiang, Y. Huang, J.A. Rogers, Mechanics of stretchable inorganic electronic materials. J. Vac. Sci. Technol. A **27**, 1107–1125 (2009)
23. J. Song, Mechanics of stretchable electronics. Curr. Opin. Solid State Mater. Sci. **19**, 160–170 (2015)
24. Z.Y. Huang, W. Hong, Z. Suo, Nonlinear analyses of wrinkles in a film bonded to a compliant substrate. J. Mech. Phys. Solids **53**, 2101–2118 (2005)
25. X. Chen, J.W. Hutchinson, Herringbone buckling patterns of compressed thin films on compliant substrates. J. Appl. Mech. Trans. ASME **71**, 597–603 (2004)
26. H.Q. Jiang, D.Y. Khang, J.Z. Song, Y.G. Sun, Y.G. Huang, J.A. Rogers, Finite deformation mechanics in buckled thin films on compliant supports. PNAS **104**, 15607–15612 (2007)
27. J. Song, H. Jiang, Z.J. Liu, D.Y. Khang, Y. Huang, J.A. Rogers et al., Buckling of a stiff thin film on a compliant substrate in large deformation. Int. J. Solids Struct. **45**, 3107–3121 (2008)
28. H.Y. Cheng, Y.H. Zhang, K.C. Hwang, J.A. Rogers, Y.G. Huang, Buckling of a stiff thin film on a pre-strained bi-layer substrate. Int. J. Solids Struct. **51**, 3113–3118 (2014)
29. H.Q. Jiang, D.Y. Khang, H.Y. Fei, H. Kim, Y.G. Huang, J.L. Xiao et al., Finite width effect of thin-films buckling on compliant substrate: experimental and theoretical studies. J. Mech. Phys. Solids **56**, 2585–2598 (2008)
30. W.M. Choi, J.Z. Song, D.Y. Khang, H.Q. Jiang, Y.Y. Huang, J.A. Rogers, Biaxially stretchable "Wavy" silicon nanomembranes. Nano Lett. **7**, 1655–1663 (2007)
31. D.H. Kim, J.H. Ahn, W.M. Choi, H.S. Kim, T.H. Kim, J.Z. Song et al., Stretchable and foldable silicon integrated circuits. Science **320**, 507–511 (2008)
32. R. Huang, S.H. Im, Dynamics of wrinkle growth and coarsening in stressed thin films. Phys. Rev. E **74**, 026214 (2006)
33. J. Song, H. Jiang, W.M. Choi, D.Y. Khang, Y. Huang, J.A. Rogers, An analytical study of two-dimensional buckling of thin films on compliant substrates. J. Appl. Phys. **103**, 014303 (2008)
34. J.F. Zang, C.Y. Cao, Y.Y. Feng, J. Liu, X.H. Zhao, Stretchable and high-performance supercapacitors with crumpled graphene papers. Sci. Rep. **4**, 6492 (2014)
35. S.P. Lacour, J. Jones, S. Wagner, T. Li, Z.G. Suo, Stretchable interconnects for elastic electronic surfaces. Proc. IEEE **93**, 1459–1467 (2005)
36. D.H. Kim, J.Z. Song, W.M. Choi, H.S. Kim, R.H. Kim, Z.J. Liu et al., Materials and noncoplanar mesh designs for integrated circuits with linear elastic responses to extreme mechanical deformations. PNAS **105**, 18675–18680 (2008)
37. H.C. Ko, G. Shin, S.D. Wang, M.P. Stoykovich, J.W. Lee, D.H. Kim et al., Curvilinear electronics formed using silicon membrane circuits and elastomeric transfer elements. Small **5**, 2703–2709 (2009)
38. J. Lee, J. Wu, J.H. Ryu, Z.J. Liu, M. Meitl, Y.W. Zhang et al., Stretchable semiconductor technologies with high areal coverages and strain-limiting behavior: demonstration in high-efficiency dual-junction GaInP/GaAs photovoltaics. Small **8**, 1851–1856 (2012)
39. J. Lee, J.A. Wu, M.X. Shi, J. Yoon, S.I. Park, M. Li et al., Stretchable GaAs photovoltaics with designs that enable high areal coverage. Adv. Mater. **23**, 986–991 (2011)
40. J. Song, Y. Huang, J. Xiao, S. Wang, K.C. Hwang, H.C. Ko et al., Mechanics of noncoplanar mesh design for stretchable electronic circuits. J. Appl. Phys. **105** (2009)
41. R. Li, M. Li, Y.W. Su, J.Z. Song, X.Q. Ni, An analytical mechanics model for the island-bridge structure of stretchable electronics. Soft Matter **9**, 8476–8482 (2013)

42. Y. Su, J. Wu, Z. Fan, K.C. Hwang, J. Song, Y. Huang et al., Postbuckling analysis and its application to stretchable electronics. J. Mech. Phys. Solids **60**, 487–508 (2012)
43. C. Chen, W.M. Tao, Y.W. Su, J. Wu, J.Z. Song, Lateral buckling of interconnects in a noncoplanar mesh design for stretchable electronics. J. Appl. Mech. Trans. ASME **80** (2013)
44. Y.H. Zhang, S. Xu, H.R. Fu, J. Lee, J. Su, K.C. Hwang et al., Buckling in serpentine microstructures and applications in elastomer-supported ultra-stretchable electronics with high areal coverage. Soft Matter **9**, 8062–8070 (2013)
45. Y.H. Zhang, H.R. Fu, Y.W. Su, S. Xu, H.Y. Cheng, J.A. Fan et al., Mechanics of ultra-stretchable self-similar serpentine interconnects. Acta Mater. **61**, 7816–7827 (2013)
46. T. Widlund, S.X. Yang, Y.Y. Hsu, N.S. Lu, Stretchability and compliance of freestanding serpentine-shaped ribbons. Int. J. Solids Struct. **51**, 4026–4037 (2014)
47. T. Li, Z.G. Suo, S.P. Lacour, S. Wagner, Compliant thin film patterns of stiff materials as platforms for stretchable electronics. J. Mater. Res. **20**, 3274–3277 (2005)
48. M. Gonzalez, F. Axisa, M.V. BuIcke, D. Brosteaux, B. Vandevelde, J. Vanfleteren, Design of metal interconnects for stretchable electronic circuits. Microelectron. Reliab. **48**, 825–832 (2008)
49. Y.H. Zhang, S.D. Wang, X.T. Li, J.A. Fan, S. Xu, Y.M. Song et al., Experimental and theoretical studies of serpentine microstructures bonded to prestrained elastomers for stretchable electronics. Adv. Funct. Mater. **24**, 2028–2037 (2014)
50. Y.A. Huang, Y.Q. Duan, Y.J. Ding, N.B. Bu, Y.Q. Pan, N.S. Lu et al., Versatile, kinetically controlled, high precision electrohydrodynamic writing of micro/nanofibers. Sci. Rep. **4**, 5949 (2014)
51. Y.A. Huang, Y.Z. Wang, L. Xiao, H.M. Liu, W.T. Dong, Z.P. Yin, Microfluidic serpentine antennas with designed mechanical tunability. Lab Chip **14**, 4205–4212 (2014)
52. N.S. Lu, C. Lu, S.X. Yang, J. Rogers, Highly sensitive skin-mountable strain gauges based entirely on elastomers. Adv. Funct. Mater. **22**, 4044–4050 (2012)
53. Y.Q. Duan, Y.A. Huang, Z.P. Yin, N.B. Bu, W.T. Dong, Non-wrinkled, highly stretchable piezoelectric devices by electrohydrodynamic direct-writing. Nanoscale **6**, 3289–3295 (2014)
54. L. Gao, Y.H. Zhang, V. Malyarchuk, L. Jia, K.I. Jang, R.C. Webb et al., Epidermal photonic devices for quantitative imaging of temperature and thermal transport characteristics of the skin. Nat. Commun. **5**, 4938 (2014)
55. J. Kim, M. Lee, H.J. Shim, R. Ghaffari, H.R. Cho, D. Son et al., Stretchable silicon nanoribbon electronics for skin prosthesis. Nat. Commun. **5** (2014)
56. L.Z. Xu, S.R. Gutbrod, Y.J. Ma, A. Petrossians, Y.H. Liu, R.C. Webb et al., Materials and fractal designs for 3D multifunctional integumentary membranes with capabilities in cardiac electrotherapy. Adv. Mater. **27**, 1731–+ (2015)
57. Y.A. Huang, W.T. Dong, T. Huang, Y.Z. Wang, L. Xiao, Y.W. Su et al., Self-similar design for stretchable wireless LC strain sensors. Sens. Actuators, A Phys. **224**, 36–42 (2015)
58. Y.W. Su, S.D. Wang, Y.A. Huang, H.W. Luan, W.T. Dong, J.A. Fan et al., Elast. Fractal Inspired Interconnects. Small **11**, 367–373 (2015)
59. Y.H. Zhang, H.R. Fu, S. Xu, J.A. Fan, K.C. Hwang, J.Q. Jiang et al., A hierarchical computational model for stretchable interconnects with fractal-inspired designs. J. Mech. Phys. Solids **72**, 115–130 (2014)
60. J.A. Fan, W.H. Yeo, Y.W. Su, Y. Hattori, W. Lee, S.Y. Jung et al., Fractal design concepts for stretchable electronics. Nat. Commun. **5**, 3266 (2014)
61. Q. Cheng, Z.M. Song, T. Ma, B.B. Smith, R. Tang, H.Y. Yu et al., Folding paper-based lithium-ion batteries for higher areal energy densities. Nano Lett. **13**, 4969–4974 (2013)
62. Z.M. Song, T. Ma, R. Tang, Q. Cheng, X. Wang, D. Krishnaraju et al., Origami lithium-ion batteries. Nat. Commun. **5**, 3140 (2014)
63. Z.Y. Wei, Z.V. Guo, L. Dudte, H.Y. Liang, L. Mahadevan, Geometric mechanics of periodic pleated origami. Phys. Rev. Lett. **110**, 215501 (2013)
64. C. Lv, D. Krishnaraju, G. Konjevod, H.Y. Yu, H.Q. Jiang, Origami based mechanical metamaterials. Sci. Rep. **4**, 5979 (2014)

65. Y. Cho, J.H. Shin, A. Costa, T.A. Kim, V. Kunin, J. Li et al., Engineering the shape and structure of materials by fractal cut. PNAS **111**, 17390–17395 (2014)
66. T.C. Shyu, P.F. Damasceno, P.M. Dodd, A. Lamoureux, L. Xu, M. Shlian et al., A kirigami approach to engineering elasticity in nanocomposites through patterned defects. Nat. Mater. **14**, 785–789 (2015)
67. Z.M. Song, X. Wang, C. Lv, Y.H. An, M.B. Liang, T. Ma et al., Kirigami-based stretchable lithium-ion batteries. Sci. Rep. **5**, 10988 (2015)
68. K.-I. Jang, H.U. Chung, S. Xu, C.H. Lee, H. Luan, J. Jeong et al., Soft network composite materials with deterministic and bio-inspired designs. Nat. Commun. **6**, 6566 (2015)

Chapter 4
Soft Power: Stretchable and Ultra-Flexible Energy Sources for Wearable and Implantable Devices

Timothy F. O'Connor, Suchol Savagatrup and Darren J. Lipomi

Abstract The development of ultra-compliant power sources is prerequisite to the realization of imperceptible biomedical systems destined to be worn or implanted in the human body. This chapter assesses the viability of conformal piezo and triboelectric, thermoelectric, and photovoltaic technologies as power sources for biomedical applications. It begins by identifying the amount of energy available to each of these modes of power conversion and then gives a brief overview on the methods of fabricating stretchable electronic devices using deterministic structures, random composites, or molecularly stretchable electronic materials. It then provides a detailed description of innovations in "soft power," where the mentioned design techniques have been employed to develop mechanically compliant power scavengers amenable to integration with stretchable medical devices. The chapter concludes with an analysis of system level power requirements and application specific compatibility, the result of which identifies piezoelectrics and triboelectrics as well suited for intermittent and implantable devices, such as low-power pacemakers for piezoelectrics or higher power wearables and neural stimulators for triboelectrics. Thermoelectrics are highly compatible with epidermal and wearable applications, and can be used as a consistent source of power for tattoo chemical or heat sensors, and photovoltaics can generate large amounts of power in full sun, for high power applications like cochlear implants, or less energy in diffuse or ambient light, for powering hearing aids.

Keywords Stretchable electronics · Stretchable biomedical devices · Wearable power sources · Wearable electronics · Organic electronics · Implantable power sources

T.F. O'Connor · S. Savagatrup · D.J. Lipomi (✉)
Department of NanoEngineering, University of California, 9500 Gilman Drive,
Mail Code 0448, La Jolla, San Diego, CA 92093-0448, USA
e-mail: dlipomi@ucsd.edu

© Springer International Publishing Switzerland 2016
J.A. Rogers et al. (eds.), *Stretchable Bioelectronics for Medical Devices and Systems*, Microsystems and Nanosystems,
DOI 10.1007/978-3-319-28694-5_4

4.1 Introduction

An attractive aspect of biointegrated electronics from the standpoint of the research community is the opportunity to reimagine the components of conventional microelectronics [1–7]. Rigid integrated circuits on planar substrates—always connected to a stable source of power—must be transformed into form factors that can conform to the curved and soft surfaces of biological tissue, and which must store or harvest their own power. Delivering, managing, and harvesting energy to power these implantable and wearable devices are critical to the development of this technology [8, 9]. Seamless integration of stretchable power sources into biomedical devices requires the development of "soft power"—highly deformable systems for harvesting and storing energy. This chapter will highlight strategies that enable the design and production of stretchable and ultra-flexible devices for energy harvesting. We define stretchability and ultra-flexibility by the capacity to withstand significant deformation without degradation in performance. We identify four relevant technologies (piezoelectric, triboelectric, thermoelectric, and photovoltaic), their potential applications based on availability of power sources, method of transduction, the performance of the devices (i.e., power output, lifetime of the device, and mechanical properties), and methods of producing them in form factors that are highly deformable.

We preface this chapter by describing the availability of the viable sources of power, Fig. 4.1. Scavenging energy from the human body could potentially be one of the most convenient methods of powering and extending the operation of

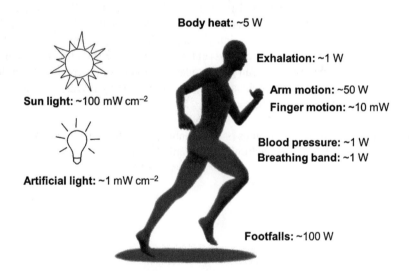

Body heat: ~5 W

Exhalation: ~1 W

Sun light: ~100 mW cm⁻²

Arm motion: ~50 W
Finger motion: ~10 mW

Blood pressure: ~1 W
Breathing band: ~1 W

Artificial light: ~1 mW cm⁻²

Footfalls: ~100 W

Fig. 4.1 Total power available from body-driven sources in comparison to sunlight and artificial light sources. Adapted from Ref. [10]

biomedical devices [8]. At rest, the human body generates roughly 7 W of power ($\sim$5 W from heat, 1 W from breathing, and 1 W from blood pressure) [10]. When in motion, the human body has an additional 100 W of kinetic power available, along with a small amount of power, on the order of 10 mW, that can be collected from small motions of extremities [10]. Physical methods of energy harvesting are usually based on transducers utilizing mechanical energy, such as heart beats [11], blood flow [12], walking [13, 14], breathing [15], and stretching of muscles [16]. Methods of harvesting additional energy from the components of sweat—i.e., bioelectrocatalytic glucose oxidation [17–19]—are exciting developments in the field [20], but our focus is on power that can be harvested from physical motions of the human body and from ambient light. In particular, the sun (which provides around 100 mW cm^{-2} under ideal conditions) and indoor light sources, which can provide power on the order of 1 mW cm^{-2}. Current wearable biomedical devices have power requirements ranging from 1 μW to 10^2 mW, thus the energy available from the human body and external light sources may be sufficient to power these devices [21]. (For the sake of space, we do not cover mechanically compliant devices for energy storage, but direct the reader to the work of others in this area [22–24]).

4.2   Approaches to Making Stretchable Electronics

Three main strategies have been identified for developing stretchable electronics (Fig. 4.2) [25]. The first involves the top-down fabrication of deterministic structures that render otherwise rigid materials (e.g., silicon, metals, and ceramics) stretchable by converting global strains into local deformations of the components on a macroscopic scale. These technologies often use an "island-bridge" approach, whereby rigid components (islands) fabricated on or in a stretchable matrix are connected by fractal or serpentine interconnects (bridges) [26, 27]. Devices can also be compressed to create sinusoidal structures through buckling instabilities [28]. These structures transfer the strain associated with elongation into a decrease in the amplitude (and corresponding increase in the wavelength) of the buckled structures. The second method—random composites—takes advantage of high aspect ratio structures (i.e., nanowires or nanotubes) that form contiguous networks when deposited on or in some elastic support [29–32]. As the device is stretched, one-dimensional structures undergo configurational changes (i.e., rotation, straightening, and sliding past each other), rather than fracturing, allowing electronic performance to be maintained while deformed. Using this method, stretchable electrodes have been fabricated that can accommodate over 400 % strain [33]. The third, complementary approach is intrinsically stretchable electronics, where the active electronic layers themselves accommodate the strain [25, 29, 34]. These

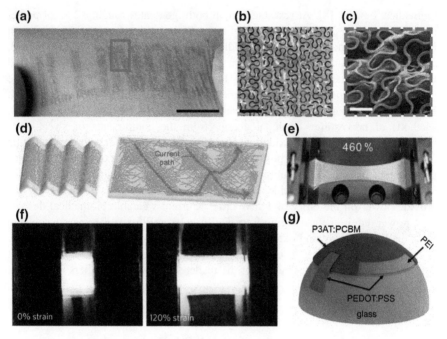

Fig. 4.2 Images of representative samples of strategies for engineering stretchable electronics: deterministic patterning of waves and fractals (**a–c**), percolation of random networks and composites (**d, e**), and use of intrinsically stretchable elastomer (**f, g**). **a** Photograph of metallic wires in a fractal pattern adhered to skin (*scale bar* 1 cm). A *blow-up* within the region indicated by the *red box* is shown in **b** the optical micrograph (*scale bar* 1 mm); and in **c** the scanning electron micrograph (*scale bar* 500 µm). **d** Schematic behavior of the very long Ag NW percolation network (VAgNPN) electrode and an Ecoflex substrate during stretching; **e** photograph of the surface morphology of an electrode on a pre-strained Ecoflex during a 460 % stretching process. **f** Intrinsically stretchable light-emitting devices, employing *Super Yellow*, a polyphenylenevinylene derivatives. **g** Schematic diagram of hemispherical solar cells using intrinsically stretchable blend of P3AT and PCBM as the active layer. **a–c** Reproduced with permission from Ref. [27]. Copyright 2014, Nature Publishing Group **d, e** Reproduced with permission from Ref. [33]. Copyright 2012, Wiley-VCH Verlag GmbH & Co. KGaA **f** Reproduced with permission from Ref. [37]. Copyright 2013, Nature Publishing Group **g** Reproduced with permission from Ref. [38]. Copyright 2014, Royal Society of Chemistry

devices are generally made from solution-processable organic materials, which in principle are amenable to high-throughput fabrication techniques [35]. The challenge of intrinsically—or "molecularly" [25]—stretchable systems is that electronic and mechanical properties are generally mutually antagonistic, and thus a material exhibiting state-of-the-art semiconducting properties with the mechanical properties of an elastomer has not yet been demonstrated [36].

4.3 Stretchable Energy Harvesting Technologies

4.3.1 Piezoelectric and Triboelectric

Flexible biomechanical energy harvesters are continuing to show their potential in converting the kinetics of the human body into usable power or signals for devices like pacemakers and pressure sensors, though many of these devices are made of thin films of rigid or brittle ceramics [8, 9, 39–41]. By nature of the transduction mechanism—which only requires physical deformation of the device structure—piezoelectric energy harvesters can be employed as power for both wearable and implantable devices. Due to the intermittency of their energy output and the fact that most of the energy put into the devices is used up to deform the crystal or material structure, piezoelectrics rarely provide the appropriate electronic outputs to continuously operate most biomedical devices without storage [8]. However, piezoelectrics can be used to extend the lifetime of implantable devices, and highly sensitive piezoelectrics can directly transduce physical stimuli for sensors that require no external power supplies [8]. Most high-performance piezoelectric materials are brittle with high-tensile moduli [42]; and the devices reviewed here are the state of the art for stretchable or soft piezoelectrics.

In work by McAlpine and co-workers, brittle lead zirconate titanate (PZT, Pb $[Zr_{0.52}Tl_{0.48}]O_3$) nanoribbons were deterministically patterned into wave-like structures, allowing for flexing and stretching operating modes by transferring the mechanical strain to the amplitudes and wavelength of the buckled structures (Fig. 4.3a–c) [42]. The buckled PZT nanoribbons exhibited nearly a two order of magnitude increase in maximum tensile strain without failure over the non-buckled counterparts, 8 % versus 0.1 %, by transferring the elongation to a reduction in the amplitude of the waves. Moreover, the structures also exhibited an enhanced electromechanical performance attributed to a flexoelectric contribution to the piezoelectric coefficient, leading to the peak power density of 2.5 W cm^{-3} under uniaxial deformation [42].

Examples of high performance devices exhibiting biaxial stretchability typically comprise functional nanocomposites in elastomeric substrates. Yao and coworkers demonstrated lead magnesium niobate-lead titanate (PMN-PT) nanoclusters (Fig. 4.3d–f) embedded in elastomeric substrates that generated voltages ranging from 4.2 to 7.8 V in an open circuit and currents ranging from 1.58 to 2.29 µA, yielding maximum instantaneous power outputs of roughly 36 µW cm^{-2} [43]. The lowest strain in these devices that produced a piezoelectric response was ~ 0.01 %, making them promising for applications in self-powered pressure sensors. Similarly, Huang et al. developed wearable triboelectric nanogenerators capable of harnessing highly available energy from walking (Fig. 4.3g–i) [44]. The all-fiber PVDF insoles were fabricated by electrospinning, which produced fibrils with nanostructured features that improved triboelectric performance. At a step

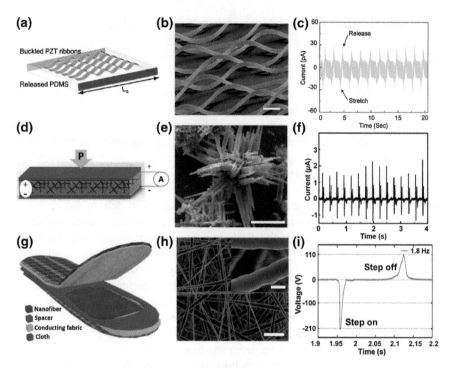

Fig. 4.3 Representative examples of piezoelectric and triboelectric devices that are possible for biointegration. **a** Schematic of lead zirconate titanate (PZT) nanoribbons and **b** scanning electron micrograph of the PZT ribbons transfer-printed to prestrained elastomeric substrate that produced the buckling morphology (*scale bar* 20 μm). **c** Short-circuit current measured from devices comprising 5 PZT ribbons under periodic stretching of 8 % strain. **d** Schematic of PMN-PT nanowire-based nanocomposite comprising PMN-PT nanowires embedded in elastomeric substrate, PDMS, and polyimide/gold electrodes. **e** High-magnification scanning electron micrograph of dispersed PMN-PT nanowires (*scale bar* 5 μm). **f** Signal generation from PMN-PT nanocomposite showing the current generation under a periodic mechanical tapping. **g** Schematic diagram of the wearable all-fiber triboelectric nanogenerator (TENG)-based insole composed of electrospun piezoelectric polyvinylidene fluoride (PVDF) nanofibers. **h** Scanning electron micrograph PVDF nanofibers (*scale bar* 10 μm; *inset scale bar* 500 nm). **i** Voltage-time curve corresponding to the mechanical stimuli. **a–c** Reproduced with permission from Ref. [42]. Copyright 2011, American Chemical Society. **d–f** Reproduced with permission from Ref. [43]. Copyright 2013, American Chemical Society. **g–i** Reproduced with permission from Ref. [44]. Copyright 2015, Elsevier Ltd

frequency of 1.8 Hz (or typical walking speed of ∼4 km h^{-1} with an average stride length), the devices produced a maximum output voltage, instantaneous power, and output current of 210 V, 2.1 mW, and 45 μA, respectively [44]. The active materials take the form of nanowoven fabrics that can be integrated into mechanoelectric textiles, including shirts, pants, and insoles [44, 45]. These devices can be potentially used to charge batteries or for neural stimulators [8].

4.3.2 Thermoelectric

Thermoelectric power generators (TEGs) can provide continuous power to wearable or subdermal biomedical devices and sensors. Unlike piezoelectric and triboelectric devices, TEGs offer a means of continuous, stable power, however they are limited by the small temperature gradients afforded by the limitations of human tolerance (between 2 and 5 K) and loss of latent heat through evaporation of sweat [13]. Notwithstanding these limitations, a lightweight, flexible TEG module was recently reported by Cho and coworkers [46, 47]. The device comprised a screen-printed inorganic porous thick film of n-type bismuth telluride (Bi_2Te_3) and p-type antimony telluride (Sb_2Te_3), infiltrated with poly(3,4-ethylenedioxythiophene):poly(styrene-sulfonate) (PEDOT:PSS) to increase the electrical conductivity and provide mechanical flexibility. With the addition of PEDOT:PSS in the composite TEGs exhibited a 10 % increase in the dimensionless figure of merit (ZT) over their screen printed, purely inorganic counterparts, and maintained high conductivity when devices were bent to 3 cm radii of curvature (Fig. 4.4a–b). At a temperature difference of 10 K, with the cold side held at 283 K, devices generated an output voltage of 19.1 mV and an output power density of 60 µW cm^{-2} [46]. To demonstrate the

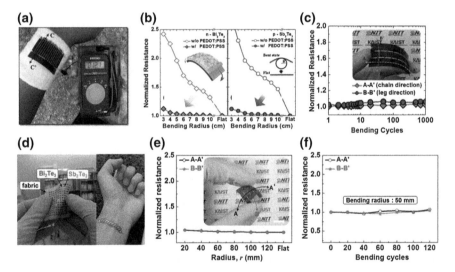

Fig. 4.4 Wearable thermoelectric devices. **a–c** Hybrid inorganic-organic composite thermoelectric modules. **a** Device in operation on the human body. **b** Normalized resistance changes before (open points) and after (closed points) PEDOT:PSS infiltration as a function of bending radius for Bi_2Te_3 (*left*) and Sb_2Te_3 (*right*). **c** Resistance changes of the module as a function of the number of bending cycles with a bending radius of 40 mm. **d–f** Wearable thermoelectric generator fabricated on a glass fabric. **d** Image of 196 Bi_2Te_3 and Sb_2Te_3 dots on glass fabric of 40 mm × 40 mm (*left*) and a complete device mounted on human skin (*right*). **e** Resistance stability of the device under bending stress along two bending axes as a function of bending radius. **f** Stability of 120 bending cycles at bending radius of 50 mm. **a–c** Reproduced with permission from Ref. [46]. Copyright 2014, Elsevier Ltd. **d–f** Reproduced with permission from Ref. [47]. Copyright 2014, Royal Society of Chemistry

mechanical endurance and potential for wearable energy applications, a TEG consisting of seven thermoelectric couples was subjected to cyclic mechanical bending (radius of curvature of 4 cm) of over 1,000 cycles. Figure 4.4c shows fatigue strength of the device, which only exhibits minimal increase in resistance.

In another study by the same group, Kim et al. improved the mechanical compliance of the TEGs by screen printing dots of Bi_2Te_3 and Sb_2Te_3 onto a woven glass fabric and subsequently sealed inside a PDMS encapsulant (Fig. 4.4d) [47]. The glass fabric, which served as a mechanically compliant substrate, also increased power generation by interrupting phonon propagation, thus reducing the thermal conductivity of the printed TE films. To demonstrate the use of this technology as a power source for wearable biomedical devices, a device comprising 11 thermocouples in the shape of a bandage was bonded to the surface of the skin. The device created an output power of 3 μW, with an open circuit voltage of 2.9 mV, on a matched external load with an air temperature of 15 °C [47]. A similar prototype consisting of eight couples was subjected to mechanical testing, whereby the device showed no significant change (<5 %) in the internal resistance with the allowed bending radius of 20 mm; furthermore, devices showed <7 % decrease in internal resistance when repeatedly bent up to 120 cycles with a radius of curvature of 50 mm (Fig. 4.4e–f) [47]. The reported power density is sufficient to activate sub-microwatt or microwatt wearable devices such as a temperature sensor or a CMOS image sensor [47].

4.3.3 Photovoltaic Devices

Unlike the two previous sections in which the energy outputs are limited by scavenging physical and thermal energy sources of the human body, photovoltaic (PV) devices have the potential to produce substantially more energy than piezoelectric and thermoelectric devices. However, the main limitations of PV devices for wearable applications will most likely be (1) the availability and the intensity of the light sources and (2) the surface area required for the photoactive components. The power output of PV devices will be significantly lower under diffuse outdoor light or ambient indoor light rather than ideal sunlight, and for devices with modest efficiency, larger active areas will be required for a viable power output. Despite the challenges related to indoor power, several photovoltaic technologies can be made in ultra-flexible or stretchable form factors while still providing useful power densities [48].

Crystalline semiconductors of which most high performing solar cells are composed are extremely brittle; however, careful engineering of the materials and creative approaches to the layout of the devices can significantly increase the deformability of whole modules. One of the first examples of stretchable solar cells was introduced by Rogers and coworkers by exploiting the "island-bridge" approach [49]. Gallium arsenide (GaAs) solar cells (roughly ∼3.6 μm thick) were transfer-printed onto prestrained elastomeric PDMS substrate with thin gold interconnects between active devices (Fig. 4.5a–b) [49]. Trenches between each active device absorbed the strains

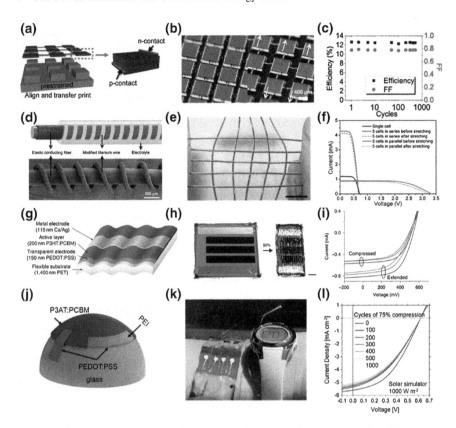

Fig. 4.5 Stretchable photovoltaic devices. **a–c** Stretchable GaAs solar cells fabricated on "island-bridge" architecture **a** Schematic diagram. **b** Unstretched cells (*left*) and 20 % biaxial strain (*right*). **c** Device performance as a function of the number of stretching cycle at 20 % biaxial strain. **d–f** Stretchable and wearable dye-sensitized solar cells comprising elastic conducting fiber, modified titanium wire, and electrolyte. **d** Schematic diagram and scanning electron micrograph of the device. **e** Photograph of a stretchable photovoltaic textile after stretching (*scale bar* 2 cm). **f** *J–V curves* of the photovoltaic textile in series and parallel before and after stretching. **g–i** Organic solar cells fabricated on ultra-thin polyester substrates. **g** Schematic diagram. **h** Prestrained substrate of ultra-thin polyester at flat (*left*) and 50 % (*right*) quasi-linear compression (*scale bar* 2 mm). **i** Device performance for 1 (*black*), 11 (*red*), and 22 (*blue*) cycles for both the fully extended and 50 % compressed states. **j** Hemispherical solar cells using intrinsically stretchable blend of P3AT and PCBM as the active layer. **k** Wearable, ultra-flexible organic solar cells comprising the composite of P3HpT:PCBM powering a digital watch in natural sunlight (98 mW cm⁻²). (l) The performance of wearable organic solar cells measured in air over 1,000 cycles of 75 % compressive strain. **a–c** Reproduced with permission from Ref. [49]. Copyright 2011, Wiley-VCH Verlag GmbH & Co. KGaA. **d–f** Reproduced with permission from Ref. [50]. Copyright 2014, Wiley-VCH Verlag GmbH & Co. KGaA. **g–i** Reproduced with permission from Ref. [51]. Copyright 2012 Nature Publishing Group. **j** Reproduced with permission from Ref. [38]. Copyright 2014, Royal Society of Chemistry. **k–l** Reproduced with permission from Ref. [52]

caused by bending or stretching of the devices, allowing biaxial stretching of 20 % strain (over 500 cycles) without degradation in performance (Fig. 4.5c) [49]. These devices performed identically in the relaxed and stretched states, providing the power conversion efficiency (*PCE*) under one sun condition (100 mW cm^{-2}) of ~ 13 % (13 mW cm^{-2}) [49]. Another example of transforming rigid materials into stretchable devices was described by Peng and coworkers; dye-sensitized solar cells were fabricated in a spring-like architecture to accommodate 30 % uniaxial strain (Fig. 4.5d–e) [50]. The solar cell consisted of two stretchable fiber electrodes (a rubber fiber wrapped with conductive multi-walled carbon nanotube sheets and a modified active titanium wire), both of which were encapsulated by a transparent polyethylene tube. The device was completed by filling the cavity of the tube with a liquid redox electrolyte and sealing the device [50]. A single cell of the wire-like solar harvester exhibited *PCE* of 7.13 % when unstretched [50]. Multiple devices were assembled into a stretchable "photovoltaic textile" comprising five cells connected in series and parallel. This assembly performed similarly when stretched and retained ~ 90 % of its original efficiency when subjected to 50 cycles of 20 % strain (Fig. 4.5f) [50].

Unlike the previous two examples, organic photovoltaics are typically much thinner (roughly ~ 200 nm); this thinness drastically increases the flexibility (provided a sufficiently thin substrate is used). In one of the most impressive demonstrations of ultra-flexibility of organic solar cells to date, Kaltenbrunner et al. fabricated ultra-thin organic solar cells based on a composite of poly (3-hexylthiophene) and [6,6]-phenyl C_{61} butyric acid methyl ester (P3HT:PCBM) on a 1.4 μm polyester foil (Fig. 4.5a) [51]. This solar cell holds the current record for specific power of organic solar cells (10 W g^{-1}) and can achieve bending radii of ~ 35 μm [51]. Using a similar approach to that described earlier by Lipomi et al. [53] the authors showed that the devices can also be reversibly compressed to 50 % of their original size when bonded to prestretched elastomeric substrates (Fig. 4.5b). Under one sun illumination (100 mW cm^{-2}), the *PCE* was measured to be around 4 % (or power density of 4 mW cm^{-2}) before deformation, and around 3 % (3 mW cm^{-2}) after 22 cycles of 50 % compression (Fig. 4.5c) [51].

Recently, our laboratory reported stretchable solar cells capable of being conformally bonded to hemispherical surfaces (Fig. 4.5j) [38]. The all-organic, fully stretchable solar cell, comprising a composite of poly(3-octylthiophene) and PCBM (P3OT:PCBM), was prefabricated onto an elastomeric substrate, then transferred to a glass hemispherical surface through contact printing. The study was a demonstration of the significant increase in compliance of the resulting devices by increasing the molecular side chain length of semiconducting polymers [54]. Devices, fabricated from a composite of P3HT:PCBM (P3HT has six carbon atoms per side chain, $n = 6$), have superior electrical performance on a flat configuration but cracked under modest strain; while devices comprising P3OT:PCBM ($n = 8$) performed worse on flat substrates but retained functionality when applied onto hemispherical substrate (conformal bonding required the solar cell to be stretched by ~ 24 % strain) [38]. Additionally, we found that poly(3-heptylthiophene) and PCBM (P3HpT:PCBM, $n = 7$) exhibited both high charge carrier mobilities and high compliance [36], enabling us to fabricate an ultra-flexible, wearable organic solar cells capable of powering wearable devices [52]. This ultra-flexible device conforms to the human body and generated uninterrupted power over repeated compressive strain of 75 % and a tensile strain of 5 % [52]. These wearable solar cells provided up to 500 μW (power density of 1 mW cm^{-2}) when measured with natural sunlight (98 mW cm^{-2}), and $\sim$5 μW (power density of 10 μW cm^{-2}) when measured indoors in diffuse artificial light [52]. These devices are promising for applications such as powering wearable applications (e.g., wearable biosensors) due to their relatively high power density, high specific power, and extreme mechanical durability.

4.4 System Level Power Requirements

While we introduced many examples of power generation devices, it is crucial to address the viability of each option by evaluating its compatibility with different biomedical applications. We will base our discussion on the power generation capability of the power sources and the feasibility of incorporating them onto the given applications. Figure 4.6 highlights the examples of several biomedical devices and the range of their typical power consumption, along with the range of energy generation for the technologies outlined previously in the chapter. Photovoltaic devices, under a full sun, could produce substantial amount of power to meet the typical power consumption of most biomedical devices [21]. For example, given an active area of 100 cm^2 (roughly the size of standard index card),

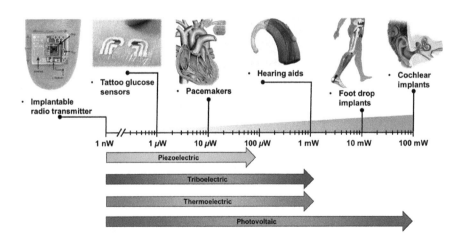

Fig. 4.6 Order of magnitude of typical power consumption of biomedical devices and the range of power generation by the four technologies outlined in this chapter. Images reproduced with permission from Ref. [58], copyright 2012, Nature Publishing Group. Images adapted from Refs. [21, 56, 57, 60]

organic solar cells can potentially produce 300 mW while GaAs solar cells upward of 1.3 W. Purely from the energy production standpoint, these PV devices would be able to power hearing aids, foot-drop implants [55, 56], and cochlear implants [57]. However, this energy production would be approximately two orders of magnitude lower in diffuse light and nonexistent without a light source. In addition, integration of PV devices and implantable devices will not be seamless and overcoming the users' barriers in terms of appeal, aesthetics, and comfort poses further challenges. Thermoelectric devices would be ideal to provide continuous power generation; despite the low efficiency arising from the low temperature gradient the body can endure, the power generated can be used to power small electronics and biosensors. For example, Mercier et al. fabricated an extremely low-power implantable chip that measures the endocochlear potential and transmits the measurement via a 2.4 GHz radio signal while consuming power on the order of 1 nW [58]. Tattoo-based technology, whose operating locations coincide with the largest temperature gradient on the human body, may also benefit from thermoelectric power generators. Epidermal electronics incorporate many electronic functionalities (e.g., temperature and strain sensors, transistors, light-emitting diodes, photodetectors, and radio frequency inductors) [59] and medical applications (e.g., pH sensor, sodium sensor, and ammonium sensor) [19, 60]. Piezoelectric and triboelectric devices provide possible sources of power for implantable devices that only consume energy intermittently. The main example is powering pacemakers by harvesting energy from the motions of the heart [61]. Improvements on current pacemakers are also aimed at reducing the power consumption by almost an order of magnitude [12, 62]. Also, the large amount of kinetic energy available in foot-falls makes triboelectric generators a viable technology for foot-drop neural stimulators, where energy generation and signal transduction could be performed simultaneously.

4.5 Conclusions and Challenges

Ultra-flexible and stretchable power sources will be essential components of future biointegrated medical devices. We have described several methods of energy harvesting that are compatible with biointegration; making devices compatible with soft, biological structures is tantamount to rendering them extraordinarily mechanically compliant. Before seamless integration between biomedical devices and power generators can be realized, more collaborative and multidisciplinary studies between the two fields of biology and electronics will be required. Understanding the transitional steps to bridge the two sides of science and engineering will most likely produce fruitful discovery and unexpected problems that could potentially bring us closer to fully functional and self-powered biomedical devices.

References

1. D.-H. Kim, J.A. Rogers, Adv. Mater. **20**, 4887 (2008)
2. J.A. Rogers, T. Someya, Y. Huang, Science **327**, 1603 (2010)
3. D.-H. Kim, J. Song, W.M. Choi, H.-S. Kim, R.-H. Kim, Z. Liu, Y.Y. Huang, K.-C. Hwang, Y. Zhang, J. Rogers, Proc. Natl. Acad. Sci. U.S.A. **105**, 18675 (2008)
4. M. Kaltenbrunner, T. Sekitani, J. Reeder, T. Yokota, K. Kuribara, T. Tokuhara, M. Drack, R. Schwödiauer, I. Graz, S. Bauer-Gogonea, S. Bauer, T. Someya, Nature **499**, 458 (2013)
5. D. Ghezzi, M.R. Antognazza, R. Maccarone, S. Bellani, E. Lanzarini, N. Martino, M. Mete, G. Perile, S. Bisti, G. Lanzani, F. Benfenati, Nat. Photon. **7**, 400 (2013)
6. L. Xu, S.R. Gutbrod, A.P. Bonifas, Y. Su, M.S. Sulkin, N. Lu, H.-J. Chung, K.-I. Jang, Z. Liu, M. Ying, C. Lu, R.C. Webb, J.-S. Kim, J.I. Laughner, H. Cheng, Y. Liu, A. Ameen, J.-W. Jeong, G.-T. Kim, Y. Huang, I.R. Efimov, J. Rogers, Nat. Commun. **5**, 3329 (2014)
7. T.F. O'Connor, K.M. Rajan, A.D. Printz, D.J. Lipomi, Mater. Chem. B **3**, 4947 (2015)
8. C.Y. Sue, N.C. Tsai, Appl. Energy **93**, 390 (2012)
9. J. Yun, S.N. Patel, M.S. Reynolds, G.D. Abowd, IEEE Trans. Mob. Comput. **10**, 669 (2011)
10. T. Starner, IBM Syst. J. **35**, 618 (1996)
11. Z. Li, G. Zhu, R. Yang, A.C. Wang, Z.L. Wang, Adv. Mater. **22**, 2534 (2010)
12. M. Deterre, E. Lefeuvre, Y. Zhu, M. Woytasik, B. Boutaud, R.D.J. Molin, Microelectromech. Syst. **23**, 651 (2014)
13. R. Riemer, A. Shapiro, J. Neuroeng. Rehabil. **8**, 22 (2011)
14. L.C. Rome, L. Flynn, E.M. Goldman, T.D. Yoo, Science **309**, 1725 (2005)
15. C. Sun, J. Shi, D.J. Bayerl, X. Wang, Energy Environ. Sci. **4**, 4508 (2011)
16. R. Yang, Y. Qin, C. Li, G. Zhu, Z.L. Wang, Nano Lett. **9**, 1201 (2009)
17. V. Coman, R. Ludwig, W. Harreither, D. Haltrich, L. Gorton, T. Ruzgas, S. Shleev, Fuel Cells **10**, 9 (2010)
18. C. Pan, Y. Fang, H. Wu, M. Ahmad, Z. Luo, Q. Li, J. Xie, X. Yan, L. Wu, Z.L. Wang, J. Zhu, Adv. Mater. **22**, 5388 (2010)
19. A.J. Bandodkar, W. Jia, J. Wang, Electroanalysis **27**, 562 (2015)
20. E. Katz, in, ed. by E. Katz, *Implantable Bioelectronics* (Wiley-VCH Verlag GmbH & Co. KGaA, 2014), p. 363
21. M. Deterre, *Toward an Energy Harvester for Leadless Pacemakers* (University of Paris-Sud, 2013)
22. M. Kaltenbrunner, G. Kettlgruber, C. Siket, R. Schwödiauer, S. Bauer, Adv. Mater. **22**, 2065 (2010)
23. S. Xu, Y. Zhang, J. Cho, J. Lee, X. Huang, L. Jia, J.a Fan, Y. Su, J. Su, H. Zhang, H. Cheng, B. Lu, C. Yu, C. Chuang, T.-I. Kim, T. Song, K. Shigeta, S. Kang, C. Dagdeviren, I. Petrov, P. V. Braun, Y. Huang, U. Paik, J.A. Rogers, Nat. Commun. **4**, 1543 (2013)
24. L. Hu, H. Wu, F. La Mantia, Y. Yang, Y. Cui, ACS Nano **4**, 5843 (2010)
25. S. Savagatrup, A.D. Printz, T.F. O'Connor, A.V. Zaretski, D.J. Lipomi, Chem. Mater. **26**, 3028 (2014)
26. S.P. Lacour, J. Jones, S. Wagner, Proc. IEEE **93**, 1459 (2005)
27. J. Fan, W.-H. Yeo, Y. Su, Y. Hattori, W. Lee, S.-Y. Jung, Y. Zhang, Z. Liu, H. Cheng, L. Falgout, M. Bajema, T. Coleman, D. Gregoire, R.J. Larsen, Y. Huang, J. Rogers, Nat. Commun. **5**, 3266 (2014)
28. D.-Y. Khang, J.a Rogers, H.H. Lee, Adv. Funct. Mater. **19**, 1526 (2009)
29. Z. Yu, X. Niu, Z. Liu, Q. Pei, Adv. Mater. **23**, 3989 (2011)
30. J. Liang, L. Li, K. Tong, Z. Ren, W. Hu, X. Niu, Y. Chen, Q. Pei, ACS Nano **8**, 1590 (2014)
31. D.J. Lipomi, M. Vosgueritchian, B.C.-K. Tee, S.L. Hellstrom, J.A. Lee, C.H. Fox, Z. Bao, Nat. Nanotech. **6**, 788 (2011)
32. S.P. Lacour, D. Chan, S. Wagner, T. Li, Z. Suo, Appl. Phys. Lett. **88**, 204103 (2006)
33. P. Lee, J. Lee, H. Lee, J. Yeo, S. Hong, K.H. Nam, D. Lee, S.S. Lee, S.H. Ko, Adv. Mater. **24**, 3326 (2012)

34. G. Kettlgruber, M. Kaltenbrunner, C.M. Siket, R. Moser, I.M. Graz, R. Schwödiauer, S. J. Bauer, Mater. Chem. A **1**, 5505 (2013)
35. F.C. Krebs, N. Espinosa, M. Hösel, R.R. Søndergaard, M. Jørgensen, Adv. Mater. **26**, 29 (2014)
36. S. Savagatrup, A.D. Printz, H. Wu, K.M. Rajan, E.J. Sawyer, A.V. Zaretski, C.J. Bettinger, D. J. Lipomi, Synth. Met. **203**, 208 (2015)
37. J. Liang, L. Li, X. Niu, Z. Yu, Q. Pei, Nat. Photonics **7**, 817 (2013)
38. T.F. O'Connor, A.V. Zaretski, B.A. Shiravi, S. Savagatrup, A.D. Printz, M.I. Diaz, D. J. Lipomi, Energy Environ. Sci. **7**, 370 (2014)
39. G.T. Hwang, H. Park, J.H. Lee, S. Oh, KIl Park, M. Byun, H. Park, G. Ahn, C.K. Jeong, K. No, H. Kwon, S.G. Lee, B. Joung, K. Lee, J. Adv. Mater. **26**, 4880 (2014)
40. Y. Qi, M.C. McAlpine, Energy Environ. Sci. **3**, 1275 (2010)
41. M. Lee, C.Y. Chen, S. Wang, S.N. Cha, Y.J. Park, J.M. Kim, L.J. Chou, Z.L. Wang, Adv. Mater. **24**, 1759 (2012)
42. Y. Qi, J. Kim, T.D. Nguyen, B. Lisko, P.K. Purohit, M.C. McAlpine, Nano Lett. **11**, 1331 (2011)
43. S. Xu, Y.W. Yeh, G. Poirier, M.C. McAlpine, R.A. Register, N. Yao, Nano Lett. **13**, 2393 (2013)
44. T. Huang, C. Wang, H. Yu, H. Wang, Q. Zhang, M. Zhu, Nano Energy **14**, 226 (2015)
45. C.K. Jeong, J. Lee, S. Han, J. Ryu, G.-T. Hwang, D.Y. Park, J.H. Park, S.S. Lee, M. Byun, S. H. Ko, K. Lee, J. Adv. Mater. **27**, 2866 (2015)
46. J.H. We, S.J. Kim, B.J. Cho, Energy **73**, 506 (2014)
47. S.J. Kim, J.H. We, B.J. Cho, Energy Environ. Sci. **2014**, 7 (1959)
48. D.J. Lipomi, Z. Bao, Energy Environ. Sci. **4**, 3314 (2011)
49. J. Lee, J. Wu, M. Shi, J. Yoon, SIl Park, M. Li, Z. Liu, Y. Huang, J.A. Rogers, Adv. Mater. **23**, 986 (2011)
50. Z. Yang, J. Deng, X. Sun, H. Li, H. Peng, Adv. Mater. **26**, 2643 (2014)
51. M. Kaltenbrunner, M.S. White, E.D. Głowacki, T. Sekitani, T. Someya, N.S. Sariciftci, S. Bauer, Nat. Commun. **3**, 770 (2012)
52. T.F. O'Connor, A.V. Zaretski, S. Savagatrup, A.D. Printz, C.D. Wilkes, M.I. Diaz, E. J. Sawyer, D.J. Lipomi, Sol. Energy Mater. Sol. Cells **144**, 438 (2016)
53. D.J. Lipomi, B.C.-K. Tee, M. Vosgueritchian, Z. Bao, Adv. Mater. **23**, 1771 (2011)
54. S. Savagatrup, A.S. Makaram, D.J. Burke, D.J. Lipomi, Adv. Funct. Mater. **24**, 1169 (2014)
55. M. Haugland, C. Childs, M. Ladouceur, J. Haase, T. Sinkjaer, in *Proceedings of the 5th Annual IFESS Conference* (2000), p. 59
56. G.M. Lyons, T. Sinkjær, J.H. Burridge, D.J. Wilcox, IEEE Trans. Neural Syst. Rehabil. Eng. **10**, 260 (2002)
57. M.W. Baker, *A Low-Power Cochlear Implant System, Massachusetts Institute of Technology* (2007)
58. P.P. Mercier, A.C. Lysaght, S. Bandyopadhyay, A.P. Chandrakasan, K.M. Stankovic, Nat. Biotechnol. **30**, 1240 (2012)
59. D.-H. Kim, N. Lu, R. Ma, Y.-S. Kim, R.-H. Kim, S. Wang, J. Wu, S.M. Won, H. Tao, A. Islam, K.J. Yu, T. Kim, R. Chowdhurry, M. Ying, L. Xu, M. Li, H.-J. Chung, H. Keum, M. McCormick, P. Liu, Y.-W. Zhang, F.G. Omenetto, Y. Huang, T. Coleman, J.A. Rogers, Science **333**, 838 (2011)
60. A.J. Bandodkar, W. Jia, C. Yard, X. Wang, J. Ramirez, J. Wang, Anal. Chem. **87**, 394 (2015)
61. C. Dagdeviren, B.D. Yang, Y. Su, P.L. Tran, P. Joe, E. Anderson, J. Xia, V. Doraiswamy, B. Dehdashti, X. Feng, B. Lu, R. Poston, Z. Khalpey, R. Ghaffari, Y. Huang, M.J. Slepian, J. Rogers, Proc. Natl. Acad. Sci. U. S. A. **2014**, 111 (1927)
62. M. Deterre, B. Boutaud, R. Dalmolin, S. Boisseau, J.-J. Chaillout, E. Lefeuvre, E. Dufour-Gergam, in *2011 Symposium on Design, Test, Integration and Packaging of MEMS/MOEMS (DTIP)* (2011), p. 387

Chapter 5
Wireless Applications of Conformal Bioelectronics

Yei Hwan Jung, Huilong Zhang and Zhenqiang Ma

Abstract Conformal bioelectronics in flexible or stretchable format that make direct contact to the skin or tissues have contributed extensively to diverse clinical applications. Wireless modules in such minimally invasive forms have developed in parallel to extend the capabilities and to improve the quality of such bioelectronics, in assurances to offer safer and more convenient clinical practice. Such remote capabilities are facilitating significant advances in clinical medicine, by removing bulky energy storage devices and tangled electrical wires, and by offering cost-effective and continuous monitoring of the patients. This chapter provides a snapshot of current developments and challenges of wireless conformal bioelectronics with various examples of applications utilizing either wireless powering or communication system. The chapter begins with near-field wirelessly powered therapeutic devices owing to the simplicity of power transfer mechanism followed by far-field powering systems which require integration of numerous electrical components. In the later sections of the chapter, sensors in conformal format that transfer clinical data wirelessly are discussed and ends by reviewing the developments of wireless bioelectronics that utilize integrated circuits for advanced capabilities in clinical applications.

Keywords Wireless medical devices · Wireless powering · Wireless sensors · Microwave antennas · Wireless communication systems · Stretchable wireless bioelectronics · High-speed electronics

Y.H. Jung · H. Zhang · Z. Ma (✉)
Department of Electrical and Computer Engineering,
University of Wisconsin-Madison, Madison WI 53706, USA
e-mail: mazq@engr.wisc.edu

5.1   Introduction

According to the nonprofit Institute for Healthcare Improvement, an unmonitored patient has a 6 % chance of surviving a cardiac arrest compared with a 48 % chance for a monitored patient [1]. As such, implantable or wearable medical electronics have been used in patients for decades to treat and monitor physiological conditions within the body [2]. Medical devices are comprised of electronic circuits which have driven advances in technology for over a half century. Wireless capability in medical devices adds a great benefit of being able to monitor and treat remotely, eliminating threats to patients from tangled electrical wires, and allowing cost-effective, continuous monitoring for medical practitioners. The history of wireless biomedical devices dates back to the late 1950s when the wireless transmission of data and power became a reality with the invention of transistors. The ability to manufacture transistors on the micro- and nanoscale have allowed medical devices to measure physiological conditions within or on the body remotely via radio frequency (RF) electronics. The first biomedical telemetry devices based on RF electronics were introduced in 1957 when two groups, Farrar et al. [3] and MacKay et al. [4], almost simultaneously reported capsule-type wireless bioelectronics, which measured physiological conditions (gastrointestinal motility in the two devices) inside a living person and sent out information remotely. The capsule form of wireless bioelectronics has not changed much since then, and conformal type devices (i.e., in an ultrathin flexible or stretchable format) have only recently started to develop. This is due to the fact that most electronics are mechanically limited to having rigid components and are confined to brittle chips and boards. While the ability to create such miniature components gave rise to a myriad of clinical capabilities, the rigid and bulky format that has potential invasive effects to surrounding tissues in the body limited many long term or reliable operations.

Recent progress in flexible or stretchable electronics presents a route to creating high-performance semiconductor electronics in forms that enable conformal contact to curvilinear surfaces, such as skin, tissues, and organs. Utilizing the fact that any material becomes flexible if it is thin enough, like a flexible sheet of paper made out of bulky wood, bending of high-performance inorganic materials such as silicon (Si) became realizable by thinning them down to the nanometer scale [5], thus called a nanomembrane (NM) . Moreover, due to their excellent biocompatibility, many inorganic materials can be used to make implantable electronics to cure and sense diseases [6]. To create such electronics, a new class of electronics manufacturing technique called the 'transfer printing' process is utilized, which transfers an exfoliated sheet of ultrathin inorganic material to a different substrate. This method enables placement of electronics on nearly any type of substrate including but not limited to rubber, plastic, fabric, and glass [7]. The most appealing feature about this approach is that the majority of the manufacturing process involves conventional fabrication technology, which already has a mature, established commercial infrastructure, thereby accelerating time towards commercialization and practical applications. This not only preserves the high-performance characteristics, but also makes the chip flexible or stretchable depending on the material of the target substrate. Introduction of such technology has already reshaped the rigid and conventional electronics into thin flexible devices that can stretch, bend, and twist seamlessly with the human body [8]. One of the promising applications of such electronics that could not be addressed with conventional technology includes advanced wireless biomedical devices, especially implantable and wearable medical electronics. As the ability to provide wireless powering or data transfer capabilities are highly desirable for many clinical applications, efforts have been made to overcome the challenges that constrain the development of implantable or wearable wireless electronics. Because high-performance components are generally required for RF electronics that enable wireless transmission, challenges such as the difficulty of designing a RF circuit in a time-dynamic media of organs and the limitation of high temperature manufacturing processes required to fabricate high-performance devices have been solved in the recent decade using 'transfer printing' technology.

This chapter discusses the techniques to create such wireless electronics in a flexible or stretchable format, which could potentially be used to create biomedical telemetry devices for either power or data transfer. Two different types of mechanisms, the near-field and far-field systems that allow wireless power or data transfer will be discussed. Near-field wireless systems utilize inductive or capacitive coupling mechanisms to communicate at a short distance. In most clinical applications, near-field systems are sufficient for either delivering power or communicating data; however, where long-distance wireless functionalities are required, far-field systems that are more complex and difficult to design and manufacture must be developed. We distinguish wireless power transfer and wireless data transfer into two sections and discuss both near-field and far-field systems using RF electronics in each section followed by existing examples of wireless conformal bioelectronics that either power or transfer data in and out of the body.

5.2 Near-Field Wireless Powering Systems for Conformal Bioelectronics

With near-field systems, powering a biomedical device via remote power transfer can be achieved by either inductive or capacitive coupling. In an inductive power-transmission system, a transmitter coil and a receiver coil form a system of magnetically coupled inductors, where the alternating current from the transmitter coil induces a voltage in the receiver coil via magnetic field generation [9]. While the inductive system is the more commonly used method for wireless transfer of power, the capacitive system, where the coupling occurs between two parallel plates like a capacitor is seldom used in wireless power transfer for biomedical devices. This is because the inductive power transfer system offers higher power transmission capacity than capacitive power transfer. Capacitive power transfer is sometimes used where the electromagnetic interference (EMI) must be reduced [10]. Here, we only discuss the importance and development of inductive power transfer. In either case, the ability to efficiently deliver the power to the device without losing energy in a time-dynamic media such as skin, blood, or tissues is significant. At the same time, the energy waves that deliver power must not have any adverse effects on the body.

Successful designs of wearable or implantable inductive coupling systems for wireless power transfer should be able to survive in time-dynamic media, withstand strain and stress due to large deformations, and be durable enough to sustain long term operation in such conditions. The materials used for the conducting lines in the design must also be biocompatible and resistive to corrosions to be minimally invasive as bioelectronics. In this section, materials and design considerations for flexible and stretchable wireless powering systems are discussed, along with clinical applications that may benefit from the technology.

5.2.1 Fabrication of Inductive Coupling Systems

Designing an inductive coil that can transfer power wirelessly involves sufficient knowledge in the electromagnetic (EM) theory. Because near-field systems utilize non-radiative fields (magnetic fields in inductive components and electric fields in capacitive components), where the energy stays within a short distance of the transmitter, the fields will not be able to couple to the receiver if it is not within the range [11–13]. The range of the fields depends on the size and shape of the transmitter; moreover, the fields also decrease exponentially with distance. Typically, ordinary inductive or capacitive wireless power transfer systems have field regions of up to about one antenna diameter:

$$D_{range} < D_{ant}$$

where D_{range} is the distance from the transmitting antenna, and D_{ant} is the diameter of the antenna. Sometimes, resonance of the two systems increases the distance 10-fold, which requires matching components on either side to match the resonance [14]. For biomedical applications, the designer must consider the media where the two coils (transmitting and receiving) will operate in. Typically, the transmitting coil is in air and is more controllable allowing the designer to tune the frequency or power level. The receiving coil will either be laminated on to the skin or implanted under the skin which requires sophisticated materials and design considerations for high durability, flexibility, and stretchability, while satisfying the biocompatibility for the surrounding tissues. The simplest approach to creating a flexible inductive coil is to deposit a thin layer of conducting metal on a plastic or rubber substrate and connect it with the device that needs to be powered wirelessly. Other methods include utilizing microfluidic alloys [15] or horseshoe shaped serpentine designs [16] to prevent cracking or breaking due to strain. Bioresorbable inductive systems using degradable metals were also developed for implantable electronics for physically transient therapeutic devices.

5.2.2 Light Emitting Systems with Wireless Capabilities

Light emitting systems can perform various clinical treatments inside the body such as accelerating wound healing [17, 18], activating photosensitive drugs [19], or performing imaging and spectroscopic characterization of internal tissues [20]. Recent advances in implantable light emitting systems offer conformal and minimally invasive operations within the body to perform such applications. For instance, various types of light emitting diodes (LEDs) such as arsenic-based red or infrared LEDs and nitride-based blue or ultraviolet LEDs are developed with advanced biocompatible packaging techniques for implantable applications [21–23]. For most of the devices that operate in the body, power transfer for turning on the LEDs has been one of the most difficult problems encountered. Because wireless power transfer can offer safe and convenient clinical applications, several techniques were developed to achieve this [24]. Figure 5.1a presents one of the techniques that considers materials and designs for contactless power transfer for the LEDs implanted under the skin, where an inductive coil is utilized. Here, an array of ultrathin GaN-based μ-LEDs is first fabricated on a sapphire substrate and subsequently released by the laser lift off technique. Selective transfer of individual LEDs using PDMS μ-stamp delivers each LED onto a temporary carrier substrate coated with a thin epoxy-based adhesive. Inductive coils and interconnects are then deposited onto the LEDs with another layer of epoxy coating to separate the coil lines from the interconnects. Finally, the entire structure, including the LEDs, the inductive coil, and the interconnects are released from the temporary carrier substrate and transfer printed onto an ultrathin flexible or stretchable substrate, as shown in Fig. 5.1b. Such fabrication techniques can create both flexible and stretchable systems. While the rectangular spiral coil and straight lines for the metal structures are sufficient to create a flexible system, the design must be modified to include horseshoe-shaped serpentine interconnects (Fig. 5.1c) in order for the entire system to be stretchable. Due to the irregular and curvilinear shapes and time-dynamic environment of the tissues, a stretchable system offers higher durability and reliability that prevents the lines from mechanical failures due to large strain (>1 %). Nevertheless, because such serpentine-based structures require larger spacing in between the lines, the spiral inductors must be designed with additional turns in order to compensate for the increase in resonance frequency due to larger spacing. As a result, with similar resonance frequency, stretchable systems have larger dimensions compared to flexible systems with straight lines as shown in Fig. 5.1d.

5.2.3 Remotely Controlled Bioresorbable Thermal Therapeutic Devices

Micro-scale therapeutic devices that can deliver controlled heat within small area are especially useful as implantable bioelectronics to manage infections by thermal

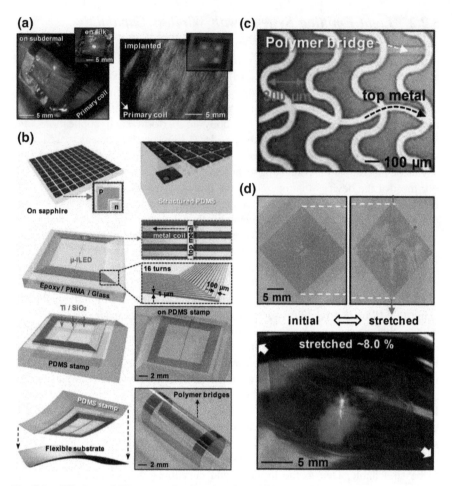

Fig. 5.1 **a** Wireless μ-LED system, laminated on the sub-dermal region of a mouse model; inset image provides the initial form of the device (*left image*). Image of an animal model with a wireless μ-LED device implanted under the skin, and on top of the muscle tissue; inset shows device before implantation (*right image*). **b** Fabrication process of the wireless μ-LED system. InGaN μ-LEDs are formed on sapphire substrate, picked up using PDMS stamp. The μ-LED is printed on to a temporary glass substrate, followed by metallization of interconnect and wireless coils. The wireless μ-LED system is retrieved with a PDMS stamp and printed onto a flexible substrate. **c** Optical micrograph showing the interconnect scheme for stretchable design. Both spiral and cross-over metal lines adopt serpentine shapes. **d** Optical images of the stretchable wireless μ-LED system. Reproduced with permission. Copyright 2012, John Wiley and Sons [24]

treatment or remote triggering of drug release. Out of many types of thermal therapeutic devices with remote triggering capabilities for implantable purposes, a physically transient form of devices that resorb in the body to avoid adverse long-term effects has been the most promising technology as it does not require any secondary surgery for removal of the electronics after use [25]. Ideally, the entire

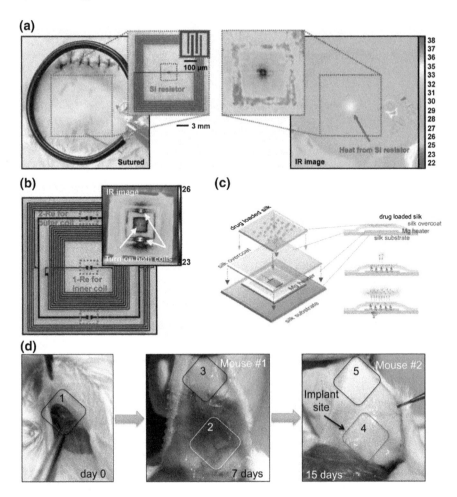

Fig. 5.2 a Wireless transient heater implanted and sutured for transient thermal therapy; inset shows the device before implantation (*left image*). Thermal image collected during the wireless operation of the device through the skin; inset shows the magnified view (*right image*). Copyright 2012, American Association for the Advancement of Science [25]. **b** Transient wireless device consisting of two resistors connected to a first wireless coil (70 MHz; *outer coil*) and a second resistor connected to a second coil (140 MHz; *inner coil*); inset shows thermal image with both resistors turned on. Reproduced with permission. Copyright 2012, American Association for the Advancement of Science [25]. **c** Schematic of the device integrated with antibiotics-doped silk film for wirelessly activated drug release. Reproduction permission required [27]. **d** Monitoring of the transient drug releasing device degradation implanted in mouse. Devices are implanted and examined after 7 and 15 days. Reproduction permission required [27]

system dissolves after operation, including the heater itself and the wireless coils that enabled remote power delivery. In Fig. 5.2a, inductive coils of Mg, combined with resistive microheaters of doped Si NMs, integrated on silk substrates, and housed in silk packages operating under skin is shown. As one of the first transient electronic devices ever made, this remotely controlled therapeutic device can provide transient thermal therapy to control surgical site infections. As shown in Fig. 5.2b, the system consists of two different inductive coils that operate at different frequencies (~ 70 and ~ 140 MHz) allowing clinicians to turn on different heaters selectively or simultaneously. Furthermore, careful design and material considerations must be made for dissolvable inductive coils because the performance of the coils will start to degrade or fail at a certain point of time as the coils dissolve gradually in the body. For most of the transient electronics, the favorable conductor material used for spiral inductors is Mg due to its excellent biocompatibility, electrical properties, and dissolving characteristics. Other dissolvable metals have also been investigated for transient electronics such as Mg alloy, Zn, Fe, W, and Mo [26]. While the feasibility of these metals as inductor coils have not been investigated except Mg, the electrical properties of the listed metals are suitable for spiral coil lines giving inductor designers a wide range of material choices for dissolvable metals. In order for a stable and long-term operation of the spiral coils, the conducting lines must be sufficiently thick. In a study of transient microheaters, the total calculated amount of Mg used for a resistor (~ 200 nm thick) and a spiral coil (~ 2 μm thick) was 0.35 and 26.43 μg, respectively [27]. While the amount used for the spiral inductor may account for most of the mass, the total amount is still a minimal quantity considering that the suggested daily intake of Mg for adults is ~ 350 mg. Further studies on the histological sections showed that the surrounding tissues were undisturbed by the dissolved Mg. Figure 5.2c shows a resistive heater with Mg coils for a therapeutic device that releases drugs using remote triggering [27]. Far more advanced than a regular heater, this device contains drugs or enzymes in the silk packaging, which is released upon thermal triggering of the wireless heater at the interface. Once the thermal triggering has released the drugs, the entire system dissolves with the body fluids as shown in Fig. 5.2d. This offers programmable remote control of release kinetics of a drug contained and stabilized within the silk material matrix. As such, bioresorbable electronics combined with wireless functionality provides safer and more advanced biomedical applications for clinicians. The development of a transient form of wireless powering system with electronics other than heaters will positively benefit implantable clinical therapeutics.

5.2.4 Stretchable Wireless Charging Batteries

Batteries are alternatives to powering many clinical systems that are fully implanted. While numerous types of flexible and stretchable batteries were developed for potential bioelectronics applications [28, 29], the batteries must be able to harvest energy to charge themselves. In many practical applications for bioelectronics, the ability to charge the battery without any physical wire connections would be extremely valuable. Innovative techniques, such as the piezoelectric energy harvester were developed to charge the battery, which harvests and converts mechanical energy to electrical energy from natural contractile and relaxation motions of organs

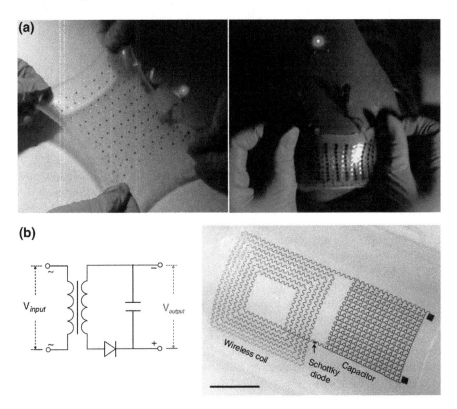

Fig. 5.3 **a** Operation of a stretchable battery connected to a red LED while biaxially stretched to 300 % (*left*) and mounted on the human elbow (*right*). **b** Circuit diagram of the stretchable system for wireless charging of the batteries (*left*) and image of the integrated system with different components labeled (*right*). Reproduced with permission. Copyright 2013, Macmillan Publishers Ltd. [29]

such as the heart [30]; but, the energy harvested from such devices are insufficient to power many clinical devices that demand high power. Moreover, the ability to integrate stretchable batteries with piezoelectric system is not yet proven. As a result, the most promising technology is to charge the battery using wireless powering systems. Where wireless powering is not available, the electronics have the option to use power from the charged battery. This allows the bioelectronics with batteries to continuously perform clinical trials. Examples of such devices are presented in Fig. 5.3, where the stretchable type of battery shown in Fig. 5.3a is integrated with wireless charging coils monolithically [29]. This system (Fig. 5.3b) includes a secondary coil which couples the electromagnetic flux from the primary coil. The embedded Schottky diode further rectifies the signals and the array of parallel capacitors smoothes oscillations in the output voltages. Overall, the entire monolithically integrated system is ultrathin and highly stretchable, which makes it a highly promising candidate for implantable or wearable powering systems.

5.2.5 MEMS Technology for Large Area Power Transfer

The main drawbacks of inductive coupling power transfer systems are that the primary and secondary coils must be well-aligned for efficient power transfer. For this reason, far-field wireless power transfer systems are preferred over inductive systems since the radiation of the power from the transmitter is well spread out which allows the receiver to receive power anywhere in a defined region. Before getting to the far-field transmission systems, one of the smart techniques that still utilize inductive power transfer for large area applications will be discussed. In 2007, Someya et al. at the University of Tokyo engineered a smart wireless powering system in a thin sheet form that combines multiple spiral inductor coils with complementary circuits integrating plastic microelectromechanical system (MEMS) switches and organic transistors [31, 32]. This smart system shown in Fig. 5.4a senses the position of any electronic object with the plastic MEMS switch using the organic transistor active matrix. The circuit schematic of both the position-sensing coil and the sender coil is presented in Fig. 5.4b. The position-sensing coil then selects one of the closest sender coils to selectively transfer power to the target electronic object via the inductive coupling system as shown in Fig. 5.4c. As a result, this entire integrated system is able to transfer a power of 40.5 W with an 81.4 % coupling efficiency over a 21×21 cm^2 large area. Photographs of the system operating at various conditions are shown in Fig. 5.4d. While no clinical medicine has seen this technology to power bioelectronics, this unique integrated system may be adapted for many clinical uses where large area power transmission is required.

5.3 Far-Field Wireless Powering Systems for Conformal Bioelectronics

In applications that require wireless power transfer over a long distance, inductive or capacitive coupling mechanism cannot be utilized, as the coils must greatly exceed in size. Therefore, far-field systems that utilize radiative fields must be used. In order for the power to be transmitted over a long distance, integrated circuits operating at microwave frequencies, together with far-field antenna must be used. Difficulties and numerous challenges arise when trying to design a microwave circuit in an ultrathin format and also on or inside a time-dynamic media. There are two major challenges that must be overcome for the successful operation of far-field wireless systems inside a body. First, the environment must be composed of invariable, solid materials. The far-field wireless system is based on EM wave propagation, where the EM waves travel between two transceivers in a medium or many media to transfer electric power or data. Propagation of EM waves depends on the dielectric properties of the medium. Thus, RF engineers design wireless circuits with consideration of the dielectric properties of every surrounding material. Organs, tissues, and blood are

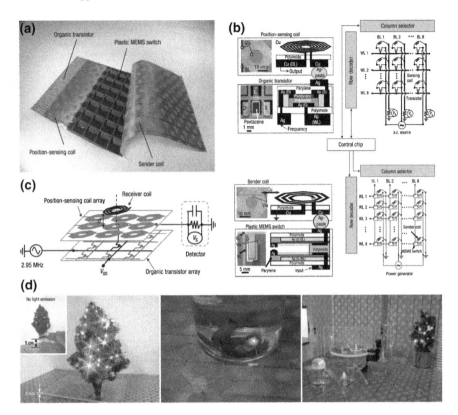

Fig. 5.4 a Photograph showing an exploded view of the wireless power-transmission sheet embedded in the floor and comprising a wireless power-transmission system and contactless position-sensing system. **b** Pictures, *cross-sectional* diagrams and *circuit* diagram of the position-sensing coil and organic transistor (*top*). Pictures, *cross-sectional* diagrams, and *circuit* diagram of the sender coil and plastic MEMS switch (*bottom*). Electrodes for electrostatic attraction are connected to the *word line* (WL) and *bit line* (BL) of the MEMS switch. Electrodes for power transmission are connected to the sender coils and a power generator. WLs and BLs of a position-sensing system and power-transmission system are connected to row decoders, column selectors, and control chips for coordination, for example, for addressing and reading-out to each other. **c** Schematic diagram of the position-sensing unit. **d** Demonstration photographs of the wireless power-transmission sheet. Reproduced with permission. Copyright 2007, Macmillan Publishers Ltd. [31]

dynamic and do not have fixed material properties; instead, it varies by conditions such as temperature and age. Therefore, this proposes a primary design challenge in making an implantable wireless system. Second, implantable wireless systems must operate at high power since the far-field transmitted power decreases rapidly with increasing distance. In an RF system, highly performing active and passive devices, such as transistors, diodes, capacitors, and inductors are needed to operate at a certain allowed frequency like the Industrial Scientific Medical (ISM) bands. Nevertheless, fabrication of high-performance devices are especially challenging on biocompatible polymers, due to the constraints in using high temperature processes generally required for fabricating high-performance electronics. Finally, all the individual key components (active devices, passive devices, and antennas) required for building the implantable wireless system must be combined carefully in order to transmit wireless power efficiently.

To effectively transmit and receive power over a long distance, a microwave power amplifier is necessary at the transmitting end in order to amplify the power to compensate for energy loss over the long distance. At the receiving end, a microwave rectifier is typically connected to the receiving antenna to convert the incoming high-frequency alternating signal into direct currents. On either side, individual components such as the transistor, diode, capacitor and inductor, must be able to perform at relatively high, yet allowed frequencies. In the following sections, individual components that are developed on either flexible or stretchable substrates that can ultimately be used for wearable and implantable conformal devices are discussed. While many of the work demonstrate only the success of individual components on such substrates, the compatibility of the fabrication techniques used for the devices with the already developed conformal bioelectronics described in this book show the feasibility of such microwave technology for far-field power transfer in biomedical applications. Following this section, several examples of wireless far-field power transfers in biomedical devices are shown that may benefit from this technology.

5.3.1 High-Speed Active Devices

Transistors and diodes are used in a variety of electronic chips to serve many purposes. While in most cases the high frequency performance of the devices are not particularly of interest, in circuits for radio frequency applications, the high frequency performance of the devices define the level of frequency of the entire circuit. For instance, an efficient power amplifier that amplifies a 2 GHz signal would normally require a transistor with maximum oscillation frequency (f_{max}) of at least 20 GHz, which is ten times higher than the operating frequency. The specific requirements for the flexible high-speed devices to achieve high frequency performance are that they must possess high carrier mobility, be integrable on foreign substrates, and be flexible with minimum performance changes. Single-crystalline semiconductors, such as Si and III–V materials, are the most favorable option that satisfies the high mobility criteria. Furthermore, with the invention of transferable single-crystal Si NMs from silicon-on-insulator (SOI) wafers and a variety of techniques to release many types of III–V materials manufacturing of high-speed devices on flexible or stretchable substrates can be accomplished [5, 7, 16, 33–35]. Nevertheless, poor heat and chemical resistances of flexible substrates have prevented the utilization of conventional manufacturing techniques. Therefore, a sophisticated process flow and design including the development of effective doping processing techniques and advanced structural designs must be considered to achieve a truly high-speed, flexible electronic device. With such active devices, more complicated/sophisticated RF circuits and systems can be realized by the integration of flexible passive components such as inductors and capacitors [36]. This type of technology provides RF electronics a platform for its unusual mechanics and for its potential as a low-cost alternative to conventional systems that require semiconductor wafers as substrates [5].

Numerous efforts to improve the high frequency performance of Si-based field effect devices have been made ever since the demonstration of the first flexible high-speed transistor that achieved a cut-off frequency of 515 MHz [37]. Engineers have developed fabrication techniques to reduce the gate length of the devices, and improved the doping characteristics and gate dielectric materials in order to improve the high frequency performance. As a result, a record-breaking transistor that is purely comprised of Si with an f_{max} of 12 GHz has been demonstrated [38]. The devices were further improved by utilizing strained Si NMs to speed up the electrons which resulted in an f_{max} of 15.1 GHz [39]. Here, a uniquely designed epitaxial layers that include SiGe induces self-sustained tensile strain to the active Si layer, which enhances the mobility of electrons [40]. While such Si-based devices may be the suitable choice for many circuits, amplifiers that require higher power and speed must utilize III–V semiconductors, such as GaN or GaAs. GaN-based high electron mobility transistors (HEMTs) have demonstrated one order magnitude higher power density and efficiency compared to other competing technologies including the Si-based transistors [41]. Thin-film GaN HEMT on a flexible plastic substrate can be fabricated by selectively removing the growth substrate (Si), while preserving the high frequency and high power characteristics [42]. The superb high frequency performance (cut-off frequency $(f_T)/f_{max} = 60/115$ GHz) on flexible plastic substrate, along with high output power density, makes the GaN HEMT the ideal transistor for power amplifying applications. GaAs-based heterojunction bipolar transistors (HBT) are widely used in low noise amplifiers [43]. GaInP/GaAs HBT has also been demonstrated on flexible substrates [44]. This device with relatively high frequency responses ($f_T/f_{max} = 38/7$ GHz) may also be utilized for many clinical applications for low noise amplifiers with minor modifications to its fabrication processes. In addition, a typical RF circuit would require not only transistors, but also high-speed diodes or switches for full functionality. Thus, it is important to realize that such devices are also developed on flexible substrates. High frequency GaAs-based Schottky diodes [44], Si-based PIN diodes [45], and Ge-based PIN diodes [46] are demonstrated on flexible substrates. Such diodes are especially useful in forming microwave switches and rectifiers that are commonly present in both the transmitter and the receiver [47, 48]. As such, many types of transistors and diodes have been developed on flexible substrates, which are process compatible with biomedical devices developed on flexible or stretchable substrates. Although the fabrication process may be compatible, for such devices to be used in clinical applications numerous clinical issues must also be resolved. Operating at high speed and high power, most of these devices generate large amounts of heat. For instance, a GaN-based HEMT that outputs 8 W could reach up to 225 °C in its hot spot [49]. Therefore, a well-designed packaging technique that includes a heat insulating layer in order to protect the surrounding tissue or skin must be developed. Furthermore, a device like GaAs is known to be toxic due to the presence of arsenic [50]. Thus, the packaging of such devices must ensure that there is no leakage of toxic materials into the body.

5.3.2 High-Speed Passive Devices

Together with the active devices, passive devices such as inductors and capacitors are essential components for fast, flexible, and stretchable electronics. Because conformal bioelectronics must be ultrathin, planar passive components, such as the spiral inductors and parallel plate capacitors are preferred. Mechanically flexible inductors and capacitors capable of handling multi-gigahertz frequencies have been reported, which are suitable for designing a microwave integrated circuit [36, 51]. Several important factors must be optimized to design a spiral inductor that has large inductance with a high quality factor, such as the type of substrate, the number of turns, the spacing between the conducting lines, and the width of the lines. By calculating the self-inductance and mutual inductance, the inductance value and the Q factor of an inductor can be obtained. In addition, because the capacitance of a parallel plate capacitor is reliant on the overlap area of two plates, the dielectric properties and the dielectric thickness, all of these parameters should be optimized for a reasonable size required capacitance and enough Q factor. State-of-the-art inductor and capacitor developed on a flexible substrate can operate up to 10 and 20 GHz, respectively [51].

5.3.3 Far-Field Receiving Antennas

Antennas for far-field differ from those used for near-field systems in that they achieve longer ranges, where the working distance is much greater than the diameter of the device. Once the radiated RF energy from the transmitting antenna reaches the bioelectronics devices, it must be received by the receiving antenna for powering. Among all other components for far-field systems, the antenna is the most difficult component to design as it is always application- and situation-specific due to the diverse types of surrounding media in the body. In wearable applications, the design criteria becomes less complex as the direction of the incoming EM waves do not have to go through any biological media, so any antenna in the flexible or stretchable format maybe suitable. Interesting antenna designs using conducting elastic materials have been presented as shown in Fig. 5.5. These antennas combine several elastic conductor materials, such as liquid metal (Fig. 5.5a) or silver nanowires (Fig. 5.5b) with advanced fabrication processes to create durable stretchable antennas [52–58]. Other forms of conformal antennas with unique conductor geometry designs, such as compact designs for flexible [59] and fractal designs for stretchable antennas [60], were also developed as shown in

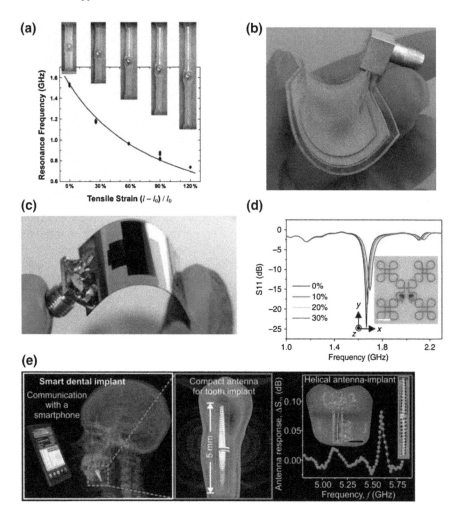

Fig. 5.5 a Sequence of optical images with corresponding resonance frequency graph upon stretching of a microfluidic radio frequency antenna. Reproduced with permission. Copyright 2010, John Wiley and Sons [55]. **b** Photograph of a stretchable microstrip patch antenna composed of Ag nanowires and PDMS conductor. Reprinted with permission from (*ACS Appl. Mater. Interfaces*, 2014, *6* (6), pp 4248–4253). Copyright (2014) American Chemical Society [57]. **c** Photograph of a compact parylene coated biocompatible flexible antenna. Reproduction permission required [59]. **d** Return loss parameters of a stretchable fractal antenna under different amounts of tensile strain; inset shows the unstrained antenna. Reproduced with permission. Copyright 2014, Macmillan Publishers Ltd [60]. **e** Compact helical antenna for implant applications, with images showing the concept of a smart implant for in-body applications on tooth. The response of the antenna implant in tooth can be accessed wirelessly, as revealed by the presence of a peak in the scattered signal at the resonant frequency of the antenna shown on the *right* graph [65]

Fig. 5.5c, d, respectively, to minimize size and increase stretchability. Typically, such antennas are designed at high frequencies (>1 GHz) to compensate for the small size requirement of wearable electronics. Implantable antennas also follow similar rules, but are far more complicated in terms of its design. To make things even more challenging, the area for the antenna is extremely limited for implanted antennas. For instance, the thickness of the antenna must be kept ultrathin and confined in a small area for it to have no adverse effects on the surrounding tissues. While designers have the choice to minimize the antenna size by increasing the operating frequency, the penetration depth of EM waves in biological tissues decreases with increasing frequency [61]. There are restrictions on using certain frequency bands for bioelectronics like the ISM bands set by the Federal Communications Commission (FCC), which poses another challenge for designers [62]. For over a decade, numerous types of antennas for implantable applications have been introduced [63–68]. Figure 5.5e represents one of the important findings for implantable antennas in helical format [65] where the operating frequencies all meet FCC regulations. In most cases, the designs started by careful simulation and modeling followed by multiple iterations of the design and real environment testing using matching and tuning techniques. Real environment tests can be done not only in the actual target tissue, but also using biological phantom material, comprised of a mixture of water, salt, sugar, etc., to facilitate the design process [62].

While combining all the active and passive devices and antennas in either the flexible or stretchable format summarized in this section can yield a complete far-field transceiver either for implantable or wearable applications, the development is still in its early stages and many engineering challenges must be resolved before such technology can be demonstrated. Stemming from the properties of being flexible and stretchable, with the added benefit of being ultrathin, such highly performing individual components will be beneficial for many applications, where wireless power transfer capabilities over a long distance are required.

5.4 Applications of Far-Field Wireless Powering Systems

As described in earlier sections, designing power transfer systems on flexible substrates require high-performance devices on every component that must operate at higher frequencies. In general, the smaller the critical dimensions for a device, the faster the device can operate. This proposes a challenge in fabrication as smaller devices are difficult to fabricate on flexible substrates. While the power transmitter normally requires such sophisticated circuit designs, the wireless receiver may be designed with less sophisticated integrated circuits and include only simple diodes to rectify the receiving signals. As most of clinical applications require only receivers, many receiver designs on flexible substrates have been developed for such applications. Two important applications are shown in the following sections: bioresorbable RF electronics for implantable applications, and miniaturized wireless power scavenger on flexible substrates for optogenetics.

5.4.1 Bioresorbable Radio Frequency Electronics

Devices that have capabilities to dissolve or degrade with time have many roles in clinical medicine, such as drug delivery and sutures. Completely water-soluble devices that have engineering capabilities derived from the integrated circuit industry may be implanted into the human body to function for a period of time and then dissolve gradually. This would eliminate the need to remove the device via surgical extraction, which would be the case for non-dissolving devices. Numerous types of circuits in this form that operate for various purposes in clinical applications have been demonstrated, all of which require some sort of power source for their functionality. Thus, RF operation of similar types of bioresorbable electronics would add great benefits for nearly all types of implanted devices. Figure 5.6a–d shows the antennas, rectifying diodes, capacitors, and inductors that are all essential components for wireless power transfer systems [69]. In addition, a wireless power scavenger in the bioresorbable form is also shown in Fig. 5.6d. Here, the Si-based diodes are combined into an integrated circuit to form a rectifier, which converts incoming AC signals to DC signals to power the LEDs. The capacitors and inductors that act to match the microwave signals for better efficiency are also used in the circuit. Combined with the Mg-based antenna, the entire receiver can receive power from a distance of up to 2 m to turn on an LED. In addition, with the Si-based rectifier, the system may efficiently rectify over a frequency range from 10 kHz to 950 MHz. In this frequency range, the antenna size must be large enough to achieve sufficient efficiency, such as the one shown in Fig. 5.6a. Nevertheless, such large antennas may not be suitable for many implantable applications as the devices must be kept small for non-invasive operations. Therefore, the frequency must be higher for practical purposes. Different types of rectifiers operating at 1.6 GHz [70], 2.45 GHz [71], and 5.8 GHz [44] built on flexible substrates have been reported. Utilizing such integrated circuits at high frequencies will eventually lead to implantable receiver designs with smaller dimensions.

5.4.2 Wireless Power Receivers for Optogenetics

Another rising technique in neuroscience that uses bioelectronics is to stimulate neurons using light, which is often referred to as optogenetics. Introduced in the early 1970s, the term optogenetics combine optics and genetics to control specific cells of living tissue [72]. A unique set of viruses that responds to light can be injected into individual neurons in the brain, where the neurons can then be controlled with external light sources. During the early development stages, optical fibers connected to external lasers or a light source was injected into the brain for stimulations. However, the technique has greatly advanced and scientists are now able to inject cellular scale optoelectronic systems, such as the μ-LEDs, which eliminates the use of

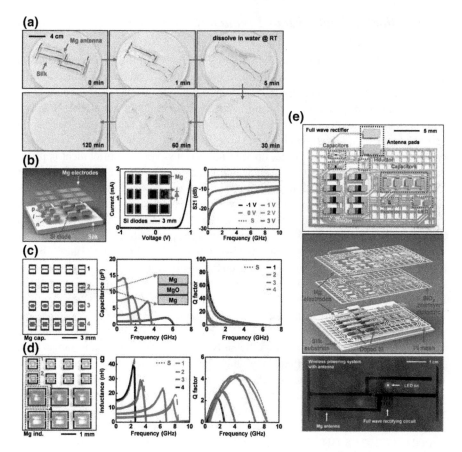

Fig. 5.6 **a** A set of images of an antenna built with Mg on silk substrate with illustrations of the process of dissolution in water. **b** Schematic illustration of Si NM PIN diodes (rectifiers) fabricated on a silk substrate with Mg electrodes (*left*). Current-voltage characteristics (*middle*) and measured (*lines*) and simulated (*dots*) RF characteristics (*right*) of a transient diode under different DC biases **c** Image of collection of capacitors of different sizes built using Mg and MgO (*left*), with measured capacitance (*middle*) and Q factor (*right*) as a function of frequency. **d** Image of collection of inductors of different size built using Mg and MgO (*left*), with measured inductance (*middle*) and Q factor (*right*) as a function of frequency. **e** Photographs and schematic illustration of transient RF power scavenging circuits, integrated with transient antenna. Reproduced with permission. Copyright 2013, John Wiley and Sons [69]

long fibers that allow experimental animals to move around more freely [23]. With development of the technology, the use of wireless scavengers has also been advanced to deliver power to the LEDs from a long distance. Figure 5.7a shows a wireless power scavenger that uses the bulky PCB board mounted on a freely moving mouse for optogenetics experiment [23]. This bulky looking device is too heavy for a mouse to freely move around with when it is placed on its head. Therefore, flexible

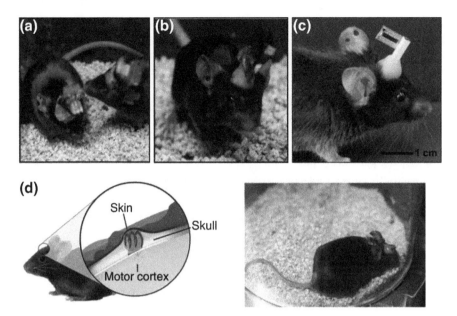

Fig. 5.7 **a** Optical image of a wireless powered optical neural device made using conventional chips and boards. Copyright 2013, American Association for the Advancement of Science [23]. **b** Optical image of a lightweight, flexible wireless power scavenger device mounted on a freely moving mouse. Copyright 2013, American Association for the Advancement of Science [23]. **c** Optical image of a miniature, lightweight energy harvester mounted on a freely moving mouse. Reproduction permission required [73]. **d** Schematic illustration of a fully implantable wireless energy harvester inserted into the brain directly above motor cortex (*left*). Freely moving mouse with the brain implant (*right*). Reproduced with permission. Copyright 2015, Macmillan Publishers Ltd. [74]

devices to minimize the substrate weight and size were developed as shown in Fig. 5.7b [23]. And with further engineering techniques, the same system is minimized to the one shown in Fig. 5.7c [23]. This ultraminiaturized version also includes photovoltaic cells to enable operation in challenging scenarios offering versatile capabilities in optogenetics. Some fully implantable wireless power receivers (Fig. 5.7d) have been developed as well, which eliminates unnecessary substrates and utilizes the smallest antenna to minimize the overall size [74].

5.5 Near-Field Wireless Sensors for Conformal Bioelectronics

Remote extraction of clinical data from wearable and implantable sensors using wireless devices greatly ensures safe and convenient data transfer in biomedical applications. Traditional wireless bio-sensing systems combine rigid sensors, commercial data processors, and transceiver chips together on a conventional printed circuit board (PCB). For sensors, such a rigid and flat format that differs from the soft nature of the human body results in lower measurement accuracy and unconformity. As biomedical sensors transform to ultrathin flexible or stretchable forms like epidermal electronics [16], which allow conformal contact to the target tissues, measurement accuracies were significantly improved. Moreover, combined with advanced conformal sensors, transmission of sensed data using either near-field or far-field communication systems in the flexible or stretchable format would overcome the issues of existing traditional rigid electronics-based systems. Efforts have been made in development of data processor and transceiver chips to match curvilinear surfaces and to create wireless data communication systems in the flexible and stretchable format. In this section, numerous types of conformal wireless sensors are discussed with applications ranging from wearable to implantable sensing stations.

5.5.1 Wireless LC Resonator-Based Flexible or Stretchable Sensors

LC resonators that include an inductor and a capacitor in parallel with or without resistors are widely used in wireless sensors due to their simplicity and their ability to change frequency response due to strain and varying dielectric properties of their surroundings. Epidermal electronics-based LC resonators were developed widely to create sensors that can sense clinical conditions such as hydration and strain. A primary coil is placed in proximity and inductively coupled to the LC resonator in the sensor. Once the value of the capacitive or inductive components change during the sensing process, the resonant frequency of the LC resonator will change accordingly which results in the change of frequency response in the primary coil [75]. The deformation of the LC resonator placed on the skin and the change of the skin dielectric property result in the change of LC values, which affect the time varying electromagnetic fields generated by the primary coil and the frequency response of the primary coil.

In Fig. 5.8a, a wireless epidermal electronic system based on LC resonators with capacitive and/or inductive electrodes is shown [76]. The sensor combines a dielectric sensor and two strain sensors for both the x- and y-axis directions, and is able to establish conformal contact with the human skin. The dielectric sensor and strain sensors are comprised of inductors made from serpentine-shaped metal lines with a planar concentric capacitor and serpentine-shaped interdigitated capacitor, respectively. The layer of serpentine interdigitated electrodes and circular electrodes are insulated from the serpentine coils by a layer of polyimide, which is also used to passivate the entire structure of the sensor. To utilize flat wafer-based fabrication technology, the sensor was completed on a temporary glass substrate and transferred onto a silicone elastomer substrate using water-soluble tape. Utilizing

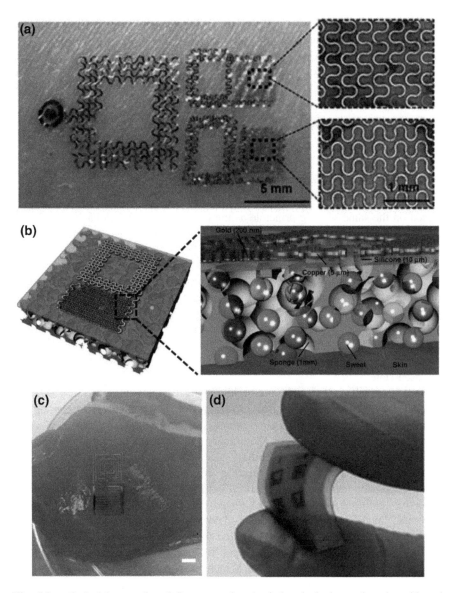

Fig. 5.8 **a** Optical image of an LC resonator-based wireless hydration and strain epidermal sensors on the skin. Reproduced with permission. Copyright 2014, John Wiley and Sons [76]. **b** Schematic illustration of a passive wireless capacitive sensor designed for sensing of sweat using porous sponge. Reproduced with permission. Copyright 2014, John Wiley and Sons [77]. **c** An optical image of a graphene-based wireless bacteria detection sensor mounted on a muscle tissue. Reproduced with permission. Copyright 2012, Macmillan Publishers Ltd. [78]. **d** An optical image of a bent 2 × 2 flexible array of 2 × 2 mm² large wireless passive pressure sensor for health monitoring. Reproduced with permission. Copyright 2014, Macmillan Publishers Ltd. [81]

serpentine-shaped metal lines and the stretchable elastomer substrate, the finished sensor can function without failing, even with large deformations of the skin. During the sensing process, a copper primary coil is placed in proximity to the sensor and its frequency response is measured. With further enhancements in LC resonating sensors, different types of biomedical sensors such as hydration sensors, bacteria sensors, or pressure sensors can be created.

5.5.2 Functional Materials for LC Resonator-Based Wireless Sensors

Based on the same sensing principle as the LC resonator, functional materials can also be incorporated as substrates to sense body fluids in wireless epidermal sensors like the one in Fig. 5.8b [77]. Here, a sponge was used as functional porous soft substrate to collect body fluids from the skin. The dielectric property of the sponge changed with the amount of absorbed body fluid, which then affects the frequency response of the serpentine-shaped LC resonator that is in contact detected by the primary coils. The ability to provide analysis of body fluids such as blood, interstitial fluid, sweat, saliva, and tears combined with thin film near-field wireless epidermal electronics provide a portable and convenient way to measure important effects on the skin, such as body temperature, fluid and electrolyte balance, and diseased state.

Graphene, which is flexible and highly conductive in nature, can also be combined with flexible LC resonator for bio-selective detection of bacteria [78]. Combined with naturally occurring antimicrobial peptides (AMPs) that serve as robust biorecognition molecules, the graphene may be functionalized with chemically synthesized bifunctional peptides, allowing it to become ultra-sensitive to pathogenic bacteria. Such functionalized graphene, grown on Ni films by chemical vapor deposition, can be transfer printed onto bioresorbable silk films. The planar inductive coil antenna and interdigitated electrodes are then simultaneously deposited on the graphene nanosensor to form an LC resonator, which is inductively coupled to primary coils. Because the silk is water soluble and biocompatible, the use of silk films allows an interface to form between the passive wireless graphene monolayers and biomaterials, such as tooth enamel or tissue, via bioresorption. The finished sensor laminated onto muscle tissue is shown in Fig. 5.8c. The resulting graphene-based sensor is capable of extremely sensitive spotting, with detection limits down to a single bacterium. Such biohazard monitoring of bacteria may provide early detection against pathogenic threats at the point of contamination.

Pressure monitoring is also an essential part of clinical practice including blood, intraocular and intracranial pressure conditions. Biomedical sensors that monitor such physiological parameters using small integrated circuits have been reported [79, 80]. Although the size of these sensors is very small (at the level of cubic millimeters), they are still in rigid packages which can pose as potential threats for soft tissues and organs. To overcome such issues, printed wireless sensor arrays with LC resonator-based passive flexible pressure sensors at the millimeter scale size have been demonstrated [81]. The beauty of this miniaturized device for implantable applications is that the sensor operates at high frequencies. In traditional implantable devices, the operating frequencies had to be kept below a few hundred MHz due to increasing loss at higher working frequencies; thus, a large antenna is needed for lower working frequencies. This study has shown that even at high frequencies, the detection could be read out by eliminating noise and unwanted signals via defining power reflection distortion (PRD) and group delay distortion (GDD). In addition, a deformable dielectric layer made from micro-structured styrene-butadiene-styrene (SBS) elastomer, which has lower loss at higher frequencies than other elastomers such as PDMS, is inserted between two layers of spiral metal on a flexible polyimide substrate. As shown in Fig. 5.8d, the resulting pressure sensor is one order magnitude smaller in volume (0.1 mm^3) than previously reported devices. This work has demonstrated that existing wireless sensing platforms with large antennas can be further reduced and lead to more opportunities in continuous wireless monitoring of physiological conditions in the body.

5.6 Integrated Circuits for Wireless Systems

Due to the simplicity of wireless flexible or stretchable sensing devices based on LC resonators, the signals that they can measure are limited. To enhance the systems' functionality and performance, either integrating multiple sensors into one device or combining flexible elements like wireless coils or stretchable interconnects with commercial rigid chips or thin-film transistors (TFT)-based circuits must be explored [16, 83–87]. Numerous integration techniques at the electrical engineering level to develop circuits for safe and convenient wireless bioelectronics and at the materials engineering level to develop complex assembly of soft and hard materials have been realized.

5.6.1 Wireless Flexible Systems with Thin-Film Transistor Circuit

Organic or inorganic TFT-based circuits are essentially mechanically flexible, which make them suitable for application in conformal biomedical electronics. Advanced versions of flexible wireless electronic systems integrated with wireless power and data transmission have been reported for applications like wireless sensors [82] and identification tags [86]. For instance, a wet sensor sheet that integrates thin-film circuits with near-field communication systems provides safer

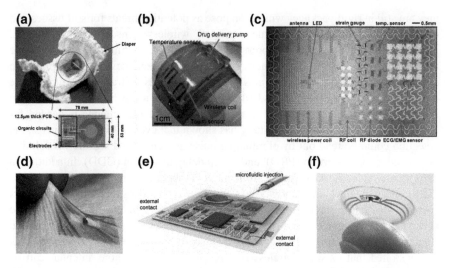

Fig. 5.9 **a** Flexible wet sensor sheet for urination detection with electrostatic discharge protection capability. Reproduction permission required [82]. **b** Photograph of the wireless wearable, human-interactive device, integrated with touch and temperature sensors, a wireless coil, and a capacitive touch panel, for multifunctional health monitoring. Reproduced with permission. Copyright 2014, John Wiley and Sons [83]. **c** Optical image of a demonstration platform for multifunctional electronics with physical properties matched to the epidermis. Reproduced with permission. Copyright 2011, American Association for the Advancement of Science [16]. **d** Photograph of an epidermal electronic systems with advanced capabilities in near-field communication on skin. Reproduced with permission. Copyright 2015, John Wiley and Sons [84]. **e** A schematic illustration of a soft, stretchable electronic system that integrates strain-isolated device components and a free-floating interconnect network in a thin elastomeric microfluidic enclosure. Reproduced with permission. Copyright 2014, American Association for the Advancement of Science [92]. **f** Photograph of a smart lens system that can measure glucose level in tears, which sends the data to a mobile device over integrated wireless system. Reproduction permission required [93]

measurements as it incorporates organic transistors-based electrostatic discharge (ESD) protection circuits to prevent shock from electrodes in contact with wet human skin, as shown in Fig. 5.9a [82]. Through detecting the resistance between two electrodes, the presence or absence of liquid is sensed and sent to the primary reader coil. In another wireless communicating device, an amorphous Indium-Gallium-Zinc-Oxide (a-IGZO) TFT-based circuit was integrated on a flexible polyester substrate [86]. In this a-IGZO-based NFC tag, a high frequency capacitor, rectifier, 12b code generator, and load modulator are all fabricated on top of the antenna foil, which allows data transmission to commercial NFC enabled smartphones or computers. In both, the wet sensor and the NFC tag, tens to hundreds of flexible TFTs were integrated together with near-field communicating system, which shows the feasibility of integration of advanced electronics with a variety of wireless biomedical devices.

5.6.2 Wireless Wearable Human-Interactive Device

Besides sensing and identification applications, wireless flexible or stretchable devices that combine both the ability to sense and interact with humans have attracted people for years with hopes of bringing a more convenient lifestyle. Although commercial wearable devices like the Apple Watch is capable of sensing several human body signals and interacting with humans, they are still limited by bulky parts and devices. Flexible or stretchable electronic devices that have similar performance to their rigid counterparts have been in rapid development. As an example, the "smart bandage", a wearable human-interactive device fabricated on flexible Kapton substrate is able to sense human body signals, interact with humans, and send out signals wirelessly [83]. The capacitive touch sensor and wireless coil in the device are printed using silver ink and the embedded temperature sensor is printed using a mixture of conductive polymer and carbon nanotube paste. A drug delivery pump with a microfluidic channel made from PDMS was also bonded into the device. All the material patterning processes done in this work used a macroscale printing method, which eliminates the need for costly traditional photolithography; hence, the reduction in cost. The printing method used to fabricate Ag electrodes and temperature sensors also play an important role in ensuring the flexibility of the device. The resulting device, shown in Fig. 5.9b, is able to monitor human body temperature, interact with humans through a touch panel, deliver drugs by pressing the bandage, and detect touch wirelessly through the combination of wireless coils and integrated capacitive touch sensor which works as a resonator.

5.6.3 Wireless Epidermal Electronic Systems

Although wireless flexible electronic systems and simple LC resonator-based epidermal sensors have been demonstrated, the need for a system level epidermal electronic device still exists due to the fact that flexible electronic systems cannot form conformal contact with skin and the function of simple LC resonator-based epidermal sensors are too limited. Rogers and colleagues have demonstrated a platform for the wireless epidermal electronics, which is shown in Fig. 5.9c [16]. Multifunctional sensors were integrated for monitoring temperature, strain, and electrophysiological signals on the platform. Circuit elements like transistors, diodes, resistors, capacitors, and inductors included in the platform allow for signal amplification or data processing. High-frequency elements like inductors, capacitor, oscillators, and antennas shown in the platform show the systems' potential of transmitting data by RF communication. Other elements like micro-scale LEDs can be used for illuminating or light therapy. All of these sensors, circuit elements, and wireless power and signal transferring elements are integrated on an ultrathin, gas-permeable elastomeric sheet with a low Young's modulus and connected by ultrathin serpentine metal lines, which provide the system the ability to form conformal contact with the skin.

5.6.4 Integration with Near-Field Communication Capabilities

Today's NFC chips appear in many convenient applications that span the banking, medical, military, transportation, and entertainment industries. To enable such benefits in epidermal electronic systems, Rogers and colleagues have reported sensors combining flexible passive coils, ultrathin NFC dies and other components like light emitting diodes (LEDs) [84, 85] which are able to communicate with NFC-enabled smartphones. The wireless coils used in both sensors are composed of two layers of copper coils; layers of polyimide were used to both insulate the two copper layers and encapsulate the coil structure. NFC dies or LED dies are flip-chip bonded to the coil structure using indium/Ag-based solder paste. After chip bonding, the whole structure is encapsulated by an elastomer, which protects the whole device. The coils combining the capacitance in NFC dies formed an LC resonator. Through changing the shape of wireless coils by stretching or bending, the resonating frequency changes accordingly and can be detected by the primary coils such as the NFC coils in smartphones. The NFC chip-enabled sensor can be laminated onto human skin and allow seamless conformal contact with the skin as shown in Fig. 5.9d. Different functionalities, such as identification, authentication, and temperature sensing, are offered by the system by selecting appropriate NFC dies. Due to the ability of forming conformal contacts with the skin, and ability to be integrated with various sensors, this design shows great potential for epidermal electronic systems and possibly for implantable bio-sensing systems.

5.6.5 Soft Microfluidic Assemblies of Sensors, Circuits, and Radios for the Skin

Besides the simple integration of the flexible coil and single commercial communication chip, various methods have been developed to fabricate the multi-chip module (MCM) onto flexible substrates [87–91]. Although the integration of MCMs onto flexible substrate enables multifunctional, flexible electronic systems, they are still bulky and cannot achieve conformal contact with the human body, which limits their application in epidermal electronics systems. To overcome this challenge, an advanced assembly technique utilizing microfluidics to assemble sensors, circuits, and radios has been developed to combine the advantages of multi-functionality of MCMs and conformability of epidermal systems [92]. The microfluidic-assembled system is fabricated by connecting lapped chips using serpentine interconnects on top of polyimide/PDMS, which is covered by silicone superstrate and sealed with additional partially cured silicone. The liquid PDMS was finally injected into the capped cavity through an edge to isolate rigid materials from elastomeric enclosures as shown in Fig. 5.9e. The serpentine interconnects and injected liquid PDMS make sure that the whole system has a low Young's modulus and offer great stretchability. Integrated with multifunctional electrodes, the device can fulfill tasks like temperature monitoring, motion tracking, electrocardiography (ECG), electroencephalography (EEG), electrooculography (EOG), and electromyography (EMG) measurements by selecting chips with different functionalities. Additionally, the device is powered wirelessly through inductive coupling with the primary coil in proximity and the data can be sent out wirelessly through an RF module in the device; thus, eliminating the complex and bulky power supply and signal transmission systems.

5.7 Conclusion

Being more convenient, accurate and safe, than existing bioelectronics, there is no doubt that next generation biomedical devices will be in conformal format. Products of such conformal bioelectronics using wireless functionalities are being commercialized rapidly and big players in the industries are investing billions of dollars into this area. For instance, Google Inc. has announced that they have built a wireless contact lens that can monitor diabetes by sensing the glucose level of the tears [93]. The smart contact lens, as shown in Fig. 5.9f, also utilizes circular shape wireless antenna, which transmits clinical data to a mobile device for continuous monitoring. The lens has a tiny hole to allow tear to be collected by the system and with ultra-miniaturized integrated circuit-based wireless controller chip connected to the sensor, the contact lens is able to read glucose level every one second. It seems that the technology of such kind is already mature enough to be commercialized; but in order for this smart lens to be get access to market, it must go through rigorous reviews, tests, and surveys, to be proven for safety. Federal institutes that regulate these products include the Food and Drug Administration (FDA), which protects and promotes public health, and the FCC, which regulates frequency bands for wireless devices. As such, there are more steps on these research-to-product process, before people start seeing them in local pharmacies and hospitals. However, with some of the already developed high technology conformal bioelectronics using wireless functionalities discussed in this chapter, people will be able to use such devices daily very soon.

References

1. K. LineBaugh, Medical devices in hospitals to go wireless (2012), http://www.wsj.com/articles/SB10001424052702304065704577422633456558976. Accessed 18 Aug 2012
2. W. Greatbatch, C.F. Holmes, History of implantable devices. IEEE Eng. Med. Biol. Mag. **10**, 38–41 (1991)
3. J.T. Farrar, V.K. Zworykin, J. Baum, Pressure-sensitive telemetering capsule for study of gastrointestinal motility. Science **126**, 975–976 (1957)
4. R.S. Mackay, B. Jacobson, Endoradiosonde. Nature **179**, 1239–1240, (1957)

5. J.A. Rogers, M.G. Lagally, R.G. Nuzzo, Synthesis, assembly and applications of semiconductor nanomembranes. Nature **477**, 45–53 (2011)
6. G. Park, H.-J. Chung, K. Kim, S.A. Lim, J. Kim, Y.-S. Kim et al., Immunologic and tissue biocompatibility of flexible/stretchable electronics and optoelectronics. Adv. Healthc. Mater. **3**, 515–525 (2014)
7. D.-Y. Khang, H. Jiang, Y. Huang, J.A. Rogers, A stretchable form of single-crystal silicon for high-performance electronics on rubber substrates. Science **311**, 208–212 (2006)
8. D.-H. Kim, J. Viventi, J.J. Amsden, J. Xiao, L. Vigeland, Y.-S. Kim et al., Dissolvable films of silk fibroin for ultrathin conformal bio-integrated electronics. Nat. Mater. **9**, 511–517 (2010)
9. G.A. Covic, J.T. Boys, Inductive Power Transfer. Proc. IEEE **101**, 1276–1289 (2013)
10. A.M. Sodagar, P. Amiri, Capacitive coupling for power and data telemetry to implantable biomedical microsystems, in *09. 4th International IEEE/EMBS Conference on Neural Engineering, 2009. NER*, pp. 411–414 (2009)
11. A. E. Umenei, Understanding low frequency non-radiative power transfer, in *Wireless Power Consortium contribution by Fulton Innovation LLC*, vol. 7575 (2011)
12. J.I. Agbinya, Wireless power transfer, in *Principles of Inductive near Field Communications for Internet of Things*, vol. 18 (2011), pp. 281–300
13. N. Shinohara, Theroy of WPT, in *Wireless Power Transfer Via Radiowaves*, pp. 21–52 (2014)
14. A. Karalis, J.D. Joannopoulos, M. Soljacic, Efficient wireless non-radiative mid-range energy transfer. Ann. Phys. **323**, 34–48 (2008)
15. S.H. Jeong, K. Hjort, Z. Wu, Tape transfer atomization patterning of liquid alloys for microfluidic stretchable wireless power transfer. Sci. Rep. **5**, 8419 (2015)
16. D.-H. Kim, N. Lu, R. Ma, Y.-S. Kim, R.-H. Kim, S. Wang et al., Epidermal electronics. Science **333**, 838–843 (2011)
17. H.L. Liang, H.T. Whelan, J.T. Eells, M.T.T. Wong-Riley, Near-infrared light via light-emitting diode treatment is therapeutic against rotenone—and 1-methyl-4-phenylpyridinium ion-induced neurotoxicity. Neuroscience **153**, 963–974 (2008)
18. J.L.N. Bastos, R.F.Z. Lizarelli, N.A. Parizotto, Comparative study of laser and LED systems of low intensity applied to tendon healing. Laser Phys. **19**, 1925–1931, (2009)
19. B.P. Timko, T. Dvir, D.S. Kohane, Remotely triggerable drug delivery systems. Adv. Mater. **22**, 4925–4943 (2010)
20. S. Waxman, Near-Infrared Spectroscopy for plaque characterization. J. Intervent. Cardiol. **21**, 452–458 (2008)
21. R.-H. Kim, D.-H. Kim, J. Xiao, B.H. Kim, S.-I. Park, B. Panilaitis et al., Waterproof AlInGaP optoelectronics on stretchable substrates with applications in biomedicine and robotics. Nat. Mater. **9**, 929–937 (2010)
22. T.-I. Kim, Y.H. Jung, J. Song, D. Kim, Y. Li, H.-S. Kim et al., High-efficiency, microscale GaN light-emitting diodes and their thermal properties on unusual substrates. Small **8**, 1643–1649 (2012)
23. T.-I. Kim, J.G. McCall, Y.H. Jung, X. Huang, E.R. Siuda, Y. Li et al., Injectable, cellular-scale optoelectronics with applications for wireless optogenetics. Science **340**, 211–216 (2013)
24. R.-H. Kim, H. Tao, T.-I. Kim, Y. Zhang, S. Kim, B. Panilaitis et al., Materials and designs for wirelessly powered implantable light-emitting systems. Small **8**, 2812–2818 (2012)
25. S.-W. Hwang, H. Tao, D.-H. Kim, H. Cheng, J.-K. Song, E. Rill et al., A physically transient form of silicon electronics. Science **337**, 1640–1644 (2012)
26. L. Yin, H. Cheng, S. Mao, R. Haasch, Y. Liu, X. Xie et al., Dissolvable metals for transient electronics. Adv. Funct. Mater. **24**, 645–658 (2014)
27. H. Tao, S.-W. Hwang, B. Marelli, B. An, J.E. Moreau, M. Yang et al., Silk-based resorbable electronic devices for remotely controlled therapy and in vivo infection abatement. Proc. Natl. Acad. Sci. U.S.A. **111**, 17385–17389 (2014)
28. M. Koo, K.-I. Park, S.H. Lee, M. Suh, D.Y. Jeon, J.W. Choi et al., Bendable inorganic thin-film battery for fully flexible electronic systems. Nano Lett. **12**, 4810–4816 (2012)

29. S. Xu, Y. Zhang, J. Cho, J. Lee, X. Huang, L. Jia et al., Stretchable batteries with self-similar serpentine interconnects and integrated wireless recharging systems. Nat. Commun. **4**, 1543 (2013)
30. C. Dagdeviren, B.D. Yang, Y. Su, P.L. Tran, P. Joe, E. Anderson et al., Conformal piezoelectric energy harvesting and storage from motions of the heart, lung, and diaphragm. Proc. Natl. Acad. Sci. U.S.A. **111**, 1927–1932 (2014)
31. T. Sekitani, M. Takamiya, Y. Noguchi, S. Nakano, Y. Kato, T. Sakurai et al., A large-area wireless power-transmission sheet using printed organic transistors and plastic MEMS switches. Nat. Mater. **6**, 413–417 (2007)
32. M. Takamiya, T. Sekitani, Y. Miyamoto, Y. Noguchi, H. Kawaguchi, T. Someya et al., Design solutions for a multi-object wireless power transmission sheet based on plastic switches, in *Solid-State Circuits Conference, 2007. ISSCC 2007. Digest of Technical Papers. IEEE International*, 2007, pp. 362–609
33. D.-H. Kim, J.-H. Ahn, W.M. Choi, H.-S. Kim, T.-H. Kim, J. Song et al., Stretchable and foldable silicon integrated circuits. Science **320**, 507–511 (2008)
34. J. Yoon, S. Jo, I.S. Chun, I. Jung, H.-S. Kim, M. Meitl et al., GaAs photovoltaics and optoelectronics using releasable multilayer epitaxial assemblies, Nature **465**, 329–333 (2010)
35. H.-S. Kim, E. Brueckner, J. Song, Y. Li, S. Kim, C. Lu et al., Unusual strategies for using indium gallium nitride grown on silicon (111) for solid-state lighting. Proc. Natl. Acad. Sci. U.S.A. **108**, 10072–10077 (2011)
36. L. Sun, G. Qin, H. Huang, H. Zhou, N. Behdad, W. Zhou et al., Flexible high-frequency microwave inductors and capacitors integrated on a polyethylene terephthalate substrate. Appl. Phys. Lett. **96**, 013509 (2010)
37. J.-H. Ahn, H.-S. Kim, K.J. Lee, Z. Zhu, E. Menard, R.G. Nuzzo et al., High-speed mechanically flexible single-crystal silicon thin-film transistors on plastic substrates. IEEE Electron Device Lett. **27**, 460–462 (2006)
38. L. Sun, G. Qin, J.-H. Seo, G.K. Celler, W. Zhou, Z. Ma, 12-GHz thin-film transistors on transferrable silicon nanomembranes for high-performance flexible electronics. Small **6**, 2553–2557 (2010)
39. H. Zhou, J.-H. Seo, D.M. Paskiewicz, Y. Zhu, G.K. Celler, P.M. Voyles et al., Fast flexible electronics with strained silicon nanomembranes. Sci. Rep. **3**, 1291 (2013)
40. K. Rim, K. Chan, L. Shi, D. Boyd, J. Ott, N. Klymko et al., Fabrication and mobility characteristics of ultra-thin strained Si directly on insulator (SSDOI) MOSFETs, in *Electron Devices Meeting, 2003. IEDM '03 Technical Digest. IEEE International*, 2003, pp. 3.1.1–3.1.4
41. U.K. Mishra, L. Shen, T.E. Kazior, Y.-F. Wu, GaN-based RF power devices and amplifiers. Proc. IEEE **96**, 287–305 (2008)
42. T.-H. Chang, K. Xiong, S.H. Park, H. Mi, H. Zhang, S. Mikael et al., High power fast flexible electronics: transparent RF AlGaN/GaN HEMTs on plastic substrates, in *Microwave Symposium (IMS), 2015 IEEE MTT-S International*, 2015, pp. 1–4
43. S.P. Voinigescu, M.C. Maliepaard, J.L. Showell, G.E. Babcock, D. Marchesan, M. Schroter et al., A scalable high-frequency noise model for bipolar transistors with application to optimal transistor sizing for low-noise amplifier design. IEEE J. Solid-State Circ. **32**, 1430–1439 (1997)
44. Y.H. Jung, T.-H. Chang, H. Zhang, C. Yao, Q. Zheng, V.W. Yang et al., High-performance green flexible electronics based on biodegradable cellulose nanofibril paper, Nat. Commun. **6**, 7170 (2015)
45. G. Qin, H.-C. Yuan, G.K. Celler, W. Zhou, Z. Ma, Flexible microwave PIN diodes and switches employing transferrable single-crystal Si nanomembranes on plastic substrates, J. Phys. D-Appl. Phys. **42**, 234006 (2009)
46. G. Qin, H.-C. Yuan, Y. Qin, J.-H. Seo, Y. Wang, J. Ma et al., Fabrication and characterization of flexible microwave single-crystal germanium nanomembrane diodes on a plastic substrate. IEEE Electron Device Lett. **34**, 160–162 (2013)

47. H.-C. Yuan, G.. Qin, G.K. Celler, Z. Ma, Bendable high-frequency microwave switches formed with single-crystal silicon nanomembranes on plastic substrates. Appl. Phys. Lett. **95**, 043109 (2009)

48. G. Qin, L. Yang, J.-H. Seo, H.-C. Yuan, G.K. Celler, J. Ma et al., Experimental characterization and modeling of the bending strain effect on flexible microwave diodes and switches on plastic substrate. Appl. Phys. Lett. **99**, 243104 (2011)

49. CGH60008D, 8 W, 6.0 GHz, GaN HEMT Die, CREE, Ed., ed

50. D.R. Webb, I.G. Sipes, D.E. Carter, In vitro solubility and in vivo toxicity of gallium arsenide. Toxicol. Appl. Pharmacol. **76**, 96–104 (1984)

51. S.J. Cho, Y.H. Jung, Z. Ma, X-Band compatible flexible microwave inductors and capacitors on plastic substrate. IEEE J. Electron Devices Soc. **3**, 435–439 (2015)

52. S. Cheng, A. Rydberg, K. Hjort, Z. Wu, Liquid metal stretchable unbalanced loop antenna. Appl. Phys. Lett. **94**, 144103 (2009)

53. S. Cheng, Z.G. Wu, P. Hallbjorner, K. Hjort, A. Rydberg, Foldable and stretchable liquid metal planar inverted cone antenna. IEEE Trans. Antennas Propag. **57**, 3765–3771 (2009)

54. J.-H. So, J. Thelen, A. Qusba, G.J. Hayes, G. Lazzi, M.D. Dickey, Reversibly deformable and mechanically tunable fluidic antennas. Adv. Funct. Mater. **19**, 3632–3637 (2009)

55. M. Kubo, X. Li, C. Kim, M. Hashimoto, B. J. Wiley, D. Ham et al., Stretchable microfluidic radiofrequency antennas, Adv. Mater., **22**, pp. 2749–2752, (2010)

56. M. Park, J. Im, M. Shin, Y. Min, J. Park, H. Cho et al., Highly stretchable electric circuits from a composite material of silver nanoparticles and elastomeric fibres. Nat. Nanotechnol. **7**, 803–809 (2012)

57. L. Song, A.C. Myers, J.J. Adams, Y. Zhu, Stretchable and reversibly deformable radio frequency antennas based on silver nanowires. ACS Appl. Mater. Interfaces **6**, 4248–4253 (2014)

58. G.J. Hayes, J.-H. So, A. Qusba, M.D. Dickey, G. Lazzi, Flexible liquid metal alloy (EGaIn) microstrip patch antenna. IEEE Trans. Antennas Propag. **60**, 2151–2156 (2012)

59. Y. Qiu, Y.H. Jung, S. Lee, T.-Y. Shih, J. Lee, Y. Xu et al., Compact parylene-c-coated flexible antenna for WLAN and upper-band UWB applications. Electron. Lett. **50**, pp. 1782–1784 (2014)

60. J.A. Fan, W.-H. Yeo, Y. Su, Y. Hattori, W. Lee, S.-Y. Jung et al., Fractal design concepts for stretchable electronics, Nat. Commun. **5**, 3266 (2014)

61. H. Bizri, F. Toameh, W. Hassan, A. Hage-Diab, L. Mustapha, Simulation of RF biological tissues response towards remote sensing ECG device, in *2nd International Conference on Advances in Biomedical Engineering (ICABME)* (2013), pp. 9–13

62. R.E. Fields, Evaluating compliance with FCC guidelines for human exposure to radiofrequency electromagnetic fields (1997)

63. R.S. Alrawashdeh, Y. Huang, M. Kod, A.A. Sajak, A broadband flexible implantable loop antenna with complementary split ring resonators, IEEE Antennas Wirel. Propag. Lett. **14**,1506–1509 (2015)

64. M.L. Scarpello, D. Kurup, H. Rogier, D. Vande Ginste, F. Axisa, J. Vanfleteren et al., Design of an implantable slot dipole conformal flexible antenna for biomedical applications. IEEE Trans. Antennas Propag. **59**, 3556–3564 (2011)

65. D.D. Karnaushenko, D. Karnaushenko, D. Makarov, O.G. Schmidt, Compact helical antenna for smart implant applications. NPG Asia Mater. **7**, e188 (2015)

66. R. Alrawashdeh, Y. Huang, P. Cao, Flexible meandered loop antenna for implants in MedRadio and ISM bands. Electron. Lett. **49**, 1515–1516 (2013)

67. Z. Duan, Y.-X. Guo, M. Je, D.-L. Kwong, Design and in vitro test of a differentially fed dual-band implantable antenna operating at MICS and ISM Bands. IEEE Trans. Antennas Propag. **62**, 2430–2439 (2014)

68. H.R. Raad, A.I. Abbosh, H.M. Al-Rizzo, D.G. Rucker, Flexible and compact AMC based antenna for telemedicine applications. IEEE Trans. Antennas Propag. **61**, 524–531 (2013)

69. S.-W. Hwang, X. Huang, J.-H. Seo, J.-K. Song, S. Kim, S. Hage-Ali et al., Materials for bioresorbable radio frequency electronics. Adv. Mater. **25**, 3526–3531 (2013)

70. N. Sani, M. Robertsson, P. Cooper, X. Wang, M. Svensson, P.A. Ersman et al., All-printed diode operating at 1.6 GHz. Proc. Natl. Acad. Sci. USA. **111**, 11943–11948 (2014)
71. J. Zhang, Y. Li, B. Zhang, H. Wang, Q. Xin, A. Song, Flexible indium-gallium-zinc-oxide Schottky diode operating beyond 2.45 GHz. Nat. Commun. **6**, 7561 (2015)
72. F.H.C. Crick, Thinking about the brain. Sci. Am. **241**, 219–232 (1979)
73. S.I. Park, G. Shin, A. Banks, J.G. McCall, E.R. Siuda, M.J. Schmidt et al., Ultraminiaturized photovoltaic and radio frequency powered optoelectronic systems for wireless optogenetics. J. Neural Eng. **12**, 056002 (2015)
74. K.L. Montgomery, A.J. Yeh, J.S. Ho, V. Tsao, S. Mohan Iyer, L. Grosenick et al., Wirelessly powered, fully internal optogenetics for brain, spinal and peripheral circuits in mice. Nature Methods, **12**, 969–974 (2015)
75. E.L. Tan, W.N. Ng, R. Shao, B.D. Pereles, K.G. Ong, A wireless, passive sensor for quantifying packaged food quality. Sensors **7**, 1747 (2007)
76. X. Huang, Y. Liu, H. Cheng, W.-J. Shin, J.A. Fan, Z. Liu et al., Materials and designs for wireless epidermal sensors of hydration and strain. Adv. Funct. Mater. **24**, 3846–3854 (2014)
77. X. Huang, Y. Liu, K. Chen, W.-J. Shin, C.-J. Lu, G.-W. Kong et al., Stretchable, wireless sensors and functional substrates for epidermal characterization of sweat. Small **10**, 3083–3090 (2014)
78. M.S. Mannoor, H. Tao, J.D. Clayton, A. Sengupta, D.L. Kaplan, R.R. Naik et al., Graphene-based wireless bacteria detection on tooth enamel. Nat. Commun. **3**, 763 (2012)
79. C. Peng, N. Chaimanonart, W.H. Ko, D.J. Young, A wireless and batteryless 130 mg 300 μW 10b implantable blood-pressure-sensing microsystem for real-time genetically engineered mice monitoring, in *Solid-State Circuits Conference - Digest of Technical Papers, 2009. ISSCC 2009. IEEE International*, 2009, pp. 428–429,429a
80. C. Po-Jui, S. Saati, R. Varma, M.S. Humayun, T. Yu-Chong, Wireless intraocular pressure sensing using microfabricated minimally invasive flexible-coiled LC sensor implant. J. Microelectromech. Sys. **19**, 721–734 (2010)
81. L.Y. Chen, B.C.-K. Tee, A.L. Chortos, G. Schwartz, V. Tse, D.J. Lipomi et al., Continuous wireless pressure monitoring and mapping with ultra-small passive sensors for health monitoring and critical care. Nat. Commun. **5**, 5028 (2014)
82. H. Fuketa, K. Yoshioka, T. Yokota, W. Yukita, M. Koizumi, M. Sekino et al., 30.3 Organic-transistor-based 2 kV ESD-tolerant flexible wet sensor sheet for biomedical applications with wireless power and data transmission using 13.56 MHz magnetic resonance, in *Solid-State Circuits Conference Digest of Technical Papers (ISSCC), 2014 IEEE International*, 2014, pp. 490–491
83. W. Honda, S. Harada, T. Arie, S. Akita, K. Takei, Wearable, human-interactive, health-monitoring, wireless devices fabricated by macroscale printing techniques. Adv. Funct. Mater. **24**, 3299–3304 (2014)
84. J. Kim, A. Banks, H. Cheng, Z.. Xie, S. Xu, K.-I. Jang et al., Epidermal electronics with advanced capabilities in near-field communication. Small **11**, 906–912 (2015)
85. J. Kim, A. Banks, Z. Xie, S.Y. Heo, P. Gutruf, J.W. Lee et al., Miniaturized flexible electronic systems with wireless power and near-field communication capabilities. Adv. Funct. Mater. **25**, 4761–4767 (2015)
86. K. Myny, B. Cobb, J.L. van der Steen, A.K. Tripathi, J. Genoe, G. Gelinck et al., 16.3 Flexible thin-film NFC tags powered by commercial USB reader device at 13.56 MHz, in *Solid- State Circuits Conference—(ISSCC), 2015 IEEE International*, 2015, pp. 1–3
87. M. Murugesan, J.C. Bea, T. Fukushima, T. Konno, K. Kiyoyama, W. C. Jeong et al., Cu lateral interconnects formed between 100-μm-thick self-assembled chips on flexible substrates, in *Electronic Components and Technology Conference, 2009. ECTC 2009. 59th*, 2009, pp. 1496–1501
88. T.-Y. Chao, Y.T. Cheng, Wafer-level chip scale flexible wireless microsystem fabrication, in *IEEE 24th International Conference on Micro Electro Mechanical Systems (MEMS)*, 2011, pp. 344–347

89. H. Rempp, J. Burghartz, C. Harendt, N. Pricopi, M. Pritschow, C. Reuter et al., Ultra-thin chips on foil for flexible electronics, in *Solid-State Circuits Conference, 2008. ISSCC 2008. Digest of Technical Papers. IEEE International*, 2008, pp. 334–617

90. P. Mostafalu, W. Lenk, M. Dokmeci, B. Ziaie, A. Khademhosseini, S. Sonkusale, Wireless flexible smart bandage for continuous monitoring of wound oxygenation, in *Biomedical Circuits and Systems Conference (BioCAS), 2014 IEEE*, 2014, pp. 456–459

91. J.G. McCall, T.-I. Kim, G. Shin, X. Huang, Y.H. Jung, R. Al-Hasani et al., Fabrication and application of flexible, multimodal light-emitting devices for wireless optogenetics. Nat. Protoc. **8**, 2413–2428, (2013)

92. S. Xu, Y. Zhang, L. Jia, K.E. Mathewson, K.-I. Jang, J. Kim et al., Soft microfluidic assemblies of sensors, circuits, and radios for the skin. Science **344**, 70–74 (2014)

93. B. Otis, B. Parviz, Introducing our smart contact lens project, in *Google Official Blog* vol. 2015, ed. by Google (2014)

Part II
Wearable Electronics Systems

Chapter 6
Ultrathin, Skin-Like Devices for Precise, Continuous Thermal Property Mapping of Human Skin and Soft Tissues

R. Chad Webb, Siddharth Krishnan and John A. Rogers

Abstract Precision thermal measurements of skin and soft tissue can provide clinically relevant information about cardiovascular health, cognitive state, hydration levels, heterogeneousvasculature changes, and many other important aspects of human physiology. In this chapter we discuss recent advances in ultrathin, compliant skin-like sensor/actuator technologies that enable forms of continuous thermal mapping, of temperature as well as transport properties, that are unavailable with other methods. We review the key mechanical and thermal properties that are fundamental to the operation of this class of devices. Further discussion of devices configured for mapping temperature, monitoring local thermal transport and skin hydration, and mapping thermal transport for blood flow analysis provides a few examples of the types of capabilities that are enabled with these technologies.

Keywords Thermal sensors · Blood flow · Skin · Circulation · Wearable electronics · Thermal transport · Mapping · Hydration · Imaging

6.1 Introduction

Thermal monitoring and manipulation of human skin or other organs of the body provide compelling application opportunities for soft, stretchable bioelectronics, due to the extraordinary thermal sensitivity of many tissue systems. The temperature and thermal transport properties of skin, in particular, depend critically on important

R. Chad Webb
3M Company, Saint Paul, MN, USA
e-mail: rwebb@mmm.com

S. Krishnan · J.A. Rogers (✉)
Department of Materials Science and Engineering, University of Illinois at Urbana-Champaign, Urbana, IL, USA
e-mail: jrogers@illinois.edu

S. Krishnan
e-mail: krishnn6@illinois.edu

© Springer International Publishing Switzerland 2016
J.A. Rogers et al. (eds.), *Stretchable Bioelectronics for Medical Devices and Systems*, Microsystems and Nanosystems,
DOI 10.1007/978-3-319-28694-5_6

physiological states such as blood flow, hydration, and metabolic rates. As a result, thermal parameters are also an indirect function of the physiological stressors that influence the aforementioned states, such as vascular disease, physical and mental exertion, body positioning, and trauma.

Numerous methods exist for assessing temperature of biological tissues each with associated strengths and weaknesses. Infrared (IR) thermography [1–3] provides an option for mapping surface temperatures with millikelvin precision and fine spatial resolution. However, such imaging tools are expensive, bulky, and require immobilization of the patient in order to minimize any relative motion between the object of interest and the imaging optics. Point contact sensors adhered directly to tissue avoid the sensitivity to motion and are inexpensive, but they do not have the ability to perform spatial mapping, which is typically required to extract meaningful information in biological studies or clinical applications. Typical sensors of this type also irritate the tissue and modify its natural physiological responses due to imposed thermal and mechanical loads associated with the physical influence of the device.

An analogous set of techniques exists for measuring thermal transport properties where the goal is typically to obtain information about blood flow. Optical measurements of blood flow rely on laser speckle contrast imaging (LSCI) [4–6] and laser Doppler flowmetry (LDF) [7–9] techniques. Such approaches are, however, more sensitive to motion than IR thermography because these measurements rely on the scattering of photons from moving blood cells and transmission via fiber optic cables. The addition of laser heating to these types of optical setups provides a means for also measuring thermal transport properties [10], via the spatiotemporal dynamics of heat diffusion, but with the disadvantage of additional complexity and cost in the optical systems.

Acoustic methods, such as ultrasound [11, 12], suffer from similar sensitivities to motion. Point contact thermal elements [13–16] that use metal heating and sensing elements applied to the tissue surface do not suffer from such limitations, but they cannot easily provide capabilities in spatial mapping, they cannot track rapid or subtle changes (due to the relatively high thermal mass of the devices compared to skin), nor can they monitor unaltered, natural signals (due to the mechanical pressure required to achieve the necessary intimate contact with the surface).

Epidermal thermal sensor systems that conform intimately to the tissue surface [17–20], allowing unrestricted motion, can now be created (select examples in Fig. 6.1) by extending many of the principles of ultrathin, stretchable electronic sensor design established during the past decade [18, 20–34]. Extremely thin metallic (bottom inset, Fig. 6.1a) and semiconducting (top inset, Fig. 6.1a) elements, with thicknesses on the order of hundreds of nanometers or less, afford the soft mechanics necessary for intimate thermal contact against heterogeneous tissue surfaces without reduction in electronic performance. These elements can be integrated as part of a hard/soft materials system consisting of 'soft' polymeric layers and 'hard' metallic/semimetallic layers to provide a device suitable for application to soft tissues (Fig. 6.1). Optimized design of open mesh layouts with serpentine interconnects [23, 24, 30, 35–39] and mechanical strain engineering can result in

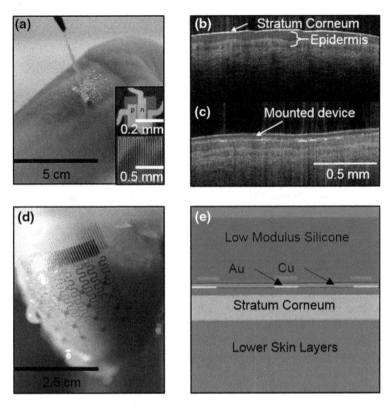

Fig. 6.1 Soft, stretchable skin-mounted electronics for precision thermography of skin. **a** Optical image of 16-channel monitoring device with sensors that rely on the temperature coefficient of resistance (TCR) of thin metallic traces, mounted on human skin. *Inset* Micrographs of a sensor that uses a p-n semiconductor junction (*above*) and a TCR metallic trace (*below*). Originally published in [18]. **b, c** Optical coherence tomography (OCT) of a human palm (**b**) with and (**c**) without an epidermal sensor. Originally published in [19]. **d** Thermal array mounted on the surface of an artificial pericardial membrane. Originally published in [20]. **e** Cross section of device mounted on skin. Originally published in [19]

devices that are highly stretchable without significant impact on device operation. Figure 6.1e shows a representative design configured for direct integration onto the surface of the skin. A thin silicone layer (~ 5 µm) between the device and skin provides robust but reversible adhesion via the action of van der Waals forces alone, without the need for penetrating pins, separated adhesives or straps/bands. Polyimide encapsulation (~ 1 µm on each side of the electronic elements, not shown) serves as electrical isolation and as a means to engineer the distribution of strain to ensure fully elastic, soft mechanical responses. The metal and/or semiconducting components can be arranged in a series of layers with polyimide as a dielectric. The entire device can be fabricated on rigid substrates using conventional microelectronics fabrication tools, and then released and transfer printed to the silicone layers. When the device exists in a mesh layout with small polyimide islands to encapsulate the sensors, the relatively low modulus silicone (~ 60 kPa), as opposed to the relatively high modulus polyimide (2–3 GPa), accommodates a majority of any applied strain.

6.2 Temperature Mapping on SoftTissues

Devices of this type provide a general platform that can be modified to perform a variety of thermal measurements on tissue locations of interest. The most straightforward operational mode involves measurement of temperature on the surface of the tissue. Numerous physiological phenomena yield measureable signals related to changes in temperature, but the focus here is on three particularly useful platforms for epidermal electronics.

The first two rely on electronic sensors that respond to changes in temperature. Device designs that exploit the temperature coefficient of resistance (TCR) of metal provide one option (bottom inset, Fig. 6.1a), while designs that rely on the temperature dependence of the charge carrier density in semiconductors provide another (top inset, Fig. 6.1a). Devices that rely on TCR in thin metal films offer simplicity of fabrication and mechanical robustness compared to those that involve semiconductor materials. Those that exploit semiconducting sensors, on the other hand, provide straightforward paths to high density arrays of elements through multiplexed addressing. Both concepts have been demonstrated in epidermal electronic formats, as shown in Fig. 6.1a–c and described in detail by Webb et al. [18, 19]. A third device option avoids electronic sensing entirely, and relies instead on skin-conformal arrays of thermochromic liquid crystal elements [17], where a colorimetric response serves as the measurement mechanism.

TCR type devices have been used extensively on skin in many systematic studies and broader human clinical investigations, to reveal the underlying physics of their operation and to examine various aspects of skin physiology [18, 19, 40]. Optimized devices achieve a measurement precision of 10 mK, which exceeds that of many of the most sophisticated IR thermography systems. Additionally, the thermal response time (the time for the device to achieve a 90 % response to a step change in temperature) can be as low as 4 ms, well below all known timescales associated with physiological changes. Analytical models yield insights into thermal design considerations for high performance operation, as detailed by Bian et al. [41]. The extremely low thermal mass of the devices, relative to skin, plays a key role. For example, the 4×4 TCR sensor array design shown in Fig. 6.1 has a thermal mass per unit area of 150 μJ cm^{-2} K^{-1}. The addition of a 50 μm thick silicone support to this device increases the thermal mass to 7.2 μJ cm^{-2} K^{-1}. These thermal masses are equivalent to those of skin with thickness <500 nm and <25 μm, respectively, for cases without and with the silicone support. In other words, the thermal mass load associated with these devices is a small fraction of the thermal mass of the dermal/epidermal matrix itself.

An important additional parameter is the strain response of the device. The resistance of a metal element is altered by strain due to induced geometrical changes. Because change in the resistance of the element is the signal used to calculate temperature changes, and the devices are intended to be functional even during motion and strain, it is important to minimize the strain response of the device. The considerations described previously provide a framework for designing devices that operate in a robust fashion, even when deformed, stretched, or compressed. In general, strain is accommodated by skin via a combination of mechanical buckling (accommodated by device flexibility) and tensile strain (accommodated by device stretchability). For example, for the 4 × 4 TCR device, a 15 % global tensile strain is absorbed almost entirely by the silicone layers and serpentine interconnects, resulting in only ∼0.005 % average strain in the sensors. This strain result is a relative temperature error (due to a shift in resistance) of <0.1 K. Although this level of error is of little concern in most cases, methods exist to further minimize such effects. For example, typical strains in the body, such as those induced by walking, appear with frequency characteristics that can be filtered electronically. Also, the use of a stacked sensor configurations, using two different metals with different responses to strain and temperature, provide an additional means to decouple the effects.

These types of stretchable sensors yield previously unavailable capabilities in the study of human tissue temperature dynamics. Examples of the types of data that can be captured from these sorts of devices appear in Fig. 6.2. Here, an array of elements generates a spatial map of temperature across the surface of the skin, which enables the visualization of thermal heterogeneity caused by the varied functional anatomy of skin. An example is in Fig. 6.2, with an IR thermograph in Fig. 6.2a, and the corresponding raw epidermal electronic spatial map in Fig. 6.2b. The coarse granularity of the spatial map can be improved by data processing techniques and/or by the use of sensor arrays with improved spatial density. The measurement sensitivity is extremely high due to the thermal mass and response time considerations described above. A comparison of thermal signals over time from the epidermal device and from an IR camera appears in Fig. 6.2c, where the signals (offset for clarity) show an almost perfect match. The data in Fig. 6.2c also reveal a compelling practical application of the devices, where the temperature on the palm drops dramatically due to prolonged mental stress. A similar set of data, this time showing temperature recordings over time from each sensor in the 4 × 4 array, is in Fig. 6.2d. In this case, the array is placed above the ulnar artery on the wrist such that externally induced changes in blood flow result in localized changes in temperature. Even with the coarse granularity in this example design, the spatial orientation of the artery can be clearly identified. The accuracy of the sensors is predominantly a function of drift in the measurement electronics, and is typically on the order of 1–2 °C. The sensitivity of these types of devices is typically orders of magnitude better than the accuracy. Clinical applications typically require precise measurement of temperature changes over time, rather than absolute accuracy. In addition, methods described in the following sections rely on thermal actuation for

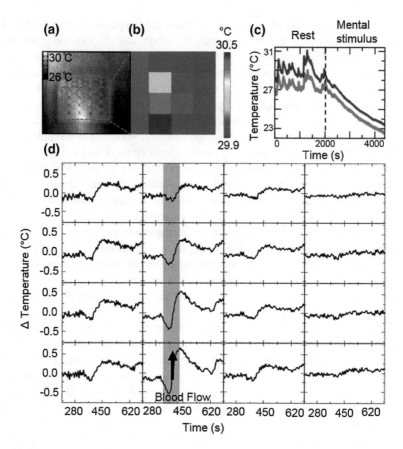

Fig. 6.2 Temperature coefficient of resistance (TCR) sensors for temperature mapping. **a** Infrared (IR) image of a 4 × 4 thermal epidermal thermal sensor array on human skin. **b** Colormap of average temperature measured by each sensor element in the array. **c** Comparison between IR data and the epidermal sensor response for the case of a patient during mental stimulus. The IR and epidermal data are offset for clarity. **d** Temperature recordings over a blood vessel on a subject's arm, during a reactive hyperaemia experiment. The data show a decline in temperature corresponding to occlusion, and a spike corresponding to reperfusion. Figure modified from originally published form in [18]

monitoring thermal transport properties, and as such, they rely on differential changes in temperature over time, rather than absolute values.

While the majority of this discussion focuses on devices designed for the skin, the same underlying concepts translate readily to other soft tissue systems. For example, similar soft TCR arrays can be applied in similar formats to the surface of the heart, as a synthetic electronic pericardium, as shown in Fig. 6.1d. In this example, the thermal sensors exist as part of a larger overall sensor network, as described in detail by Xu et al. [20].

6.3   Local Thermal Transport Property Measurements for Monitoring Tissue Hydration and Structure

An additional benefit of metallic resistive sensors is their straightforward application as thermal actuators, enabled by Joule heating associated with the passage of controlled amounts of current through the element. An example appears in the thermal image of Fig. 6.3a, where the Joule heating process results in a bright spot at the associated element. The use of thermal actuation (heating), together with sensing, provides a mean to measure the thermal transport properties of soft tissues, in vivo. In fact, metallic elements can simultaneously act as thermal actuators and temperature sensors. For devices of this sort, the thermal dynamics of heating at a given power level are dictated almost entirely by the surrounding tissue, due to the extremely small size and thermal mass/high diffusivity of the elements. Without significant loss in measurement precision, an individual element can be treated as a point heat source on the surface of skin, with air as a vacuum. Here, the change in temperature near the heat source can be written [42] as

$$T = T_\infty + \frac{Q}{2\pi r k} \operatorname{erfc}\left(\frac{r\sqrt{\rho c_p}}{\sqrt{4kt}} \right) \tag{6.1}$$

Where t is the time, k is the thermal conductivity (of skin for our case), ρc_p is the volumetric heat capacity (of skin), r is the distance from the heat source, Q is the power of the heat source, and T_∞ is the initial temperature of the medium. Although this expression ignores the finite geometry and complex multilayered structures of the device and the skin, it provides a basic understanding of the overall principles. Details associated with various approximations for the skin and exact numerical solutions appear elsewhere [19, 43, 44].

In practice, these sorts of devices allow measurement of local thermal conductivities and heat capacities of the upper layers of skin in a noninvasive, wearable format. These types of capabilities are particularly valuable for monitoring skin hydration, which has an approximately linear effect on the thermal conductivity of skin and is an important parameter in dermatology and cosmetics. In this application, the lack of mechanical loading on skin is particularly important. Conventional devices for assessing skin hydration use a rigid electrical impedance probe pressed against the tissue to obtain good contact, although in a manner that can alter the natural skin state via compression. Additionally, the degree of contact, although important, has significantly less impact on thermal measurements than on electrical impedance measurements. The epidermal devices described here also offer formats that allow long term, continuous monitoring in a wearable mode. They can also be easily integrated with ultrathin impedance sensors or other devices with identical mechanics, to yield multimodal operation.

An example of the concept appears in Fig. 6.3b, which shows the temperature rise of a 1 mm^2 element (the same type as in the TCR arrays described previously) on skin with 10 mW of applied power, on both the cheek and the heel pad. The cheek has a very thin stratum corneum layer (~ 10 μm), supported by a relatively well hydrated dermal layer, while the heel pad has a very thick stratum corneum layer (~ 1 mm) with low water content. As a result of these differences, the temperature rise on the cheek is lower than that on the heel. Figure 6.3c shows similar results, where multiple measurements are performed at the same body location throughout the day, and compared to hydration measurements performed using a commercial moisture meter. The thermal conductivity and diffusivity (the ratio of thermal conductivity to volumetric heat capacity) can be calculated using algorithms similar to those that rely on Eq. (6.1), but which account for corrections due to finite geometries. The results show that the conductivity and diffusivity of the cheek are higher than those of the heel due to the increased water content. The optimal parameters for these types of measurements relate to the general field of transient thermal measurements [43]. In general, the penetration depth increases with increased measurement timescales. This dependence is important for evaluation of skin due to its complex layered structure. Using a 1-D heat transfer model, the measurement depth is given by

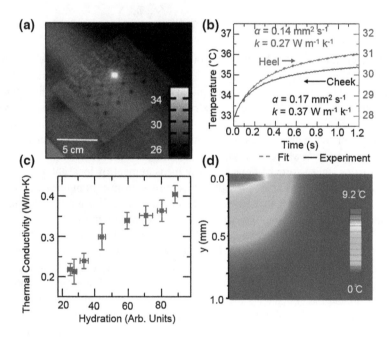

Fig. 6.3 Thermal actuation for measurement of thermal transport properties of human skin. **a** IR image of thermal actuation of an individual element in a 4 × 4 array, for the case of a ~ 4 °C rise in local temperature. Originally published in [19]. **b** Temperature rise of individual heater/sensor elements when sequentially activated. Originally published in [19]. **c** Comparison between measured thermal conductivity and hydration determined using a commercial tool (Delfin MoistureMeter®) on human skin. Originally published in [18]. **d** Finite Element Model showing the penetration depth of the thermal signal into the skin

$$\Delta_p = \sqrt{\alpha t_{max}} \qquad (6.2)$$

Where α is the thermal diffusivity of the skin and t_{max} is the total time of the measurement (2 s in this case). The 1-D assumption is valid as long as the heater area is significantly larger than the area being probed. However, if the heater reduces in size, relative to the area of interest, 2-D and 3-D effects become significant and other analytical or numerical approaches are required to predict the measurement depth. One such approach, a finite element solution to the transient, 3-D heat equation, yields the depth profile shown in Fig. 6.3d. Here, for the case of a measurement time of 2 s, and for circular sensor/actuators of 1 mm radius, the penetration depth is ~ 100 μm. Ultimately, both measurement time and device geometry, therefore affect the measurement depth. These types of devices and operational modes have now been used in clinical studies with >100 subjects, as described in part by Webb et al. [19].

In a more sophisticated embodiment, thermal sensor/actuator arrays and impedance sensors can be integrated onto the same device platform to provide capabilities in multimodal hydration sensing, as shown in Fig. 6.4a. The application of a glycerol-based humectant compound on human skin increases the epidermal water content resulting in an increased thermal conductance and a reduced impedance. These effects appear clearly in the data of Fig. 6.4a, b. When monitored over typical timescales associated with the action of such compounds, both measurement modalities show peak values followed by a subsequent return to baseline values. A representative data set that displays these trends from a patient in a clinical setting is in Fig. 6.4d. Statistical analysis of data collected from >20 patients shows the relationships between thermal conductivity, thermal diffusivity, impedance and hydration, as measured by a state-of-the-art commercial tool (MoistureMeter®, Courage Khazaka Gmbh), and summarized in Fig. 6.4e.

6.4 Local Thermal Transport Property Measurements for Monitoring Near Surface Blood Flow

In an alternative configuration, individual elements can be heated at low levels for extended periods of time, while the remaining elements in an array provide a high precision map of the time dependent distributions of temperature. These distributions depend directly on the thermal transport properties of the underlying tissue. While the previously discussed methods provide measurements of local thermal transport properties (and at relatively higher precision), they do not allow for assessment of spatial anisotropies in heat transfer. Any vasculature near the surface of the skin, for example, provides a source of anisotropic convective transport, and depending on the details of the tissue, these anisotropies can be readily assessed noninvasively using the device platforms described here.

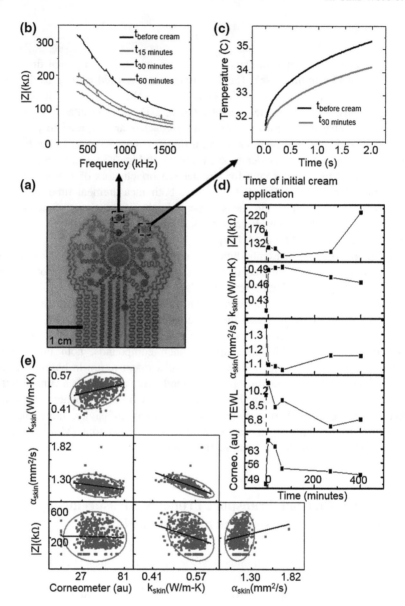

Fig. 6.4 Clinical thermography for hydration measurements. **a** Optical image of an integrated thermal/impedance sensor system. **b** Impedance frequency sweep on a patient at a range of time points, from immediately before application of a humectant cream (15 %-glycerine) to 60 min after. Increased water content in the skin results in a reduction in impedance. **c** Raw data for the rise in temperature associated with thermal actuation. An increase in thermal conductivity reduces the temperature at long times and the rate of change in temperature at short times. These changes suggest an increased level of hydration. **d** Representative time series data on a single subject, collected using epidermal impedance and thermal sensors, and commercial tools (corneometer and transepidermal water loss meter). **e** Scatterplot matrix, showing correlations between thermal conductivity, diffusivity, impedance and corneometer readings

The same principles as those described previously apply here, but with one additional complication. In particular, as relatively small thermal signals propagate through the skin, they quickly decrease in amplitude, even over relatively short distances. Such effects result in relatively low detectable signals at the sensor locations. For application to many non-biological materials, it is possible to increase the heating power to achieve the necessary signal-to-noise ratio (SNR) and avoid this problem. For evaluation of skin, or other tissues in living systems, however, the maximum temperatures are constrained to values that avoid any discomfort or sensation by the user, and of any physiological responses of the tissue itself. One can avoid excess heating levels and, at the same time, recover acceptable SNR values by increasing the overall size of the thermal actuator element. In this way, the same amount of total thermal energy can be transferred to the blood, while maintaining a lower temperature of the actuator. Such design strategies appear in Webb et al. [40], where a central actuator with 1.5 mm radius is surrounded by

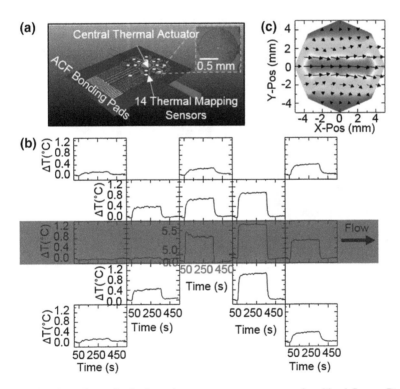

Fig. 6.5 Mapping anisotropies in thermal transport to assess near-surface blood flow. **a** Device construction, consisting of a central thermal actuating/sensing element, surrounded by 14 radially arranged temperature sensing elements designed to map anisotropies in thermal transport. **b** Temperature map determined from all sensing elements in response to thermal actuation. The data reveal significant flow directly beneath the sensors, in the direction indicated by the arrow. **c** Vector map showing net distribution of heat, which corresponds to the net direction of flow. Figure modified from original version published in [40]

14 smaller temperature sensing elements (Fig. 6.5a). With this device, maximum temperature increases can be significantly less than ~8 °C, below the sensation of the subject, while providing adequate SNR for precise blood flow measurements. The details of the thermal transport, for an example where the device is centered over a near surface vessel, are shown in Fig. 6.5b. Here, the temperature of the actuator is indicated in red in the center of the array. The vessel is beneath the actuator, flowing roughly from left to right, causing a significant anisotropy in the temperature rise of the surrounding sensors, as shown by their relative signals (i.e., the signal labeled P2 compared to P1). This data can be used to generate a vector map of blood flow, with an example in Fig. 6.5c.

In practice, these devices can be used to monitor both macrovascular flow (veins and arteries) and microvascular flow (arterioles and capillaries) over extended periods of time without disturbing the subject. An example of monitoring

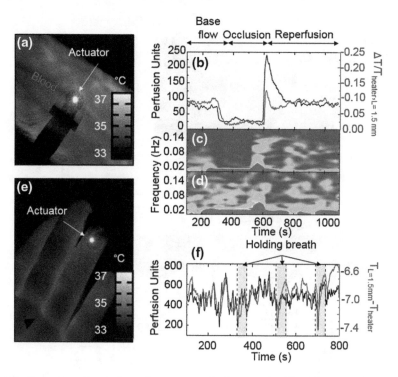

Fig. 6.6 Thermal sensing to determine macro- and microvascular blood flow. **a** IR image of a thermal sensing/actuating array above a blood vessel on the human forearm, during thermal actuation, to measure changes in macrovascular flow. **b** Changes in flow measured using a thermal array, and a laser contrast speckle imager (LSCI) during a hyperemia experiment. Spectrograms of **c** Thermal sensor data and **d** LSCI data, showing a strong correlation between the positions and relative magnitudes of the peaks. **e** IR image of thermal sensing/actuating array on a human finger, with thermal actuation, to measure changes in microvascular flow. **f** Changes in microvascular flow measured using a thermal array and LSCI, where the subject holds his breath to diminish blood flow to the extremities. Figure modified from originally published form in [40]

macrovascular flow, in this case the flow of blood through a vein in the forearm, is presented in Fig. 6.6a–d. LSCI data is provided for comparison, but cannot be assumed to be an exact assessment of the flow in a single vessel due to the details of operation (see Webb et al. [40] for more discussion). The graphs in Fig. 6.6b show the temperature differential between a pair of sensors, one downstream from the actuator and one upstream. This difference, as discussed above, is related to the anisotropy in thermal transport induced by the directional blood flow. Figure 6.6c, d shows a comparison of the spectrograms generated from the epidermal device and LSCI, respectively.

Microvascular flow can be considered in two contexts. For measurement of individual capillaries, the principles of anisotropic flow assessment can be applied in device platforms with length scales comparable to the sizes of the capillaries. Alternatively, large scale devices can probe effective properties of the overall capillary bed, as an effective medium, with an isotropic thermal diffusivity. Here, the responses of sensors on opposing sides of an actuator are nearly identical, but the temperature difference between the actuator and a ring of sensors at a given radial distance depends on the magnitude of microvascular flow. As flow increases, the overall heat transfer coefficient increases, and the temperature differences between the actuator and sensors decrease. For an example with the device placed on the fingertip (Fig. 6.6e), this straightforward analysis yields a remarkably good match to LSCI data (Fig. 6.6f).

6.5 Conclusion

Extension of the foundational principles of flexible and stretchable bioelectronics to systems for mapping the thermal properties of tissue at high precision provides a powerful new set of tools for physiological studies and clinical diagnostics. Through an optimized selection of materials, device geometries and mechanical designs, it is possible to realize a previously unattainable combination of capabilities in intimate thermal contact, minimized thermal and mechanical loading, and thermal response properties, with significant consequences in measurement capabilities and versatility in modes of use. Further technology development will provide options in wireless operation. Even in existing wired formats, the thermal mapping systems can be configured in user friendly, low cost forms for use in hospital or laboratory settings, as shown in Fig. 6.7. With minor modifications, the basic device platforms and measurement techniques can be adapted for use on internal organs and tissues, either as subsystems in implantable devices to monitor inflammatory responses, for example, or as advanced sensing nodes on surgical/diagnostic tools. This broad range of possibilities creates many opportunities for basic and applied research in stretchable, bio-integrated electronics.

Fig. 6.7 Data acquisition electronics for thermal arrays. Mobile platform for clinical thermography measurements. Originally published in [40]

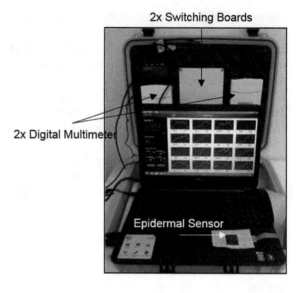

References

1. A.M. Gorbach, H.C. Ackerman, W.M. Liu, J.M. Meyer, P.L. Littel, C. Seamon, E. Footman, A. Chi, S. Zorca, M.L. Krajewski, M.J. Cuttica, R.F. Machado, R.O. Cannon, G.J. Kato, Infrared imaging of nitric oxide-mediated blood flow in human sickle cell disease. Microvasc. Res. **84**, 262–269 (2012)
2. M.B. Ducharme, P. Tikuisis, In vivo thermal conductivity of the human forearm tissues. J. Appl. Physiol. **70**, 2682–2690 (1991)
3. T. Togawa, H. Saito, Non-contact imaging of thermal properties of the skin. Physiol. Meas. **15**, 291–298 (1994)
4. D.A. Boas, A.K. Dunn, Laser speckle contrast imaging in biomedical optics. J. Biomed. Opt. **15** (2010)
5. M. Draijer, E. Hondebrink, T. Van Leeuwen, W. Steenbergen, Review of laser speckle contrast techniques for visualizing tissue perfusion. Lasers Med. Sci. **24**, 639–651 (2009)
6. A.K. Dunn, H. Bolay, M.A. Moskowitz, D.A. Boas, Dynamic imaging of cerebral blood flow using laser speckle. J. Cereb. Blood Flow Metab. **21**, 195–201 (2001)
7. G.E. Nilsson, T. Tenland, P.A. Oberg, Evaluation of a laser Doppler flowmeter for measurement of tissue blood flow. IEEE Trans. Biomed. Eng. **27**, 597–604 (1980)
8. P.A. Oberg, Laser-Doppler flowmetry. Crit. Rev. Biomed. Eng. **18**, 125–161 (1990)
9. K. Wardell, A. Jakobsson, G.E. Nilsson, Laser Doppler perfusion imaging by dynamic light scattering. IEEE Trans. Biomed. Eng. **40**, 309–316 (1993)
10. C. Jin, Z.Z. He, S.S. Zhang, M.C. Qi, Z.Q. Sun, D.R. Di, J. Liu, A feasible method for measuring the blood flow velocity in superficial artery based on the laser induced dynamic thermography. Infrared Phys. Technol. **55**, 462–468 (2012)
11. J.R. Lindner, Microbubbles in medical imaging: Current applications and future directions. Nat. Rev. Drug Discovery **3**, 527–532 (2004)
12. B.A. Schrope, V.L. Newhouse, Second harmonic ultrasonic blood perfusion measurement. Ultrasound Med. Biol. **19**, 567–579 (1993)
13. M. Nitzan, S.O. Anteby, Y. Mahler, Transient heat clearance method for regional blood-flow measurements. Phys. Med. Biol. **30**, 557–563 (1985)

14. M. Nitzan, S.L.E. Fairs, V.C. Roberts, Simultaneous measurement of skin blood flow by the transient thermal-clearance method and laser Doppler flowmetry. Med. Biol. Eng. Comput. **26**, 407–410 (1988)

15. M. Nitzan, Y. Mahler, Theoretical-analysis of the transient thermal clearance method for regional blood-flow measurement. Med. Biol. Eng. Comput. **24**, 597–601 (1986)

16. W.J.B.M. van de Staak, A.J.M. Brakker, H.E. de Rijke-Herweijer, Measurements of thermal conductivity of skin as an indication of skin blood flow. J. Invest. Dermatol. **51**, 149–154 (1968)

17. L. Gao, Y. Zhang, V. Malyarchuk, L. Jia, K.I. Jang, R.C. Webb, H. Fu, Y. Shi, G. Zhou, L. Shi, D. Shah, X. Huang, B. Xu, C. Yu, Y. Huang, J.A. Rogers, Epidermal photonic devices for quantitative imaging of temperature and thermal transport characteristics of the skin. Nat. Commun. **5**, 4938 (2014)

18. R.C. Webb, A.P. Bonifas, A. Behnaz, Y.H. Zhang, K.J. Yu, H.Y. Cheng, M.X. Shi, Z.G. Bian, Z.J. Liu, Y.S. Kim, W.H. Yeo, J.S. Park, J.Z. Song, Y.H. Li, Y.G. Huang, A.M. Gorbach, J.A. Rogers, Ultrathin conformal devices for precise and continuous thermal characterization of human skin. Nat. Mater. **12**, 938–944 (2013)

19. R.C. Webb, R.M. Pielak, P. Bastien, J. Ayers, J. Niittynen, J. Kurniawan, M. Manco, A. Lin, N.H. Cho, V. Malyrchuk, G. Balooch, J.A. Rogers, Thermal transport characteristics of human skin measured in vivo using ultrathin conformal arrays of thermal sensors and actuators. PLoS ONE **10**, e0118131 (2015)

20. L. Xu, S.R. Gutbrod, A.P. Bonifas, Y. Su, M.S. Sulkin, N. Lu, H.J. Chung, K.-I. Jang, Z. Liu, M. Ying, C. Lu, R.C. Webb, J.-S. Kim, J.I. Laughner, H.Y. Cheng, Y. Liu, A. Ameen, J.W. Jeong, G.-T. Kim, Y. Huang, I.R. Efimov, J.A. Rogers, 3D multifunctional integumentary membranes for spatiotemporal cardiac measurements and stimulation across the entire epicardium. Nature Commun. **5**, 10 (2014)

21. X. Huang, H. Cheng, K. Chen, Y. Zhang, Y. Zhang, Y. Liu, C. Zhu, S.C. Ouyang, G.W. Kong, C. Yu, Y. Huang, J.A. Rogers, Epidermal impedance sensing sheets for precision hydration assessment and spatial mapping. IEEE Trans. Biomed. Eng. **60**, 2848–2857 (2013)

22. M. Kaltenbrunner, T. Sekitani, J. Reeder, T. Yokota, K. Kuribara, T. Tokuhara, M. Drack, R. Schwodiauer, I. Graz, S. Bauer-Gogonea, S. Bauer, T. Someya, An ultra-lightweight design for imperceptible plastic electronics. Nature **499**, 458–463 (2013)

23. D.H. Kim, J.H. Ahn, M.C. Won, H.S. Kim, T.H. Kim, J. Song, Y.Y. Huang, Z. Liu, C. Lu, J. A. Rogers, Stretchable and foldable silicon integrated circuits. Science **320**, 507–511 (2008)

24. D.H. Kim, N.S. Lu, R. Ma, Y.S. Kim, R.H. Kim, S.D. Wang, J. Wu, S.M. Won, H. Tao, A. Islam, K.J. Yu, T.I. Kim, R. Chowdhury, M. Ying, L.Z. Xu, M. Li, H.J. Chung, H. Keum, M. McCormick, P. Liu, Y.W. Zhang, F.G. Omenetto, Y.G. Huang, T. Coleman, J.A. Rogers, Epidermal electronics. Science **333**, 838–843 (2011)

25. S.P. Lacour, J. Jones, Z. Suo, S. Wagner, Design and performance of thin metal film interconnects for skin-like electronic circuits. IEEE Electron Device Lett. **25**, 179–181 (2004)

26. T. Li, Z.Y. Huang, Z. Suo, S.P. Lacour, S. Wagner, Stretchability of thin metal films on elastomer substrates. Appl. Phys. Lett. **85**, 3435–3437 (2004)

27. S.C. Mannsfeld, B.C. Tee, R.M. Stoltenberg, C.V. Chen, S. Barman, B.V. Muir, A.N. Sokolov, C. Reese, Z. Bao, Highly sensitive flexible pressure sensors with microstructured rubber dielectric layers. Nat. Mater. **9**, 859–864 (2010)

28. T. Someya, Y. Kato, T. Sekitani, S. Iba, Y. Noguchi, Y. Murase, H. Kawaguchi, T. Sakurai, Conformable, flexible, large-area networks of pressure and thermal sensors with organic transistor active matrixes. Proc. Natl. Acad. Sci. USA **102**, 12321–12325 (2005)

29. J.Y. Sun, N.S. Lu, J. Yoon, K.H. Oh, Z.G. Suo, J.J. Vlassak, Inorganic islands on a highly stretchable polyimide substrate. J. Mater. Res. **24**, 3338–3342 (2009)

30. S. Wang, M. Li, J. Wu, D.-H. Kim, N. Lu, Y. Su, Z. Kang, Y. Huang, J.A. Rogers, Mechanics of epidermal electronics. J. Appl. Mech. **79**, 031022 (2012)

31. M. Drack, I. Graz, T. Sekitani, T. Someya, M. Kaltenbrunner, S. Bauer, An imperceptible plastic electronic wrap. Adv. Mater. **27**, 34–40 (2015)

32. J.A. Rogers, T. Someya, Y. Huang, Materials and mechanics for stretchable electronics. Science **327**, 1603–1607 (2010)
33. G. Schwartz, B.C.K. Tee, J. Mei, A.L. Appleton, D.H. Kim, H. Wang, Z. Bao, Flexible polymer transistors with high pressure sensitivity for application in electronic skin and health monitoring. Nat. Commun. **4** (2013)
34. C. Wang, D. Hwang, Z. Yu, K. Takei, J. Park, T. Chen, B. Ma, A. Javey, User-interactive electronic skin for instantaneous pressure visualization. Nat. Mater. **12**, 899–904 (2013)
35. Y. Zhang, H. Fu, Y. Su, S. Xu, H. Cheng, J.A. Fan, K.C. Hwang, J.A. Rogers, Y. Huang, Mechanics of ultra-stretchable self-similar serpentine interconnects. Acta Mater. **61**, 7816–7827 (2013)
36. Y. Zhang, H. Fu, S. Xu, J.A. Fan, K.C. Hwang, J. Jiang, J.A. Rogers, Y. Huang, A hierarchical computational model for stretchable interconnects with fractal-inspired designs. J. Mech. Phys. Solids **72**, 115–130 (2014)
37. Y. Zhang, Y. Huang, J.A. Rogers, Mechanics of stretchable batteries and supercapacitors. Curr. Opin. Solid State Mater. Sci. **19**, 190–199 (2015)
38. Y. Zhang, S. Xu, H. Fu, J. Lee, J. Su, K.C. Hwang, J.A. Rogers, Y. Huang, Buckling in serpentine microstructures and applications in elastomer-supported ultra-stretchable electronics with high areal coverage. Soft Matter **9**, 8062–8070 (2013)
39. Y.H. Zhang, S.D. Wang, X.T. Li, J.A. Fan, S. Xu, Y.M. Song, K.J. Choi, W.H. Yeo, W. Lee, S.N. Nazaar, B.W. Lu, L. Yin, K.C. Hwang, J.A. Rogers, Y.G. Huang, Experimental and theoretical studies of serpentine microstructures bonded to prestrained elastomers for stretchable electronics. Adv. Funct. Mater. **24**, 2028–2037 (2014)
40. R.C. Webb, Y. Ma, S. Krishnan, Y. Li, S. Yoon, X. Guo, X. Feng, Y. Shi, M. Seidel, N.H. Cho, J. Kurniawan, J. Ahad, N. Sheth, J. Kim, J.G.t. Taylor, T. Darlington, K. Chang, W. Huang, J. Ayers, A. Gruebele, R.M. Pielak, M.J. Slepian, Y. Huang, A.M. Gorbach, J.A. Rogers, Epidermal devices for noninvasive, precise, and continuous mapping of macrovascular and microvascular blood flow. Sci. Adv. **1**, p. e1500701 (2015)
41. Z.G. Bian, J.Z. Song, R.C. Webb, A.P. Bonifas, J.A. Rogers, Y.G. Huang, Thermal analysis of ultrathin, compliant sensors for characterization of the human skin. Rsc Adv. **4**, 5694–5697 (2014)
42. H.S. Carslaw, J.C. Jaeger, *Conduction of Heat in Solids*, 2d edn. (Clarendon Press, Oxford, 1959)
43. S.E. Gustafsson, Transient plane source techniques for thermal conductivity and thermal diffusivity measurements of solid materials. Rev. Sci. Instrum. **62**, 797–804 (1991)
44. D.G. Cahill, Analysis of heat flow in layered structures for time-domain thermoreflectance. Rev. Sci. Instrum. **75**, 5119–5122 (2004)

Chapter 7
Soft Biosensor Systems Using Flexible and Stretchable Electronics Technology

Tsuyoshi Sekitani

Abstract We will review the recent progresses of large-area, ultraflexible, and ultrasoft electronic sensors. This chapter focuses on integration technologies of thin-film, ultraflexible electronics comprising ultrasoft gel electrodes, thin-film amplifier, Si-LSI wireless platform, thin-film battery, and information engineering, which are imperceptible active sensors. Here we would like to demonstrate the applications of the patch-type wearable biosignal sensors including brain wave (Electroencephalogram: EEG) monitoring from a forehead. Furthermore, this chapter will review the technologies for realizing the new era of electronic system, that is, Internet of Things (IoT.)

Keywords Large-area ultraflexible sensors · Internet of Things (IoT) · Thin-film transistors · Integration technologies

7.1 Outline and Background

The era of the Internet of Things (IoT), networks of objects connected wirelessly to produce new values, has begun. One of the famous examples is "Industry 4.0", promoted through the industry–academia–government collaboration in Germany as an attempt to place various types of sensors in a factory, to control an entire factory using the numerous sensors distributed in production and manufacturing lines, and to optimize manufacturing. In Japan, similar efforts have also been intensively carried out. Information in a physical space is obtained using sensors, and the information is visualized using an algorithm and fed back to the physical space for optimization. Such systems, called cyber-physical systems (CPSs), are attracting interest (Fig. 7.1). Sufficient information in a physical space cannot be obtained by conventional Si large-scale integration (LSI) based sensors alone because the

T. Sekitani (✉)
The Institute of Scientific and Industrial Research, Osaka University,
8-1, Mihogaoka, Osaka, Ibaraki 567-0047, Japan
e-mail: sekitani@sanken.osaka-u.ac.jp

© Springer International Publishing Switzerland 2016
J.A. Rogers et al. (eds.), *Stretchable Bioelectronics for Medical Devices and Systems*, Microsystems and Nanosystems,
DOI 10.1007/978-3-319-28694-5_7

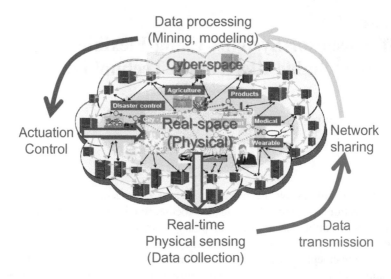

Fig. 7.1 Schematics of IoT and CPS. Cyber-physical system involves real-time physical sensing (data collection), data transmission, networksharing processing, and actuation between cyber and real spaces. The CPS aims to make every social system efficient and optimized. First-generation CPSs can be used in areas as diverse as home security, automotive systems, civil infrastructure, energy, agriculture, healthcare, manufacturing, transportation, robotics, and consumer electronics. Seamless connection between cyberspace and a physical space will optimize the systems of the physical space

objects in the physical space in which we live often have a complicated shape and occupy a large area. The measurement of physical spaces to obtain information on human conditions or bioinformation is considered to be especially important. In other words, one of the ultimate goals of the IoT is the precise acquisition and utilization of bioinformation. In recent years, many researchers have attempted to realize wearable electronics, and their attempts have been accelerating. Typical examples include watches that monitor heart rate, acceleration sensors, sphygmomanometers, and wearable computing glasses. Watches and glasses are hard objects that people have worn since early times. They also comprise the majority of current wearable devices. The monitoring of heart rates in everyday life by directly attaching a sensor on the chest and the monitoring of brain waves by wearing headgear have recently become more widespread. A ripple effect of wearable electronics is expected in various fields from medicine to entertainment. However, the use of wearable devices in daily life is not yet common. The main reasons for this are considered to be:

1. Discomfort felt when wearing sensors
2. Difficulty in accurate positioning of sensors, which differs among individuals
3. Misalignment of sensors during activity.

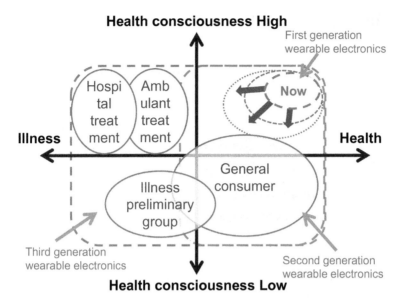

Fig. 7.2 Schematic of future applications of wearable electronics. Applications of wearable electronics. This figure has been described from private communication with Prof. S. Izumi (Kobe University)

For these three reasons, the target of current wearable electronics is being limited to healthy people with a high health consciousness (first-generation wearable electronics). The application of these sensors to the medical field has been hampered by their low measurement precision and the large difference in physical shape among individuals. Most people are healthy but have only a moderate level of health consciousness, and it is important for wearable electronics to be targeted at these people (second-generation wearable electronics), as well as at those who have health problems and require a hospital visit or hospitalization (third-generation wearable electronics). Figure 7.2 shows a schematic of future applications of wearable electronics.

We have been involved in the development of wearable bioinstrumentation (biosensor) systems using flexible and stretchable electronics technology that can be placed on human skin without causing discomfort to the user. We have succeeded in simultaneously solving the problems of differences among individuals and the misalignment of the sensor by placing a large number of sensor nodes on a flexible and/or stretchable large-area sheet, rather than a single sensor node, and covering the entire target region to be measured. Detailed examples are introduced in the following section.

7.2 Development of Soft Wearable Sensors

Five device technologies, as shown in Fig. 7.3, are required to transfer the information from flexible biosensors to external PCs and other devices.

- Flexible biocompatible electrodes and various sensor materials
- A signal amplifier
- A Si-LSI platform (a term used to collectively describe an analog-to-digital (AD) converter, a central processing unit (CPU), a power regulating circuit, and a wireless module)
- Small thin-film batteries
- Information processing technology.

By integrating and systematizing the five device technologies, sensors can transmit meaningful information to a cyberspace that can analyze the information. Namely, the simultaneous development of these device technologies and a framework of system integration are required. In the following, each device technology is first introduced and examples of bioinstrumentation obtained through system integration are described.

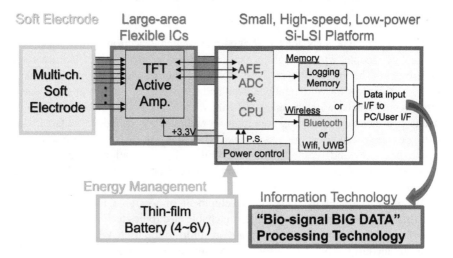

Fig. 7.3 Device technologies required to implement flexible sensors in social systems. Device technologies required to introduce flexible multichannel sensors in society/flexible electrode technology/analog front-end circuit technology/ultralow-power-consumption wireless communication technology, information processing circuit technology (Si platform technology)/flexible sheet multichannel electrode/TFT amplifier/power circuit/flexible thin-film battery/lightweight thin-film power source technology/AD converter and CPU (signal processing)/memory/wireless communication/data output/low-power-consumption control/transmission of results/logging/extraction of waveform/signal processing/analysis technology of big biodata/integration of all technologies required

7.2.1 Flexible, Stretchable, and Biocompatible Electrodes

Human skin (muscle) has a Young's modulus of elasticity of ~ 100 kPa. Therefore, flexible materials with a smaller modulus of elasticity, such as gels and rubber, are preferable for realizing sensors that are imperceptible to the user. The metals and plastics generally used as electrode materials have a modulus of elasticity of 2 MPa–3 GPa, meaning that the development of new flexible electrodes is essential.

Our research group has developed the rubber-like stretchable electrode technology for uniformly dispersing a conductive nanomaterial on a substrate of a different material, and we have succeeded in developing a stretchable conductor that is as stretchable as rubber and as conductive as a metal [1–4].

Using this original technology, we further developed a technology to uniformly disperse Ag conductive nanowires in polyurethane rubber or a halogel, realizing electrodes with both high conductivity ($\geq 10,000$ S/cm) and flexibility (stretchability ≥ 100 %) [4]. As shown in Fig. 7.4, the electrical and mechanical characteristics of the electrodes are maintained even when they are stretched by ≥ 100 %, enabling the electrodes to flexibly follow the movement of human skin during activity. In conformity with the International Organization for Standardization (ISO) standards ISO10993-5 and ISO10993-6, we carried out cytotoxicity tests and implantation tests and confirmed that the electrodes are not toxic and cause little inflammatory response.

7.2.2 Flexible Signal Amplifier

A large signal-to-noise ratio (SNR) is required for biosensors because biosignals are extremely weak. Monitoring bioinformation over a wide area using many sensor nodes on a plane, rather than a single high-sensitivity sensor node, and visualizing the bioinformation through mapping are important for obtaining sensitive bioinformation that is less affected by noise. However, when a plane sensor is fabricated on a hard substrate, it is difficult for the sensor to be physically attached to human skin and to follow its movement. In practice, in the case of multisensory systems using conventional hard Si-based technology, flexible electrodes are in contact with living tissues; however, the signal amplifier connected to the electrodes is hard, preventing the direct application of these electrodes on living tissues over a wide area. As a result, long cables connecting living tissues and the amplifier are required, which makes it difficult to develop a multichannel system because of the cross talk between cables or wirings. For this reason, the development of a multichannel brain–machine interface (BMI) is difficult in practice (whose number of channel is 64–128 at most). The development of a large-area flexible plane sensor equipped with an amplifier is required.

Our research group has realized the development and integration of the most flexible thin-film transistor (TFT) [5–10]. Using this technology, we realized a

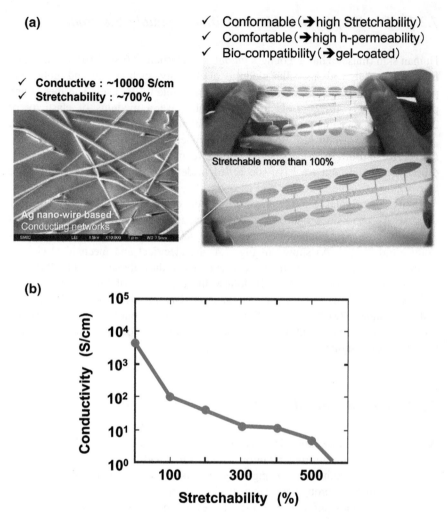

Fig. 7.4 Stretchable electrode. **a** High conductivity and flexibility are realized using a composite material of Ag nanowires and rubber. Conducting networks can be formed using Ag nanowires embedded in stretchable rubber substrates. **b** shows conductivity as a function of stretchability. Adapted from [4]. Copyright 2011, IEEE

thin-film flexible sensor that detects the distribution of pressure and can be wrapped around a 1-mm-diameter medical catheter [5–7]. In addition, we fabricated a contact-type sensor [8] and a signal amplifier [9, 10] using organic TFT integration technology. Using these devices, weak biosignals are instantly amplified at the surface of living tissues to enable the detection of high-quality biosignals.

Combining these technologies and using the technology for fabricating organic TFTs at low temperatures, we realized a large-area plane sensor on an ultrathin polymer film with a thickness of 1 μm (Fig. 7.5) with a high flexibility, high SNR,

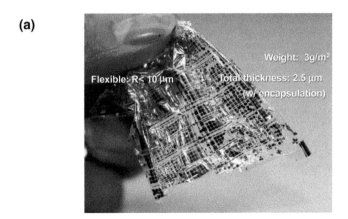

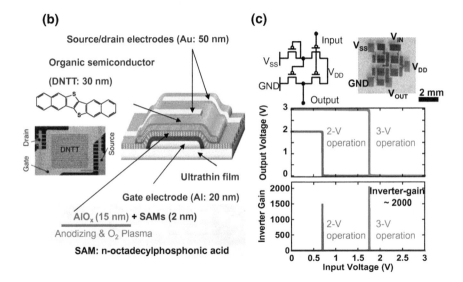

Fig. 7.5 Thin-film biosignal amplifier fabricated using organic semiconductor technology. **a** picture of ultrathin organic amplifier consisting of organic TFTs. **b** Schematic cross section of the organic TFTs. **c** Circuit diagram of organic pseudo-CMOS inverter and the characteristics [11]

and a low impact on living bodies [8–10]. Figure 7.5b, c shows the schematic cross section of our organic TFTs and the inverter performance taken from organic pseudo-CMOS inverter comprising four organic TFTs [11]. This inverter can operate within 3 V and shows very high gain greater than 1,500. The manufacturing process and technical details can be seen in our previous reports [5–11].

Such TFTs using organic semiconductors are flexible and thin. However, the electron mobility of these TFTs is only ~ 1 cm^2/Vs, which is 3–4 orders of magnitude smaller than that of conventional Si-based semiconductor transistors

	InGa As/In AlAs [20]	Single-Si nanomembrane [21]	Metal-oxide IGZO [22]	MoS$_2$ [23]	Graphene [24]	CNT [25]	a-Si TFTs [26]	Organic DNTT [8]
Mobility (cm^2/Vs)	7300	400	76	39	800	20	1.6	>1.0
Voltage (V)	0.7	5	0.5	4	2	10	15	3
Critical bending (mm)	12.5	15.5	10	1	1.3	1	0.5	0.005
Substrate	PI	PET 125 μm	PEN PET	PI 76 μm	PI 100 μm	PET 100 μm	PI 50 μm	PEN 1 μm

Fig. 7.6 Comparison of performance of various flexible TFTs. Comparison of performance of various flexible TFT

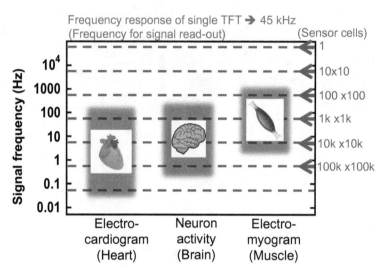

Fig. 7.7 Frequencies of biosignals. The frequencies of most biosignals are ≤1 kHz. Comparison of response rate of organic transistor (∼50 kHz) with frequencies of biosignals

(Fig. 7.6). Low mobility affects the frequency response of TFTs. The frequency response of the TFTs used in our study was ∼50 kHz [11]. With this frequency response, it is not realistic to fabricate CPUs requiring high-speed operation. In contrast, the frequencies of biopotentials obtained from living bodies are ∼200 Hz for electroencephalography (EEG) and electrocardiography (ECG) signals and ∼1 kHz for electromyography (EMG) signals (Fig. 7.7). This means that even a low-frequency response of the organic TFTs is sufficient to monitor biosignals. We succeeded in monitoring infarcted myocardial tissues by applying a flexible sensor on the surface of a beating heart, showing the possibility of the sensor when used as a new medical device.

Regarding the stretchable and flexible electronics technologies, D.H. Kim and J. Rogers groups have developed sophisticated biosensors including ultrathin Si-TFT-based active matrix neural interface, skin-like epidermal electronic system, and electrocardiogram monitoring [12–15]. These high-performance flexible and stretchable sensors have been already used in animal experiments and demonstrated the excellent feasibility of the new sensors to biomedical and wearable applications. Z. Bao and B. Tee group reported flexible, stretchable electrical functional materials and demonstrated to monitor biosignals [16, 17]. Although organic TFT-based sensors have low frequency response, it has several advantages, such as scalability, disposability, and ultimate mechanical flexibility. These flexible and stretchable electronics with combining inorganic and organic TFTs will open up new era of electronics that cannot be achieved only with rigid TFT technologies. Besides the TFT technologies, highly conformable conducting poly(3,4-ethylenedioxythiophene) poly(styrenesulfonate) (PEDOT:PSS) electrodes have been reported for in vivo electrocorticography [18]. Au-coated poly(dimethylsiloxane) has been demonstrated for stretchable in vivo neural interfaces [19].

7.2.3 Si-LSI Platform

The information in a physical space measured using multichannel sensors is analog and the quantity of information is enormous. The processing of such analog information and its analysis using external devices require (1) an AD converter, (2) a CPU used to process information, (3) a wireless module to transmit the information to external devices, and (4) a power regulation circuit to drive these modules. A system integrating these functions on a substrate is collectively referred to as a Si-LSI platform.

We succeeded in integrating our modules on a plastic film with a thickness of ≤ 100 μm to form a system. By devising a circuit for canceling noise, the noise level of the system developed was suppressed to as low as ≤ 1 μV.

7.2.4 Small Thin-Film Batteries

As the power source to operate the Si-LSI platform, we developed a thin-film Li-ion battery with a thickness of 0.5 mm and dimensions of 4×4 cm^2. The battery has a capacity of ~ 200 mAh, a current of 20 mA, and a discharge voltage of 3.75 V, and is sufficient to operate the thin-film amplifier and Si-LSI platform. The wireless module of the Si-LSI platform requires a high current of ~ 10 mA. Therefore, the electrode structure was optimized to reduce the internal resistance in the battery and ensure a current of ~ 10 mA. We confirmed that continuous monitoring for ~ 10 h is possible using a small thin-film battery.

7.2.5   Information Processing Technology

The most noteworthy feature of this research is the comprehensive approach using information processing technology in addition to the development of measurement devices. We developed a process to extract meaningful information from the huge amount of bioinformation obtained by multichannel sensors through classification of the bioinformation by frequency analysis using an algorithm. The information in a physical space obtained by the sensors is processed and fed back to the physical space as meaningful information using an algorithm. This basic technology used in research on the integration of cyberspace and physical space was demonstrated. A flowchart of bioinstrumentation data from devices to information processing is shown in Fig. 7.8.

The following are examples of bioinformation obtained by integrating the above five components into a system.

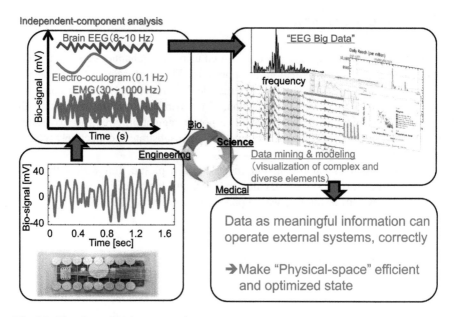

Fig. 7.8 Flowchart of bioinstrumentation

7.2.6   Examples of Bioinstrumentation

7.2.6.1   EEG

There are many diseases caused by brain dysfunction, such as Alzheimer's disease, Parkinson's disease, mental disorders including depression, and pediatric developmental disorders. The biopotential obtained from the activities of the brain or neurons is very weak; the magnitude of EEG signals is ~ 2–$3\ \mu V$. Therefore, medical devices with a large amplifier are required for precise EEG measurements. In recent years, the importance of EEG measurements has been recognized. In addition, the amplification technology has advanced and the development of EEG sensors that users can wear on their heads is underway. Most of these sensors are headgear type and adopt a comb-shaped electrode structure that can separate the hair on the top of the head, where no myopotential is detected. Such headgear-type sensors can measure brain waves without interference from the myopotential. However, while increasing the duration of contact of the comb-shaped electrodes with the scalp, either the sense of discomfort increases or the user develops a headache. At medical institutions, EEG measurements are carried out by inserting a conductive gel between the electrode and the scalp instead of using a comb-shaped electrode. However, the conductive gel must be rinsed off after each measurement, which is a burden on patients. As a consequence, there has been no easy method for EEG measurement.

We succeeded in developing a patch-type EEG sensor using flexible electrodes, a thin-film circuit, and applying our previously developed technologies. Figure 7.9

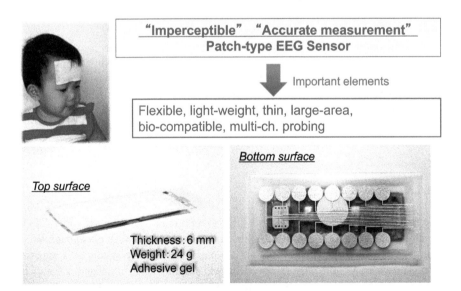

Fig. 7.9 Photographs of patch-type EEG sensor. Development of patch-type EEG sensors with a precision comparable to that of existing medical devices

shows photographs of the patch-type EEG sensor. It is an EEG sensor with eight-channel flexible electrodes (16 electrodes are used for a differential readout) on the gel surface, although at first glance it has the appearance of a cool gel patch used to reduce fever. Simply by attaching the sensor to the forehead, brain waves are measured by the multichannel sensor and the information is transmitted to an external PC in real time. When brain waves were simultaneously measured using a commercially available EEG system and the patch-type EEG sensor, the precision of the two sensors was confirmed to be comparable. Using the patch-type EEG sensor, users can measure brain waves easily at home. Therefore, the range of application of patch-type EEG sensors is considered to be wide, including the observation of patients with dementia at home, the EEG measurement of patients with sleep apnea syndrome, the observation of people in need of nursing care, and the early detection of pediatric developmental disorders.

7.2.6.2 ECG

The heart is a muscular organ that moves constantly. There are quite a few diseases related to the heart including myocardial infarction caused by ischemia and cardiomegaly. A surgical procedure is available to recover the function of the heart by removing an affected area with weak activity. However, it is generally difficult to identify such an area, although experienced surgeons can determine it from the movement and color of the heart. An area with a weak potential or deteriorated activity can be identified by covering the entire heart with a flexible and soft biopotential sheet sensor. In practice, the ischemic areas of the hearts of rats and large animals (pigs) have been successfully identified during surgery by applying a biopotential sheet sensor to their hearts.

A conventional hard potential sensor places a burden on the heart when it is pressed against the heart. This problem has been solved with our biopotential sheet sensor because of its flexibility. In addition, the biopotential sheet sensor is disposable and can be discarded after the measurement. Therefore, the lifetime of the sensor is not a concern, in contrast to organic devices. The biopotential sheet sensor has been attracting attention as a new medical device with the advantages of a large area, flexibility, and disposability of the organic flexible sensors.

7.2.6.3 EMG

The movement of muscles produces a large myopotential. Its frequency is ~ 1 kHz, which is relatively high compared with the frequencies of other biosignals. When brain waves with a potential of 2–3 μV are measured, the myopotential is regarded as noise, which is generally removed by an antiphase algorithm and a bandpass filter that rejects frequencies outside a certain range. Myopotential is an extremely important biosignal for the precise control of prosthetic limbs and hands. A change in myopotential is expressed as the movement of muscles. New human–machine

(a)

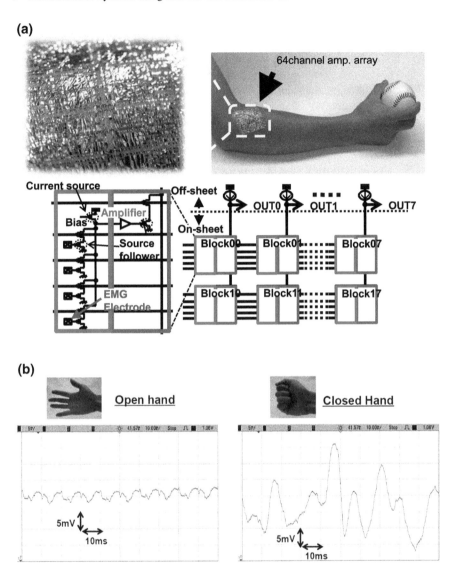

(b)

Fig. 7.10 **a** Pictures of surface electromyogram system and **b** the measurements. Adapted from [10]. Copyright 2014, IEEE

interfaces for controlling electronics using the myopotential are expected to be developed. In addition, the myopotential is expected to be used as a quantitative index to observe the movements of top athletes (Fig. 7.10).

The measurement of biosignals (e.g., EEG, ECG, and EMG signals) as a change in biopotential associated with the activity of living bodies has been explained. In addition to biopotential sensors, physical sensors detecting pressure, temperature, strain, pH, and vibration are indispensable to develop next-generation medical devices.

Our research group developed a thin pressure sheet sensor by integrating (1) a thin sheet of a piezoelectric polymer (polyvinylidene difluoride) that produces a voltage upon the application of pressure or strain and (2) a thin-film amplification circuit. This thin pressure sheet is described below as a typical example of such physical sensors.

7.2.6.4 Implementation of Thin-Film Pressure Sensor on the Surface of Catheters

We succeeded in fabricating a flexible pressure sensor sheet by integrating a piezoelectric film and a thin-film amplification circuit. Figure 7.11 shows a photograph of the thin-film pressure sensor, which was implemented as a spiral on a medial catheter with a diameter of 1 mm taking advantage of its flexibility and thinness. As shown in the upper part of Fig. 7.11, the thin-film pressure sensor can produce a voltage of ~ 1 mV when a pressure is applied to the surface of the piezoelectric film alone. By integrating the piezoelectric sheet and the thin-film amplification circuit, the output voltage could be amplified to 545 mV (lower part of Fig. 7.11).

As explained, sensors obtained by integrating a functional film that can generate a voltage and a thin-film amplification circuit can be used as thin-film physical sensors.

7.3 Final Remarks

In this article, we have introduced the basic technologies of biopotential sheet sensors taking advantage of the flexibility, large area, and low cost of organic materials. In particular, it is demonstrated that the issues with conventional wearable sensors, such as the misalignment of the sensor and the difficulty in adapting to individual differences, are resolved using multichannel sensors. In addition, multichannel sensors are found to contribute to high-sensitivity measurements. This is comparable to a multichannel surface readout function of human skin that can detect even a hair on a surface from the differential readout at multiple points (or by mapping using a plane sensor) even though individual spots sensitive to heat and pain on the human skin are not highly sensitive. Compared with experimental devices placed under ideal conditions, wearable sensors are used under very severe conditions because noise caused by sweat, chemical components, and changes in temperature, exists in signals associated with daily activities. Therefore, high

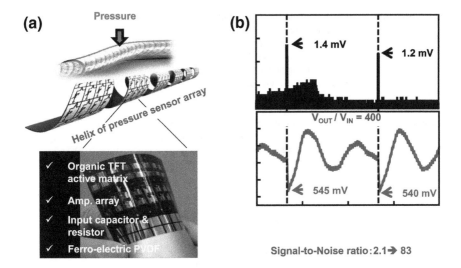

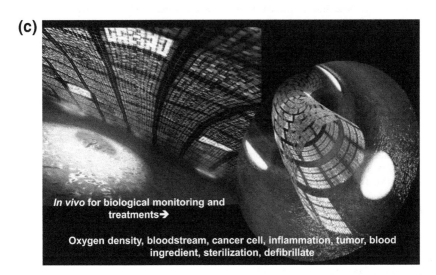

Fig. 7.11 Development of next-generation medical devices by implementing a thin-film pressure sensor on the surface of small-diameter catheter. Application of thin-film pressure sensor to disposable intelligent catheter that realizes intravascular treatment. Adapted from [9]. Copyright 2012, IEEE

measurement precision is required. However, thin-film flexible sensors with a large area, multiple channels, and a high ability to follow the shape of a surface can be used to measure biosignals that have been difficult to measure using conventional sensors. I believe that the research achievements summarized in this article are a significant step toward realizing next-generation sensors in the fields of healthcare, medicine, and welfare.

Acknowledgments Part of our studies was supported by the Center of Innovation (COI) Program of Osaka University. The research was carried out in collaboration with Takafumi Uemura (specially appointed associate professor), Teppei Araki (assistant professor), and Shusuke Yoshimoto (assistant professor), Katsuaki Suganuma (professor), Kazuhiko Matsumoto (professor), Mototsugu Ogura (specially appointed professor) of the Institute of Scientific and Industrial Research of Osaka University, and members of Sekitani's laboratory. The research on organic TFTs and their integrated circuits was carried out with the support of the Exploratory Research for Advanced Technology (ERATO) research funding program of Japan Science and Technology Agency (JST) when the author was at the University of Tokyo. I am grateful to Takao Someya (professor) and Tomoyuki Yokota (specially appointed assistant professor) of the Graduate School of Engineering of the University of Tokyo, and to Takayasu Sakurai (professor) and Makoto Takamiya (associate professor) of the Institute of Industrial Science of the University of Tokyo.

References and Published Papers

1. T. Sekitani, Y. Noguchi, K. Hata, T. Fukushima, T. Aida, T. Someya, Science **321**, 1468–1472 (2008)
2. T. Sekitani, H. Nakajima, H. Maeda, T. Fukushima, T. Aida, K. Hata, T. Someya, Nat. Mater. **8**, 494–499 (2009)
3. T. Sekitani, T. Someya, Adv. Mater. **22**, 2228–2246 (2010)
4. T. Araki, M. Nogi, K. Suganuma, M. Kogure, O. Kirihara, IEEE Electron Device Lett. **32**, 1424 (2011)
5. T. Sekitani, T. Yokota, U. Zschieschang, H. Klauk, S. Bauer, K. Takeuchi, M. Takamiya, T. Sakurai, T. Someya, Science **326**, 1516–1519 (2009)
6. T. Sekitani, M. Takamiya, Y. Noguchi, S. Nakano, Y. Kato, T. Sakurai, T. Someya, Nat. Mater. **6**, 413–417 (2007)
7. T. Sekitani, U. Zschieschang, H. Klauk, T. Someya, Nat. Mater. **9**, 1015–1022 (2010)
8. M. Kaltenbrunner, T. Sekitani, J. Reeder, T. Yokota, K. Kuribara, T. Tokuhara, M. Drack, R. Schwödiauer, I. Graz, S. Bauer-Gogonea, S. Bauer, T. Someya, Nature **499**, 458–463 (2013)
9. T. Yokota, T. Sekitani, T. Tokuhara, N. Take, U. Zschieschang, H. Klauk, K. Takimiya, T.-C. Huang, M. Takamiya, T. Sakurai, T. Someya, IEEE Trans. Electron Devices **59**, 3434–3441 (2012)
10. H. Fuketa, K. Yoshioka, Y. Shinozuka, K. Ishida, T. Yokota, N. Matsuhisa, Y. Inoue, M. Sekino, T. Sekitani, M. Takamiya, T. Someya, T. Sakurai, IEEE Trans. Biomedical Circuits and Systems **8**, 824 (2014)
11. K. Fukuda, T. Sekitani, T. Yokota, K. Kuribara, T.-C. Huang, T. Sakurai, U. Zschieschang, H. Klauk, M. Ikeda, H. Kuwabara, T. Yamamoto, K. Takimiya, T. Someya, IEEE Electron Device Lett. **32**, 1448–1450 (2011)
12. D.H. Kim et al., Science **320**, 507–511 (2008)
13. D.H. Kim et al., Science **333**, 838–843 (2011)
14. J. Viventi et al., Nat. Neurosci. **14**, 1599–1605 (2011)
15. D. Son et al., Nat. Nanotechnol. **9**, 397–404 (2014)
16. B.C.-K. Tee et al., Science **350**, 313–316 (2015)
17. G. Schwartz, Nat. Comm. **4**, 1859 (2013)
18. D. Khodagholy et al., Nat. Comm. **4**, 2133 (2013)
19. O. Graudejus, B. Morrison, C. Goletiani, Z. Yu, S. Wagner, Adv. Funct. Mater. **22**, 640–651 (2012)
20. J. Shi et al., Appl. Phys. Lett. **102**, 243503 (2013)
21. H. Zhou et al., Nature Sci. Rep. **3**, 1291 (2013)

22. H. Chang et al., ACS Nano **7**, 6 (2013)
23. H. Hsu et al., IEEE Electron Dev. Lett. **34**, 6 (2013)
24. J. Lee et al., Appl. Phys. Lett. **101**, 252109 (2012)
25. P.H. Lau et al., Nano Lett. **13**, 8 (2013)
26. S. Wagner et al., Appl. Phys. Lett. **96**, 042111 (2010)

Chapter 8
High-Performance Wearable Bioelectronics Integrated with Functional Nanomaterials

Donghee Son, Ja Hoon Koo, Jongsu Lee and Dae-Hyeong Kim

Abstract As nanotechnology has advanced, deformable nanoscale materials with superb electrical, chemical, and optical properties have made possible the development of high-performance multifunctional electronic devices with flexible and stretchable form factors. Deformability in electronics is achieved mainly by replacing rigid bulk materials (e.g., a silicon wafer) with various promising nanomaterials (e.g., silicon/oxide nanomembranes, carbon nanotubes, graphene, and metal nanoparticles/nanowires). These ultrathin, lightweight, and deformable electronics have attracted widespread interest and offer new opportunities in personalized healthcare, such as wearable bioelectronics. Their deformability, in particular, helps overcome the mechanical mismatch between the conventional bioelectronics, which are flat and rigid, and the soft, curvilinear human skin and internal organs. It resolves prevalent problems in conventional biomedical devices, such as inaccurate biosignal sensing, low signal-to-noise ratio, and user discomfort. Here, we provide an overview of recent developments in wearable bioelectronics integrated with functional nanomaterials with a focus on mobile personal healthcare technologies. The devices introduced in this chapter include wearable sensors, actuators, memory units, and nanogenerators dedicated to healthcare applications. Detailed descriptions of such integrated systems and their uses in clinical medicine are also presented.

Keywords Soft bioelectronics · Functional nanomaterials · Flexible electronics · Stretchable electronics · Wearable electronics

D. Son · J.H. Koo · J. Lee · D.-H. Kim (✉)
Center for Nanoparticle Research, Institute for Basic Science (IBS), Seoul 151-742, Republic of Korea
e-mail: dkim98@snu.ac.kr

D. Son · J. Lee · D.-H. Kim
School of Chemical and Biological Engineering and Institute of Chemical Processes, Seoul National University, Seoul 151-742, Republic of Korea

J.H. Koo · D.-H. Kim
Interdisciplinary Program for Bioengineering, Seoul National University, Seoul 151-742, Republic of Korea

© Springer International Publishing Switzerland 2016
J.A. Rogers et al. (eds.), *Stretchable Bioelectronics for Medical Devices and Systems*, Microsystems and Nanosystems,
DOI 10.1007/978-3-319-28694-5_8

151

8.1   Introduction

Wearable smart devices capable of performing multiple functions including sensing [1–26], actuation [27–31], data storage [32–40], logic control [41–46], light emission [47–51], and energy supply [52–57], have recently attracted significant attention, primarily in relation to rapid progress in flexible and stretchable electronics technology. Of their many applications, mobile healthcare systems based on wearable bioelectronic devices that enable real-time monitoring of physiological and electrophysiological signals, such as those related to body temperature, motion-induced strain, breathing, pulse, electrocardiogram (ECG), and electroencephalogram (EEG), are of particular interest. These systems offer the possibility of providing clinically valuable patient data to primary healthcare providers, and can thereby prevent fatal diseases and emergency situations through point-of-care treatments and ubiquitous healthcare systems.

Several challenges persist, however, in relation to wearable applications of bioelectronic devices, mainly due to the rigidity of prevalent electronic devices. The poor interfacial contact between such devices and the human body prevents accurate detection of physiological/electrophysiological conditions. The mechanical mismatch between the soft, curvilinear human body and rigid, planar electronic devices renders the latter susceptible to delamination and mechanical fracture due to stress from repeated deformations associated with daily movements. Low signal-to-noise ratio (SNR), high power consumption, and low operating speeds are also important issues that need to be addressed in the ongoing development of wearable bioelectronic devices.

One of the most effective research approaches intended to overcome the aforementioned problems has been the use of nanomaterials that have intrinsic flexible characteristics and excellent electrical properties, along with stretchable device design. In particular, single-walled carbon nanotubes (SWNTs) [4, 6, 36, 41, 42, 44], graphene [58, 59], molybdenum disulfide nanosheets [40, 60], Pt-coated nanofibers [61], zinc oxide (ZnO) nanowires (NWs) [55, 56, 62], pentacene nanofilm (PNF) [1, 32–34, 46], silicon nanomembranes (SiNMs) [11, 14, 15, 17, 21, 25, 29, 30, 35, 45, 63], and titanium dioxide (TiO$_2$) NMs [35, 64, 65] have been highlighted as candidates for replacing the bulky and rigid materials used in conventional planar devices. The latest research on wearable bioelectronics focuses on the monolithic integration of these nanomaterials with one another to develop a multifunctional system. This chapter covers some of the latest developments in wearable bioelectronics based on nanomaterials and their heterogeneous hybrids.

8.2   Wearable Bioelectronics Integrated with Functional Nanomaterials

8.2.1   Multifunctional Wearable Devices Based on Silicon Nanomembranes

Inorganic nanomaterials have attracted increasing research interest for their application in high-performance electronic devices because they are reliable under ambient conditions and compatible with conventional micro/nanofabrication processes. Among such nanomaterials, single-crystalline SiNMs have shown considerable potential for high-performance wearable electronic devices including various types of sensors and/or logic devices, because they exhibit highly uniform electrical properties (e.g., electron mobility) along with intrinsically superior piezoresistive properties. Ultrathin doped SiNMs can be transferred to flexible and stretchable substrates by transfer printing processes. Subsequent microprocessing techniques lead to high-performance deformable electronics. The details of the fabrication process are available elsewhere [66–68]. In this section, the discussion focuses on high-density, SiNM-based sensor arrays with multiplexers for temperature, strain, and pressure sensing.

Figure 8.1a shows a multiplexed array of SiNM-based PIN diodes (each has dimensions of approximately 100 μm × 200 μm in an 8 × 8 layout) with serpentine-shaped interconnections formed using an ultrathin metal film [14]. The SiNM-based PIN temperature sensor measures shifts in the diode's *I-V* relationship resulting from external temperature changes. The device array can be mounted directly onto the skin with a thin backing substrate of polydimethylsiloxane (PDMS, ∼ 30 μm thick, ∼ 50 kPa modulus) that makes conformal contact through the van der Waals force. Figure 8.1b presents a temperature map (right) of a thin film copper heater obtained using a SiNM-based PIN diode array (left). The SiNM-based PIN temperature sensor array provides a precise spatial map of the thermal distribution around the copper heater. SiNMs can also be used as strain sensors (Fig. 8.1c–f). Figure 8.1c, d show digital (left) and optical microscopy (right) images of the SiNM-based strain sensor [66]. The SiNM-based strain sensor embedded in polyimide (PI) layers is mechanically stable because it is located near the neutral mechanical plane and is used with stretchable electrodes (Fig. 8.1e). These strain sensors precisely monitor the movement of the chest during breathing (Fig. 8.1f).

In addition to these applications, Fig. 8.1g–j show smart prosthetic skin integrated with ultrathin SiNM-based strain, pressure, and temperature sensor arrays [63]. The site-specific design of SiNM sensor arrays is shown in Fig. 8.1g. In low-deformation region (below ∼ 5 % strain), the S1 design (top right) is used, whereas in medium- and high-deformation regions (∼ 10 and ∼ 16 % strain), the system adopts, respectively, the S3 and the S6 design (middle and bottom right). This site-specific design for strain sensors maximizes both sensitivity and dynamic sensing range. The SiNM-based electronic skin can encounter complex situations,

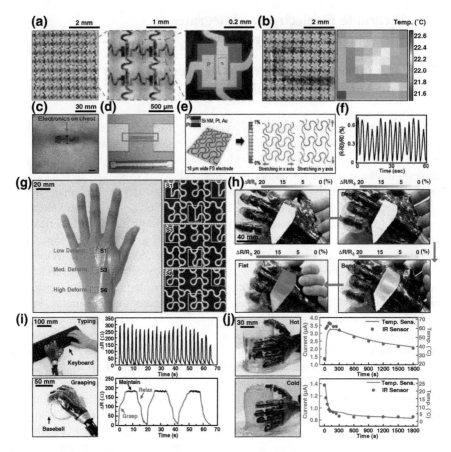

Fig. 8.1 **a** Optical microscopic images of an 8 × 8 SiNM PIN diode array integrated onto a stretchable substrate. The *boxes* bordered by *red dashes* enclose magnified images of the local PIN diode sensors. **b** Optical image and corresponding temperature distribution map for an 8 × 8 SiNM PIN diode sensor array mounted on a heated object. **c** Photograph of a multifunctional epidermal electronic device on a human chest. **d** Image of a mechanical strain gauge sensor based on a Si nanomembrane resistor. **e** FEA results for an epidermal electronic device incorporating a fully serpentine mesh, under stretching modes along the *x* and *y* directions. The neutral mechanical plane of the device is shown in the inset. **f** Sensitivity of the strain sensor in terms of change in resistance versus time. **g** Photograph of a SiNM strain sensor array with site-specific design, conformally laminated onto the back of a human hand. On the *right side* of the frames are magnified views of different designs suited to each deformation mode. **h** Images of a prosthetic hand with integrated functional sensors. Spatiotemporal maps for the resistance change of the device array are shown according to various motions. **i** Images of the prosthetic limb tapping a keyboard (*top left*) and catching a baseball (*bottom left*). Plots for the temporal resistance change of the SiNM pressure sensor (*top right*), indicating the dynamics of the prosthetic hand in grasping, maintaining, and relaxing modes (*bottom right*). **j** Photographs of the prosthetic limb touching a cup of hot (*top left*) and cold (*bottom left*) water. Plots (*right*) of the corresponding temporal current change of the SiNM-based temperature sensor (PIN diode) and the actual temperature data measured by an IR sensor

such as a handshake, typing on a keyboard, grasping a ball, or holding a cup of hot or cold liquid (Fig. 8.1h–j). The strain map shows some changes in the strain distribution near the index finger and related joints (Fig. 8.1h). Pressure sensors exhibit rapid and reliable responses to external stimuli in both situations (Fig. 8.1i, top: tapping a keyboard; bottom: catching a ball). Temporal temperature sensing is verified when a hand touches a cup containing hot or cold liquid (Fig. 8.1j). These results suggest potential for the application of soft bioelectronics to prosthetics.

8.2.2 Wearable Nonvolatile Memory Devices Based on TiO_2 Nanomembranes and an Au Nanoparticle Assembly

Nonvolatile memory technologies for wearable bioelectronics applications involve the employment of lightweight, ultrathin, and stretchable components in devices. Stretchable memory devices integrated with wearable sensors are a good example. Resistive random-access memory (RRAM) constructed from thin nanoscale oxide films is widely regarded as an emerging class of nonvolatile memory. It exhibits extraordinary electrical performance—low power consumption, fast resistive switching speed, large on/off ratio, and reliable retention/endurance. Although oxide thin films used in RRAM devices are rigid, which limits their ability to interface with curvilinear, dynamically deforming soft tissues, stretchable designs allow RRAM to be wearable. This enables the implementation of an integrated system of stretchable sensors and memory devices that incorporate such unique features as sensing, data storage, and pattern analysis on the human skin. Co-integrated drug delivery and feedback actuation devices on board complete the closed-loop for wearable diagnosis and therapy applications.

Figure 8.2a, b show a schematic illustration and corresponding image, respectively, of a representative example of a multifunctional wearable electronic device for movement disorders [35]. The system is equipped with an RRAM module with stretchable sensors and actuators. To minimize strain-induced damage in inorganic layers, the active region of the memory is located in the neutral mechanical plane. The inset in Fig. 8.2b shows a 10 × 10 wearable RRAM array. Gold nanoparticles (AuNPs) (~ 12 nm diameter) are used as charge trap layers in the RRAM. Uniform AuNP layers on a TiO_2 NM/Al substrate can be prepared over a large area by using the Langmuir–Blodgett (LB) assembly technique (Fig. 8.2c). The number of AuNP layers can also be controlled through the LB method, as confirmed by transmission electron microscopy (TEM) analysis (Fig. 8.2d) and energy-dispersive X-ray spectroscopy (EDS) (Fig. 8.2e). Figure 8.2f shows bipolar I-V characteristics for wearable RRAM devices with varying configurations. The inset in Fig. 8.2f shows the operational sequence: a negative voltage ("set") switches the RRAM to the low resistance state, and a positive voltage ("reset") switches it back to the high resistance state. Uniform-sized AuNPs increase the number of charge trap sites, thus reducing the set and reset currents by nearly two orders of magnitude relative

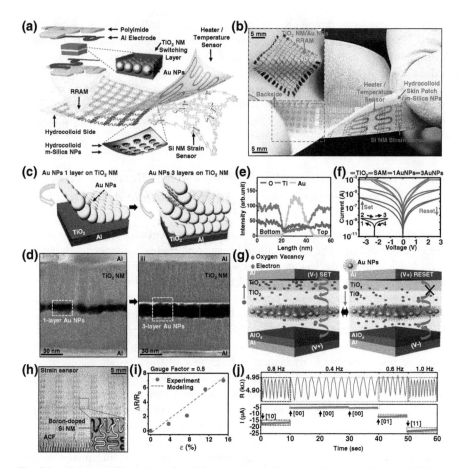

Fig. 8.2 **a** Schematic illustration of multifunctional wearable electronic devices based on RRAM, Si nanomembrane strain sensors, thermistors, and functional nanoparticles. **b** Image of wearable electronic devices integrated into an elastomeric substrate composed of hydrocolloids. The *inset* shows a magnified view of a 10 × 10 RRAM array. **c** Schematics for the LB assembly of AuNPs on the TiO₂ substrate. The number of AuNP layers can be controlled by adjusting the number of operating cycles. **d** High-resolution TEM images of an RRAM array with 1 (*left*) and 3 (*right*) AuNP layers. **e** EDS analysis of the RRAM cell with 3 AuNP layers. **f** *I-V* characteristics of the RRAM device with different traps. The *inset* shows sequential directions. **g** Schematic illustrations of the resistive switching mechanism in the RRAM cell. The "set" (*left*) and "reset" (*right*) operations are based on the trap-controlled space charge-limited conduction. **h** Image of the Si nanomembrane strain sensor array. An *inset* (area bordered by *red dashes*) shows a magnified view of the sensor array. **i** Plot of change in the resistance of the strain sensor as a function of strain for modeling the GF. **j** Plot of time-dependent change in resistance in the Si nanomembrane strain sensor caused by simulated hand tremors at frequencies of 0.8, 0.4, 0.6, and 1 Hz (*top of the frame*). A plot illustrating the multi-level cell operation of RRAM cells (*bottom of the frame*)

to those without charge trap layers. These levels are further reduced by four orders of magnitude in the three-AuNP-layer structure. This memory device follows the trap-controlled space charge-limited conduction mechanism schematically illustrated in Fig. 8.2g.

This memory can be integrated with other stretchable electronic devices. Figure 8.2h shows an array of wearable SiNM-based strain gauges as a representative example of wearable motion sensors interconnected with collocated memory modules. The strain sensor has an effective gauge factor (GF) of ~ 0.5 (Fig. 8.2i) consistent with analytically calculated GF values. This particular demonstration emulates tremor modes that manifest as part of the symptoms of epilepsy and Parkinson's disease, and which lead to tremors in the hand occurring at various frequencies (Fig. 8.2j, top frame). The varying tremor frequencies provide an important tracking factor for diagnosing and monitoring these movement disorders. The data captured from the movements are stored in separate memory cells every 10 s using the multi-level cell RRAM operation (Fig. 8.2j, bottom frame).

8.2.3 Transdermal Drug Delivery Using a Wearable Electronic Patch

By analyzing the data stored in memory devices, clinically relevant key factors, such as vital signs or tremor frequencies can be tracked. This makes it possible to determine the patient's state in real time. An appropriate response to diagnosis using stored information is to trigger controlled transdermal drug delivery as feedback through thermal and/or electrical actuations (Fig. 8.3a) [35]. For this purpose, mesoporous silica nanoparticles (m-silicaNPs) have been employed as a vehicle for drug loading/delivery in conjunction with diagnostic, data storage, and actuation elements. These m-silicaNPs (~ 40 nm diameter) have many small pores (~ 2 nm diameter). Thus, m-silicaNPs have a large surface area that can be used to store drug molecules (Fig. 8.3b). M-silicaNPs loaded with drug molecules are transfer printed onto the sticky surface of a hydrocolloid patch using a structured stamp (Fig. 8.3c). When applied to the skin, nanoparticles loaded with pharmacological agents are in direct contact with the skin and transdermally deliver drugs at appropriate energy for desorption and diffusion. Transdermal drug delivery is attained by using a resistive heater integrated on the top surface of a wearable electronic patch. The heater supplies thermal energy to accelerate the desorption/diffusion of drugs from m-silicaNPs through the skin and into the bloodstream. Figure 8.3d shows infrared (IR) camera measurements of the thermistor on the wearable patch. Finite element modeling (FEM) is used to simulate Joule heating and estimate heat transfer around the device (Fig. 8.3e). Heat weakens the physical bonding between the m-silicaNPs and the drug molecules inducing diffusion into the skin. The thermal actuation further increases the diffusion rate, which accelerates transdermal drug delivery (Fig. 8.3f). Transdermal drug delivery was

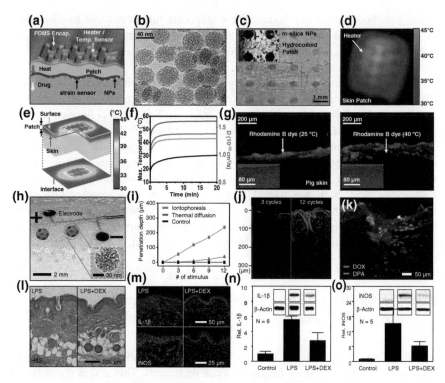

Fig. 8.3 **a** Schematic illustration of thermally controlled transdermal drug delivery system using a hydrocolloid patch and m-silicaNPs. **b** TEM image of the m-silica NPs. **c** Digital photographic image of the m-silica NP array, with *inset* showing an optical microscope image. **d** IR camera image showing the temperature distribution of the skin patch. **e** FEA of the thermal profile of the skin patch attached to human skin. **f** Maximum temperature of the heater in the skin patch (*red*), the interface between the skin and the patch (*orange*), and the heater without heating enabled (*black*) as a function of time. The y-axis of the right hand shows the diffusion coefficient with increasing temperature. **g** Cross-sectional fluorescence images of the skin of a pig showing diffusion profiles following the activation of the thermal release of Rhodamine B dyes at 25 °C (*left*) and 40 °C (*right*). **h** Wireless transdermal drug delivery system using iontophoresis attached to the skin. **i** Drug penetration depth into the skin of the mouse with the number of applied stimuli using different actuators (*black*: none; *blue*: thermal actuator; *red*: iontophoresis electrode). **j** Cross-sectional fluorescence images showing diffusion profiles for different stimuli with iontophoresis: 3 cycles (*left*) and 12 cycles (*right*). **k** Cross-sectional confocal fluorescence image showing sequential diffusion of two dyes into the mouse's skin (*red*: DOX; *green*: 9, 10-diphenylanthracene). **l** H&E stain histology showing the LPS-injected kins of the mice prior to (*left*) and following (*right*) treatment. **m** Immunohistochemistry of IL-1β (*green*) and iNOS (*red*) prior to (*left*) and following (*right*) treatment. **n** and **o** Western blot immunoassay of IL-1β (**n**) and iNOS (**o**)

confirmed by a fluorescence microscope image (Fig. 8.3g) of a dye (Rhodamine B) diffusing into the skin of a pig at room temperature (25 °C, left). The delivery rate is higher at higher temperature (40 °C, right). The depth of penetration of the skin at room temperature is three times less than that at higher temperature indicating that controlled drug diffusion was achieved through thermal actuation.

An effective alternative to a thermal actuation-based drug delivery is iontophoresis, which uses charge repulsion between electrodes and drugs within the electric field generated by a pair of electrodes [25]. A wearable smart patch equipped with iontophoresis electrodes has been shown to more efficiently control the rate of drug delivery (Fig. 8.3h). Wireless control of iontophoresis has also been demonstrated. Figure 8.3i shows a plot of the diffusion depth of doxorubicin (DOX) into the skin of a BALB/c mouse as a function of the number of applied stimuli. The penetration rate and depth with iontophoresis were higher than with thermal diffusion (Fig. 8.3j). The diffusion of the drug molecules depends on the number of cycles of iontophoresis. Another advantage of this patch-based system with iontophoresis electrodes is that the pharmacological agents (red: DOX; green: 9,10-diphenylanthracene) can easily be reloaded (Fig. 8.3k). The therapeutic effects of iontophoresis-based drug delivery were confirmed through in vivo animal experiments (Fig. 8.3l–o). A wearable smart patch was laminated onto the animal model (a mouse injected with the inflammatory mediator lipopolysaccharide) and drugs were delivered transdermally. Histopathological (H&E staining; Fig. 8.3l), immunofluorescence (Fig. 8.3m), and western blot (Fig. 8.3n and o) analyses showed that the dexamethasone treatment effectively suppressed inflammatory mediators, interleukin-1β (IL-1β) and inducible nitric oxide synthase (iNOS) compared with the control.

8.2.4 Carbon Nanotube-Based Wearable Bioelectronics

CNTs are another excellent electronic material for use in wearable electronics due to their high conductivity, carrier mobility, mechanical flexibility, and low production cost. Moreover, recent developments have enabled the facile separation and sorting of metallic and semiconducting CNTs from as-synthesized CNT bundles [67, 68]. These CNTs can be integrated with flexible/stretchable substrates as a channel material in field effect transistors and/or as conducting electrodes in electronic systems.

Figure 8.4a shows a relatively simple fabrication process for a piezoresistive pressure sensor based on porous pressure-sensitive rubber (PPSR) [23]. PPSR is composed of multiwalled carbon nanotubes (MWNTs) homogeneously mixed with porous silicone rubber. This porous structure was introduced using a reverse-micelle solution in a mixture with MWNTs and PDMS to enhance the sensitivity of the pressure-sensitive rubber. The PPSR solution was nozzle jet printed onto a flexible substrate as a patterned array. The FEA simulations for stretching and bending in Fig. 8.4b reveal that the induced strain was almost identical to the applied strain in the direction of the applied strain, whereas the normal stretching direction experienced almost no strain. The induced strain caused a change in the resistance of the PPSR patterns with no significant material degradation as shown in

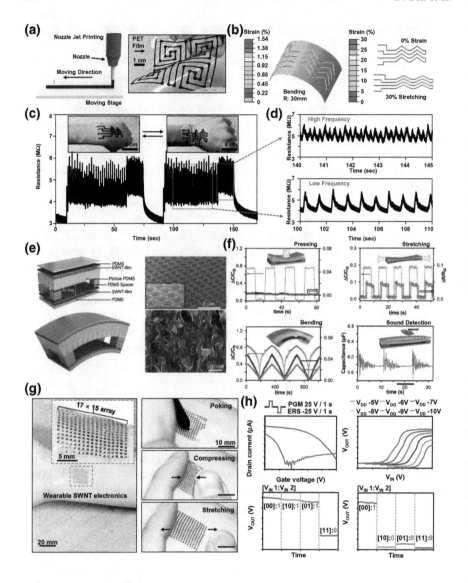

Fig. 8.4c. High- and low-frequency measurement demonstrations are presented in Fig. 8.4d.

While resistive-type sensors are beneficial in regard to the simplicity of their fabrication processes and designs, capacitive sensors generally exhibit higher sensitivity. Figure 8.4e is a schematic illustration of an SWNT-based capacitive sensor with the capability of differentiating multiple mechanical stimuli by virtue of the porous structure of its dielectric and the air gap between the two SWNT-coated plates [69]. Porous and structured PDMS are used to enhance the sensitivity in the higher pressure region (>1 kPa). The introduction of an air gap aided pressure

◄ **Fig. 8.4** **a** *Left* Schematic illustration of the nozzle jet printing method for PPSR pattern formation. *Right* Digital image of the nozzle jet-printed PPSR on a flexible PET film. **b** FEA of strain distribution in the patterned PPSR on stretchable substrate when bent with 30 mm radius of curvature (*left*) and subjected to approximately 30 % stretching (*right*). **c** Resistance change with respect to deformations caused by wrist movements in strain gauges attached to the skin. The *inset* shows the deformed states of the patterned PPSR strain gauge laminated onto a human wrist. **d** Measured signals for high-frequency movements (*top*) and low-frequency movements (*bottom*). **e** *Left* Schematic illustration of electronic skin showing device architecture in a layer-by-layer format. *Right* SEM images of porous PDMS with the surface coated with SWNTs. **f** Detection of mechanical variations caused by pressing, stretching, bending, and exposure to sound waves based on changes in capacitance. **g** Digital images of the wearable SWNT-based electronics attached to the skin showing conformal contact after poking, compressing, and stretching. **h** Electrical responses of a charge trap flash memory device (*top left*), inverter (*top right*), NAND gate (*bottom left*), and NOR gate (*bottom right*)

sensing in the low-pressure region (<1 kPa). The resulting device shows high sensitivity, such that even subtle sound vibrations are detectable, as well as the ability to differentiate modes of external stimuli—pressing, stretching, bending, etc. —with varying magnitudes (Fig. 8.4f).

Other applications of CNTs in wearable electronics include their use as channel materials in various devices and circuits including transistors, logic gates, and memory modules. The intrinsically flexible nature of CNTs is beneficial for alleviating accumulated stress in the electronic device generated by human movements. Figure 8.4g depicts an SWNT-based wearable electronic system composed of arrays of capacitors, charge trap flash memory devices, transistors, and various logic gates. These arrays exhibit consistent and excellent performance under different deformation modes [36]. The representative electrical performances of a charge trap flash memory device, an inverter, a NAND gate, and a NOR gate are shown in Fig. 8.4h, which demonstrate that replacing conventional inorganic materials with SWNTs is highly advantageous in terms of mechanical stability and endurance.

8.2.5 Transparent Nanomaterials for Wearable Interactive Human–Machine Interfaces

Another important concern for wearable electronic systems, apart from the SNR issue and discomfort resulting from the mechanical mismatch between the human body and such devices, is their unnatural appearance on the body. Although reducing overall thickness has been effective in overcoming the rigidity of the materials, the opacity of conventional electronic materials is regarded as a significant impediment to develop wearable electronics that look natural on the skin. Thus, transparent materials with superior electrical properties have been highlighted as candidates for next-generation wearable electronics that will provide pleasing esthetics while maintaining high performance and utility.

One such promising material is zinc oxide (ZnO), which is a wide-bandgap semiconductor with good transparency, high electron mobility, and piezoelectric properties. A semitransparent wearable electronic system composed of ZnO nanomembrane hybrids has been implemented; it was attached to the skin, monitored the user's motions, and was used to control the motion of a wheelchair. Figure 8.5a shows a schematic illustration of the overall operation of a system of semitransparent wearable devices—a ZnO-based piezoelectric strain sensor and a thermistor-like temperature sensor—that enabled interactive human–machine interface (iHMI) [70]. An image of the devices on the skin is shown in Fig. 8.5b. The active components of the piezoelectric strain sensor that controlled the wheelchair motion are illustrated in a magnified view in Fig. 8.5c. The piezoelectric strain sensor is composed of consecutive layers of 5 nm Cr, 700 nm ZnO, networks of semiconducting SWNTs, and 20 nm MoO_x embedded in PI/indium tin oxide (ITO) layers. The use of transparent active materials in conjunction with ultrathin and stretchable designs enables conformal lamination of the devices on the skin resulting in accurate monitoring of the user's motion, which in turn allows for proper control of the movement of the wheelchair. Figure 8.5d depicts the electrical response of the piezoelectric strain sensors to the applied strain of ~ 6 %. The Cr and SWNT layers improve the piezoelectric performance of the ZnO nanomembrane by improving the crystallinity of the ZnO and by passivating the native surface defects of the ZnO, respectively. The piezoelectric strain sensors exhibited linear voltage response to applied strains between 0.6 and 7.0 % (Fig. 8.5e). The semitransparent piezoelectronic devices were effective as wearable iHMIs for controlling the wheelchair.

A fully transparent wearable iHMI system was also demonstrated by using a combination of a piezoelectric motion sensor and an electrotactile stimulator based on transparent nanomaterials, such as graphene, SWNT networks, silver nanowire (AgNW) networks, and a piezoelectric polymer (polylactic acid, PLA). Stretchable designs enable the conformal lamination of devices onto the skin. Figure 8.5f provides an image of a transparent iHMI system conformally attached to the wrist [71]. As shown in Fig. 8.4g, the piezoelectric motion sensor consists of heterostructures of graphene/PLA-SWNT composite film (70 μm thick)/graphene between layers of polymethyl methacrylate (PMMA). The electrotactile stimulator is composed of graphene/AgNW/graphene heterostructures between epoxy layers. SWNTs are incorporated into the PLA to enhance the piezoelectric properties of the polymer; this enhancement is evident in the generated piezoelectric voltage signals shown in Fig. 8.5h. The finite element analysis (FEA) in Fig. 8.5j confirms this result. When strain is applied to the system, stress is concentrated near the SWNTs, which leads to maximum charge generation. Feedback stimulation based on the sensed signals can be employed to complete the interactive system. Figure 8.5i is a plot of the boundary condition for the stimulation current required for perception at different frequencies; it indicates an inversely proportional relationship. Figures 8.5k, l confirm bending-induced strain sensing with linear sensitivity (~ 0.12 mV/cm^{-1})

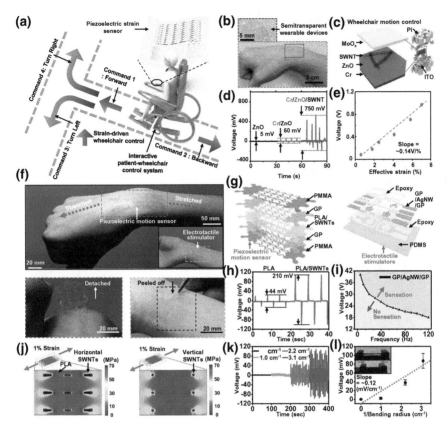

Fig. 8.5 **a** Schematic illustration of a wheelchair control system that incorporates a semitransparent piezoelectric strain sensor. **b** Digital image of the transferred semitransparent devices on human skin. **c** Diagram showing layer information of the strain sensor. **d** Output voltage curves of the piezoelectric strain sensor with ZnO (*black*), Cr/ZnO (*red*), and Cr/ZnO/SWNT (*blue*) layers. **e** Calibration curve of the strain sensor showing linear response with increasing applied strain. **f** Digital images of the transparent piezoelectric motion sensor (*top*) and electrotactile stimulator (*inset, top*) for an interactive human–machine interface attached to the *top* and *bottom* of a human wrist, respectively. The images at the *bottom* show the partially detached motion sensor (*left*) and the electrotactile stimulator (*right*). **g** Diagram of layer information of the patterned heterostructures used in the motion sensor (*left*) and the electrotactile stimulator (*right*). **h** Output voltages of sensors with PLA (*blue*) and SWNT-embedded PLA (*red*) as a function of time. **i** Voltage versus frequency curve showing the minimum stimulation current required for perception. **j** FEA of the strain distribution on the PLA film with horizontally (*left*) and vertically (*right*) aligned SWNTs. **k** Output voltage of the PLA/SWNT composite film with respect to time with variations in the bending radii. **l** Corresponding output voltage curve for different bending radii. The *inset* images show film bending in the horizontal direction (*red arrow*)

over wide strain ranges. The materials employed and the device strategy developed for these heterostructure-based devices lead to unprecedented wearability and performance allowing for elaborate interactive robot control.

8.2.6 Inorganic Nanowires for Wearable Thermotherapy

Inorganic nanowires are another important candidate material in the context of wearable electronics, especially for electrodes. Their high aspect ratio is advantageous for superior electrical percolations and they can maintain good electrical performance under various deformation conditions. Silver nanowires (AgNWs), in particular, have received significant attention owing to their intriguing electrical and optical properties. Ag has the highest electrical conductivity among all metals, and the one-dimensional structure of AgNWs provides enhanced optical transparency. Although fabrication/processing difficulties, such as achieving uniform/homogeneous networks or film without morphological irregularity, have limited their use in large-scale production; recent advances have delivered promising results and AgNW has thus been integrated into wearable device applications. In this subchapter, a wearable heater using AgNWs is introduced.

Figure 8.6a illustrates the fabrication process of a AgNW-based heater that incorporates serpentine mesh layouts to maximize stretchability for conformal

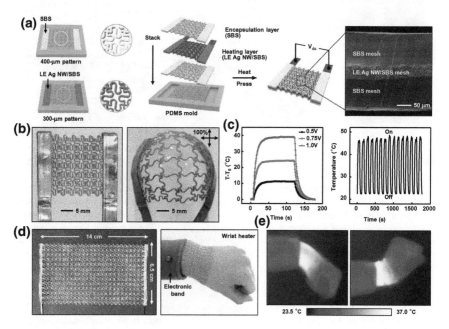

Fig. 8.6 **a** Schematic illustration of the fabrication process of the mesh heater and SEM images of the welded interfaces. **b** Photograph of the mesh heater prior to (*left*) and following (*right*) stretching under ∼100 % biaxial strain. **c** *Left* Temperature response of the mesh heater under applied voltages of 0.5, 0.75, and 1.0 V with respect to time. *Right* Cyclic temperature responses produced by turning the mesh heater on and off by applying a voltage of ∼0.5 V. **d** *Left* Digital image of a large stretchable heater designed for the human wrist. *Right* Image of the same heater worn on the wrist. **e** IR camera images of the mesh heater attached to the wrist following downward and upward motion of the wrist

integration onto movable joints, user comfort, and heat transfer efficiency by intimate contact [27]. The fabrication procedure is as follows: The polyvinylpyrrolidone (PVP) ligands in the as-synthesized AgNWs are first exchanged with hexylamine for homogeneous dispersion in an organic phase styrene-butadiene-styrene (SBS) solution. This is followed by molding and curing in a prefabricated PDMS mold with serpentine patterns. The ligand-exchanged (LE) AgNW/SBS nanocomposite serves as the heating layer. It is securely encapsulated between regular SBS insulation layers. The good interface between the insulating layers and the heating layer is confirmed by the scanning electron microscope (SEM) image at the right side of Fig. 8.6a. Wiring for the power source and the controller completes the fabrication process. Figure 8.6b shows the result of a stretching test using the fabricated heater with a biaxial strain of ∼ 100 %. The wearable heater also exhibited rapid response to small applied voltages as depicted in Fig. 8.6c. Specifically, an increase in the temperature to 40 °C within 120 s was observed with an applied voltage of 1 V. A repeated cyclic test revealed consistent temperature changes.

The possibility of using the fabricated wearable heater as an orthopedic therapeutic tool in physical therapy was assessed through large-area heating over the entire surface of an adult wrist (Fig. 8.6d). The AgNW/SBS nanocomposite-based wearable heater is integrated with a custom-designed electronic band containing the control and power supply units. The IR camera image in Fig. 8.6e shows homogeneous heat generation. The wearable heater maintains conformal contact with the wrist under different deformation states during large joint movements. As a result, effective heat transfer to the skin and underlying tissues is realized. The integrated system comprising a AgNW-based stretchable heater and a custom-made electronics band enables portable, wearable, point-of-care articular thermotherapy.

8.2.7 Wearable Triboelectric Nanogenerators

Wearable bioelectronics require an autonomous power supply system in order to continuously monitor physiological and electrophysiological signals from the human body. Triboelectric nanogenerators (TENGs) have recently been demonstrated as promising energy harvesting components for wearable devices. TENGs convert mechanical friction or pressure to electrical energy based on the coupling of the triboelectric effect and electrostatic induction [72]. When two materials with different triboelectric series repeatedly come into contact and then separate, changes in their surface charges drive alternating current through conductive materials. The generated electrical energy can supply power to other wearable electronic components, such as temperature/pressure sensors or a heartbeat monitor.

Figure 8.7a shows the fabrication process for and photographs of a fiber-based TENG [73]. A cotton thread was dipped into a carbon nanotube (CNT) ink and dried to make a conducting electrode followed by coating with polytetrafluoroethylene (PTFE) NPs of highly negative triboelectric series. Then,

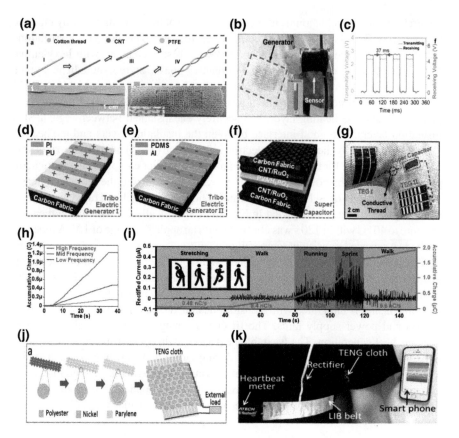

Fig. 8.7 **a** The fabrication process of cotton fiber-based TENGs and photographs of the TENGs. **b** Photograph showing the integrated system with the TENG and temperature sensor on a human wrist. **c** Modulated and demodulated signals from the temperature sensor. **d–f** Schematic illustrations of a CF-based (**d**) TENG 1 with PI and polyurethane (*PU*), (**e**) TENG 2 with Al and polydimethylsiloxane (PDMS), and (**f**) supercapacitor. **g** Photograph showing the TENGs and supercapacitor attached to clothes. **h** Accumulated charge stored in the supercapacitor at different TENG energy generation frequencies. **i** Rectified current from a TENG and accumulated charge in the supercapacitor in relation to the wearer's movements. **j** Schematic illustration of the fabrication process for the polyester fabric-based TENG cloth. **k** Photograph showing the integrated system with a TENG, a lithium-ion battery, and a heartbeat monitor transmitting signals to a smartphone

CNT-coated fiber (without PTFE coating) and PTFE-coated fiber were entangled to complete the fiber-based TENG, which can be woven into fabric; the TENG was integrated with a temperature sensor (Fig. 8.7b). The mechanical friction occurring between the two threads generated electrical energy that powered the temperature sensor (Fig. 8.7c).

Carbon fabric (CF) has also been used as an electrode for wearable TENGs; four different materials were aligned in alternating fashion to utilize sliding friction occurring under the armpit (Fig. 8.7d, e) [52]. An all-solid-state wearable supercapacitor based on CFs was incorporated into the system to store the energy generated by the TENG (Fig. 8.7f). A vertically aligned CNT forest was directly synthesized on the CF to maximize the surface area of electrodes. It was then covered with RuO_2 NPs to achieve pseudocapacitive performance. Figure 8.7g shows CF-based TENGs and an integrated supercapacitor attached to clothing in the armpit region and the chest region, respectively. The charge accumulation speed exhibited by the supercapacitor depends on the sliding frequency between the TENGs (Fig. 8.7h). In addition to generating power, TENGs can thus be used to monitor bodily activity. The accumulated charge from the TENG stored in the supercapacitor is proportional to the degree of activity, and thus the total charge accumulation can be analyzed to assess the wearer's activity level (Fig. 8.7i).

The energy harvesting performance of TENGs can be enhanced by the segmentation design shown in Fig. 8.7j [74]. Polyester fabric was plated with nickel to function as a flexible conductive electrode for textile TENG with a positive triboelectric series. Another polyester fabric was coated with a parylene film for the negative triboelectric series electrode of the TENG. The Ni-plated fabric and parylene-coated fabric were woven together to create a TENG cloth with this segmentation design. A flexible lithium-ion battery was also fabricated based on the Ni-plated polyester fabric and integrated with the TENG cloth to collect the generated electricity and supply the stored energy to a heartbeat monitor (Fig. 8.7k). The obtained heartbeat data were wirelessly transmitted to a smartphone.

8.3   Conclusion

The nanomaterial-based wearable bioelectronic devices described in this chapter introduce various new techniques and strategies for designing and fabricating physiological/electrophysiological sensors, therapeutic functional actuators, nonvolatile data storage devices, and energy supply devices for use in integrated systems. Employing nanomaterials as device components enhances performance and makes it possible to achieve hitherto nonexistent optical, mechanical, electrical, and biological properties and functions. These multifunctional wearable bioelectronic devices, independently or as part of an integrated system, can provide unprecedented advantages in personal healthcare applications. With the continuing research and development dedicated to the use of soft bioelectronics in clinical medicine, it is time to explore new scientific and commercial opportunities in the wearable healthcare industry.

Acknowledgments   This work was supported by IBS-R006-D1.

References and Notes

1. T. Someya, T. Sekitani, S. Iba et al., A large-area, flexible pressure sensor matrix with organic field-effect transistors for artificial skin applications. Proc. Nat. Acad. Sci. **101**, 9966–9970 (2004)
2. M. Kaltenbrunner, T. Sekitani, J. Reeder et al., An ultra-lightweight design for imperceptible plastic electronics. Nature **499**, 458–463 (2013)
3. C. Wang, D. Hwang, Z. Yu et al., User-interactive electronic skin for instantaneous pressure visualization. Nat. Mater. **12**, 899–904 (2013)
4. C. Yeom, K. Chen, D. Kiriya et al., Large-area compliant tactile sensors using printed carbon nanotube active-matrix blackplanes. Adv. Mater. **27**, 1561–1566 (2015)
5. S.C.B. Mannsfeld, B.C.-K. Tee, R.M. Stoltenberg et al., Highly sensitive flexible pressure sensors with microstructured rubber dielectric layers. Nat. Mater. **9**, 859–864 (2010)
6. D.J. Lipomi, M. Vosgueritchian, B.C.-K. Tee et al., Skin-like pressure and strain sensors based on transparent elastic films of carbon nanotubes. Nat. Nanotechnol. **6**, 788–792 (2011)
7. B.C.-K. Tee, C. Wang, R. Allen et al., An electrically and mechanically self-healing composite with pressure- and flexion-sensitive properties for electronic skin applications. Nat. Nanotechnol. **7**, 825–832 (2012)
8. M. Ramuz, B.C.-K. Tee, J.B.-H. Tok et al., Transparent, optical, pressure-sensitive artificial skin for large-area stretchable electronics. Adv. Mater. **24**, 3223–3227 (2012)
9. G. Schwartz, B.C.-K. Tee, J. Mei et al., Flexible polymer transistor with high pressure sensitivity for application in electronic skin and health monitoring. Nat. Commun. **4**, 1859 (2013). doi:10.1038/ncomms2832
10. L.Y. Chen, B.C.-K. Tee, A.L. Chortos et al., Continuous wireless pressure monitoring and mapping with ultra-small passive sensors for health monitoring and critical care. Nat. Commun. **5**, 5028 (2014). doi:10.1038/ncomms6028
11. D.-H. Kim, N. Lu, R. Ma et al., Epidermal electronics. Science **333**, 838–843 (2011)
12. D.-H. Kim, N. Lu, R. Ghaffari et al., Materials for multifunctional balloon catheters with capabilities in cardiac electrophysiological mapping and ablation therapy. Nat. Mater. **10**, 316–323 (2011)
13. J.-W. Jeong, W.-H. Yeo, A. Akhtar et al., Materials and optimized designs for human-machine interfaces via epidermal electronics. Adv. Mater. **25**, 6839–6846 (2013)
14. R.C. Webb, A.P. Bonifas, A. Behnaz et al., Ultrathin conformal devices for precise and continuous thermal characterization of human skin. Nat. Mater. **12**, 938–944 (2013)
15. Y.M. Song, Y. Xie, V. Malyarchuk et al., Digital cameras with designs inspired by the anthropod eye. Nature **497**, 95–99 (2013)
16. S. Xu, Y. Zhang, L. Jia et al, Soft microfluidic assemblies of sensors, circuits, and radios for the skin. Science **344**, 70–74 (2014)
17. Y. Hattori, L. Falgout, W. Lee et al., Multifunctional skin-like electronics for quantitative clinical monitoring of cutaneous wound healing. Adv. Health. Mater. **3**, 1597–1607 (2014)
18. L. Gao, Y. Zhang, V. Malyarchuk et al., Epidermal photonic devices for quantitative imaging of temperature and thermal transport characteristics of the skin. Nat. Commun. **5**, 4938 (2014). doi:10.1038/ncomms5938
19. C. Dagdeviren, Y. Su, P. Joe et al., Conformal amplified lead zirconate titanate sensors with enhanced piezoelectric response for cutaneous pressure monitoring. Nat. Commun. **5**, 4496 (2014). doi:10.1038/ncomms5496
20. H.-J. Chung, M.S. Sulkin, J.-S. Kim et al., Stretchable, multiplexed pH sensors with demonstrations on rabbit and human hearts undergoing ischemia. Adv. Health. Mater. **3**, 59–68 (2014)
21. J.-W. Jeong, J.G. McCall, G. Shin et al., Wireless optofluidic systems for programmable in vivo pharmacology and optogenetics. Cell **162**, 1–13 (2015)

22. L. Xu, S.R. Gutbrod, Y. Ma et al., Materials and fractal designs for 3D multifunctional integumentary membranes with capabilities in cardiac electrotherapy. Adv. Mater. **27**, 1731–1737 (2015)
23. S. Jung, J.H. Kim, J. Kim et al., Reverse-micelle-induced porous pressure-sensitive rubber for wearable human-machine interface. Adv. Mater. **26**, 4825–4830 (2014)
24. S.J. Kim, H.R. Cho, K.W. Cho et al., Multifunctional cell-culture platform for aligned cell sheet monitoring, transfer printing, and therapy. ACS Nano **9**, 2677–2688 (2015)
25. M.K. Choi, O.K. Park, C. Choi et al., Cephalopod-inspired miniaturized suction cups for smart medical skin. Adv. Health. Mater. (2015). doi:10.1002/adhm.201500285
26. D.-H. Kim, Y. Lee, Bioelectronics: injection and unfolding. Nat. Nanotechnol. **10**, 570–571 (2015)
27. S. Choi, J. Park, W. Hyun et al., Stretchable heater using ligand-exchanged silver nanowire nanocomposite for wearable articular thermotherapy. ACS Nano **9**, 6626–6633 (2015)
28. D.-H. Kim, R. Ghaffari, N. Lu et al., Electronic sensor and actuator webs for large-area complex geometry cardiac mapping and therapy. Proc. Nat. Acad. Sci. **109**, 19910–19915 (2012)
29. D.-H. Kim, S. Wang, H. Keum et al., Thin, flexible sensors and actuators as 'instrumented' surgical sutures for targeted wound monitoring and therapy. Small **8**, 3263–3268 (2012)
30. M. Ying, A.P. Bonifas, N. Lu et al., Silicon nanomembranes for fingertip electronics. Nanotechnology **23**, 344004 (2012)
31. T.-I. Kim, J.G. McCall, Y.H. Jung et al., Injectable, cellular-scale optoelectronics with applications for wireless optogenetics. Science **240**, 211–216 (2013)
32. T. Sekitani, T. Yokota, U. Zschieschang et al., Organic nonvolatile memory transistors for flexible sensor arrays. Science **326**, 1516–1519 (2009)
33. S.-T. Han, Y. Zhou, Z.-X. Xu et al., Microcontact printing of ultrahigh density gold nanoparticle monolayer for flexible flash memories. Adv. Mater. **24**, 3556–3561 (2012)
34. S.-T. Han, Y. Zhou, C. Wang et al., Layer-by-layer-assembled reduced graphene oxide/gold nanoparticle hybrid double-floating-gate structure for low-voltage flexible flash memory. Adv. Mater. **25**, 872–877 (2013)
35. D. Son, J. Lee, S. Qiao et al., Multifunctional wearable devices for diagnosis and therapy of movement disorders. Nat. Nanotechnol. **9**, 397–404 (2014)
36. D. Son, J.H. Koo, J.-K. Song et al., Stretchable carbon nanotube charge-trap floating-gate memory and logic devices for wearable electronics. ACS Nano **9**, 5585–5593 (2015)
37. Y. Ji, B. Cho, S. Song et al., Stable switching characteristics of organic nonvolatile memory on a bent flexible substrate. Adv. Mater. **22**, 3071–3075 (2010)
38. Y. Ji, D.F. Zeigler, D.S. Lee et al., Flexible and twistable non-volatile memory cell array with all-organic one diode-one resistor architecture. Nat. Commun. **4**, 2707 (2013). doi:10.1038/ncomms3707
39. S. Kim, J.H. Son, S.H. Lee et al., Flexible crossbar-structured resistive memory arrays on plastic substrates via inorganic-based laser lift-off. Adv. Mater. **26**, 7480–7487 (2014)
40. A.A. Bessonov, M.N. Kirikova, D.I. Petukhov et al., Layered memristive and memcapacitive switches for printable electronics. Nat. Mater. **14**, 199–204 (2015)
41. C. Wang, J.-C. Chien, K. Takei et al., Extremely bendable, high-performance integrated circuits using semiconducting carbon nanotube networks for digital, analog, and radio-frequency applications. Nano Lett. **12**, 1527–1533 (2012)
42. D.-M. Sun, M.Y. Timmermans, Y. Tian et al., Flexible high-performance carbon nanotube integrated circuits. Nat. Nanotechnol. **6**, 156–161 (2011)
43. D.-M. Sun, M.Y. Timmermans, A. Kaskela et al., Mouldable all-carbon integrated circuits. Nat. Commun. **4**, 2302 (2013). doi:10.1038/ncomms3302
44. Q. Cao, H.-S. Kim, N. Pimparkar et al., Medium-scale carbon nanotube thin-film integrated circuits on flexible plastic substrates. Nature **454**, 495–500 (2008)
45. D.-H. Kim, J.-H. Ahn, W.-M. Choi et al., Stretchable and foldable silicon integrated circuits. Science **320**, 507–511 (2008)

46. T. Sekitani, U. Zschieschang, H. Klauk et al., Flexible organic transistors and circuits with extreme bending stability. Nat. Mater. **9**, 1015–1022 (2010)
47. T. Sekitani, H. Nakajima, H. Maeda et al., Stretchable active-matrix organic light-emitting diode display using printable elastic conductors. Nat. Mater. **8**, 494–499 (2009)
48. M.S. White, M. Kaltenbrunner, E.D. Gtowacki et al., Ultrathin, highly flexible and stretchable PLEDs. Nat. Photon. **7**, 811–816 (2013)
49. R.-H. Kim, D.-H. Kim, J. Xiao et al., Waterproof AlInGaP optoelectronics on stretchable substrates with applications in biomedicine and robotics. Nat. Mater. **9**, 929–937 (2010)
50. B.H. Kim, M.S. Onses, J.B. Lim et al., High-resolution patterns of quantum dots formed by electrohydrodynamic jet printing for light-emitting diodes. Nano Lett. **15**, 969–973 (2015)
51. M.K. Choi, J. Yang, K. Kang et al., Wearable red-green-blue quantum dot light-emitting diode array using high-resolution intaglio transfer printing. Nat. Commun. **6**, 7149 (2015). doi:10.1038/ncomms8149
52. S. Jung, J. Lee, T. Hyeon et al., Fabric-based integrated energy devices for wearable activity monitors. Adv. Mater. **26**, 6329–6334 (2014)
53. J. Yoon, S. Jo, I.S. Chun et al., GaAs photovoltaics and optoelectronics using releasable multilayer epitaxial assemblies. Nature **465**, 329–333 (2010)
54. S. Xu, Y. Zhang, J. Cho et al., Stretchable batteries with self-similar serpentine interconnects and integrated wireless recharging systems. Nat. Commun. **4**, 1543 (2013). doi:10.1038/ncomms2553
55. Z. Li, G. Zhu, R. Yang et al., Muscle-driven in vivo nanogenerator. Adv. Mater. **22**, 2534–2537 (2010)
56. G. Zhu, R. Yang, S. Wang et al., Flexible high-output nanogenerator based on lateral ZnO nanowire array. Nano Lett. **10**, 3151–3155 (2010)
57. Y. Yang, H. Zhang, Z.-H. Lin et al., Human skin based triboelectric nanogenerators for harvesting biomechanical energy and as self-powered active tactile sensor system. ACS Nano **7**, 9213–9222 (2013)
58. K.S. Kim, Y. Zhao, H. Jang et al., Large-scale pattern growth of graphene films for stretchable transparent electrodes. Nature **457**, 706–710 (2009)
59. S. Bae, H. Kim, Y. Lee et al., Roll-to-roll production of 30-inch graphene films for transparent electrodes. Nat. Nanotechnol. **5**, 574–578 (2010)
60. J.N. Coleman, M. Lotya, A. O'Neill et al., Two-dimensional nanosheets produced by liquid exfoliation of layered materials. Science **311**, 568–571 (2011)
61. C. Pang, G.-Y. Lee, T.-I. Kim et al., A flexible and highly sensitive strain-gauge sensor using reversible interlocking of nanofibers. Nat. Mater. **11**, 795–801 (2012)
62. W. Wu, X. Wen, Z.L. Wang, Taxel-addressable matrix of vertical-nanowire piezotronic transistors for active and adaptive tactile imaging. Science **340**, 952–957 (2013)
63. J. Kim, M. Lee, H.J. Shim et al., Stretchable silicon nanoribbon electronics for skin prosthesis. Nat. Commun. **5**, 5747 (2014). doi:10.1038/ncomms6747
64. D.-H. Kwon, K.M. Kim, J.H. Jang et al., Atomic structure of conducting nanofilaments in TiO2 resistive swiching memory. Nat. Nanotechnol. **5**, 148–153 (2010)
65. S.J. Song, J.Y. Seok, J.H. Yoon et al., Real-time identification of the evolution of conducting nano-filaments in TiO2 thin film ReRAM. Sci. Rep. **3**, 3443 (2013). doi:10.1038/srep03443
66. W.-H. Yeo, Y.-S. Kim, J. Lee et al., Multifunctional epidermal electronics printed directly onto the skin. Adv. Mater. **25**, 2773–2778 (2013)
67. H.W. Lee, Y. Yoon, S. Park et al., Selective dispersion of high purity semiconducting single-walled carbon nanotubes with regioregular poly(3-alkylthiophene)s. Nat. Commun. **2**, 541 (2011). doi:10.1038/ncomms1545
68. M.S. Arnold, A.A. Green, J.F. Hulvat et al., Sorting carbon nanotubes by electronic structure via density differentiation. Nat. Nanotechnol. **1**, 60–65 (2006)
69. S. Park, H. Kim, M. Vosgueritchian et al., Stretchable energy-harvesting tactile electronic skin capable of differentiating multiple mechanical stimuli modes. Adv. Mater. **26**, 7324–7332 (2014)

70. M. Park, K. Do, J. Kim et al., Oxide nanomembrane hybrids with enhanced mechano- and thermos-sensitivity for semitransparent epidermal electronics. Adv. Health. Mater. **4**, 992–997 (2015)

71. S. Lim, D. Son, J. Kim et al., Transparent and stretchable interactive human machine interface based on patterned graphene heterostructures. Adv. Funct. Mater. **25**, 375–383 (2015)

72. C. Zhang, W. Tang, C. Han et al., Theoretical comparison, equivalent transformation, and conjunction operations of electromagnetic induction generator and triboelectric nanogenerator for harvesting mechanical energy. Adv. Mater. **26**, 3580–3591 (2014)

73. J. Zhong, Y. Zhang, Q. Zhong et al., Fiber-based generator for wearable electronics and mobile medication. ACS Nano **8**, 6273–6280 (2014)

74. X. Pu, L. Li, H. Song et al., A self-charging power unit by integration of a textile triboelectric nanogenerator and a flexible lithium-ion battery for wearable electronics. Adv. Mater. **27**, 2472–2478 (2015)

Chapter 9
Sensor Skins: An Overview

Jennifer Case, Michelle Yuen, Mohammed Mohammed
and Rebecca Kramer

Abstract Sensor skins can be broadly defined as distributed sensors over a surface to provide proprioceptive, tactile, and environmental feedback. This chapter focuses on sensors and sensor networks that can achieve strains on the same order as elastomers and human skin, which makes these sensors compatible with emerging wearable technologies. A combination of material choices, processing limitations, and design must be considered in order to achieve multimodal, biocompatible sensor skins capable of operating on objects and bodies with complex geometries and dynamic functionalities. This chapter overviews the commonly used materials, fabrication techniques, structures and designs of stretchable sensor skins, and also highlights the current challenges and future opportunities of such sensors.

Keywords Sensor skins · Wearables · Soft robotics · Stretchable sensors · Pressure sensors · Liquid metals · Sensor fabrication · Flexible materials · Ionic liquids · Conductive ink · Conductive composites · Microchannels

9.1 Introduction

With growing interest in soft systems and wearable technology comes a need to develop deformable sensing components. Traditional sensors and electrical components are often rigid and are best suited to well-defined systems that have discrete motions and confined trajectories. In contrast, soft structures generally have more degrees of freedom than rigid systems. These degrees of freedom come from the deformability of the soft structures themselves. State information of these soft systems may be obtained by populating the surface of the structure with sensor skins, which are stretchable planar structures with embedded sensing components.

J. Case · M. Yuen · M. Mohammed · R. Kramer (✉)
Purdue University, West Lafayette, IN, USA
e-mail: rebeccakramer@purdue.edu

© Springer International Publishing Switzerland 2016
J.A. Rogers et al. (eds.), *Stretchable Bioelectronics for Medical Devices and Systems*, Microsystems and Nanosystems,
DOI 10.1007/978-3-319-28694-5_9

The design of these highly deformable sensory skins has been guided by the flexible and stretchable characteristics of elastomers and human skin.

Let us define what we mean by sensor skins and wearable systems. There are a number of different ways that we can limit our definition of a sensor skin by including stipulations like stretchability, placement on a flexible host, proprioceptive feedback, etc.; however, in this chapter, we will discuss sensor skins that are mechanically compatible with human skin, meaning that they can undergo at least 20 % strain. Wearable systems will be defined here as systems designed to be worn by a human, but we will focus on wearable technology that targets efficient interaction with the host. Other reviews on flexible sensors that do not match this criteria are available for further reading [1–3].

Figure 9.1 shows examples of sensor skins used in different applications. Figure 9.1a (http://ieeexplore.ieee.org/xpls/abs_all.jsp?arnumber=7352295) is a sensor skin module containing three resistance-based strain gauges capable of measuring deformation. Multiple modules can be combined into an array to measure the state of deformation of a host. Figure 9.1b shows an example of a tactile sensor skin used in a surgical environment. This device is composed of an array of pressure sensors to probe the environment without damaging tissues during neuroendoscopy and gives the surgeon another tool to help operate safely on a patient [4]. Two examples of wearable sensor skins can be seen in Fig. 9.1c, d. These sensors can detect pose of the lower limbs (Fig. 9.1c) [5] and of the hand (Fig. 9.1d) [6].

In the following sections, we discuss the materials and processing approaches of substrates and conductors, the structures and designs of elements, the systems used in sensor skins, and conclude with potential future directions of sensor skins.

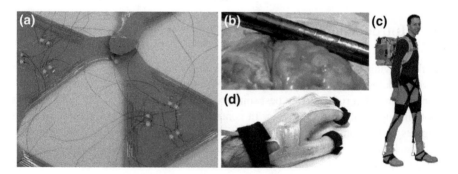

Fig. 9.1 Examples of sensors skins. **a** Sensor skin modules composed of silicone elastomer with three embedded liquid metal sensing strain gauges each. **b** Sensor skin composed of liquid metal pressure sensors for detecting tissue damage [4]. **c** Wearable system with liquid metal strain sensors for detecting pose [5]; and **d** wearable system with ionic liquid strain sensors for detecting hand pose [6]. http://ieeexplore.ieee.org/xpls/abs_all.jsp?arnumber=7352295

9.2   Materials and Processing

Sensor skins are generally composed of at least two types of materials: substrates and conductors. These materials give the skins stretchability as well as the ability to interact in some capacity with itself or a host. This section overviews common substrates and conductors that have been used in sensor skin fabrication and their processing techniques.

9.2.1   Substrate

A sensor skin is, in essence, a substrate onto which or into which sensors are integrated. Substrates can be defined in terms of attributes such as stretchability, breathability, toughness, tear resistance, weight, and compatibility with the host or existing manufacturing techniques. Due to the wide range of substrates available and the complex interactions between the substrate and the host, the substrate should be carefully chosen so that it matches the target properties of the sensor and the fabrication process. Furthermore, in some wearable applications there is a need to use two substrates: the sensor substrate and the garment that is meant to be worn. The integration of two substrates into a single device adds another layer of complexity, since the compatibility between substrate materials must be considered in addition to the substrate–host compatibility.

In the following subsections, we discuss two common types of substrate materials: elastomers and woven fabrics.

9.2.1.1   Elastomers

Elastomers are the most common substrates in soft sensor applications. They are capable of supporting structures and encasing functional elements. In addition, elastomers are compatible with other types of substrates, such as woven fabrics [7–9]. There are many commercially available low-cost elastomers with a wide range of stretchability (from 40 % [10] up to 700 % [11]). Because elastomers are highly stretchable, they are conformable to human skin, which can strain up to 30 %, and thus are less likely to limit natural motion than nonstretchable substrates. However, this high flexibility and stretchability comes at a cost; elastomers are subject to viscoelastic behaviors, such as the Mullins effect [12], creep, and stress relaxation [13, 14].

Commercial elastomers usually are sold as two parts (free chains and crosslinker). Polymerization starts by mixing both parts at a specific ratio, and can be triggered by heating [10] or exposure to UV light [15]. Uncured elastomer can be easily cast in a pre-made mold, which is convenient for creating specific geometries for different part functions and material properties. Researchers use different techniques to fabricate molds such as lithography [16] and 3D printing [17].

It is possible to use methods to shape the elastomer other than replica molding, such as spin coating or other coating methods to make thin films. For example, elastomer films cast onto polyethylene terephthalate (PET) sheets can be integrated into roll-to-roll machines. Elastomers can be both physically and chemically altered through processes such as laser ablation [18] and plasma treatment [19].

Elastomers were originally used to coat and support the structure of conductive solid substrates [20–23]. However, the concept of building microchannels into elastomers [24] allowed researchers to build sensor components by filling elastomeric microchannels with functional materials, such as liquid metals [4, 16, 18, 25–37] or ionic liquids [6, 38, 39].

9.2.1.2 Woven Fabrics

Woven fabric is most recognizable as the material of which our garments are comprised. More generally, woven fabrics are composed of two sets of fibers interlaced together. This construction gives rise to the tensile strength and tear resistance of fabrics. Properties of the fabric such as the elasticity, stiffness, chemical resistance, and thermal properties can be tuned based on the choice of the constituent fibers and the pattern with which they are woven.

The first attempts to integrate sensors and fabrics were by means of simple attachment, such as sewing. This concept was improved upon by the invention of conductive fibers that can be woven the same way as conventional fibers and act as sensory elements without additional components [40]. Woven fabrics using conductive threads and fibers have been employed as strain, pressure, respiratory, heart rate, and electrochemical sensors, as well as gesture-input devices [41–43]. Conductive threads woven with a known spacing can act as capacitive pressure sensors by measuring the change in capacitance between fibers due to thread shifting from applied pressure. Conductive fibers can act as resistive sensors when they are woven as single or multiple threads that gain contact with each other and reduce resistance when strain or pressure is applied to them [43, 44].

Woven skin sensors can achieve higher strains by sewing the components onto pre-wrinkled fabrics. This technique allows the devices to be stretched beyond the stretch limit of the fabric itself [45]. Alternately, conductive coatings can be applied to the same pre-wrinkled construct to form stretchable electronics-compatible fabrics [46, 47].

9.2.2 Conductor

A conductor is a material that allows the flow of electrons or ions through it. In sensor skins, conductors have two major functions: conveying information (i.e. a trace on a circuit board) and collecting information (i.e. a sensor). In many cases, conductors can be used for both purposes. The stability of the conductor and its conformal contact with the interface are essential to ensure efficient sensor performance. In wearable applications, it is also important to consider the biocompatibility (i.e. toxicity) of a sensor.

The conductors that we highlight in this section are thin metal films, liquid metals, ionic liquids, conductive polymers, and conductive inks.

9.2.2.1 Thin Metal Films

Conductive materials are, in most cases, rigid. They show flexible behavior when they are shaped as thin films. The first attempt to use thin metal films in flexible electronics was in 1967 to produce the first flexible solar cell [48, 49]. Advanced electronics typically consist of insulators, conductors, and semiconductors, and it is possible to use processes like roll-to-roll to produce electronics at larger scales [50]. Metallic thin films have been fabricated to accommodate moderate strains using clever geometries (waves and nets) made from both highly conductive metals (such as gold, silver, copper, and aluminum) and semiconductors (such as silicon) [51]. Thin-film electronics enable other applications such as displays [9, 52, 53], electrodes [54, 55], LEDs [56] and wearable electronics [57–59].

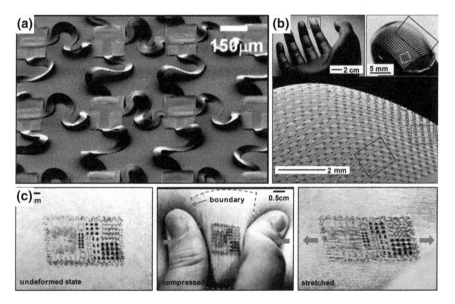

Fig. 9.2 **a** Extremely stretchable metal films with serpentine-design bridges [51]. **b** Stretchable circuit with noncoplanar metal films on a model finger tip [63]. **c** A multifunctional epidermal electronic circuit mounted on human skin in undeformed, compressed, and stretched states [69]

The devices fabricated using thin metal films are flexible. However, they lack stretchability due to the very low fracture strain of most metals. Researchers have used geometry to cause this nonstretchable material to behave elastically. For instance, curved and wavy metal films can undergo strains that are impossible to achieve using flat films [21–23, 60–64]. Figure 9.2a, b shows two examples of thin film structures. This approach enables stretchable interconnects [65, 66], integrated circuits [67], batteries [68] and epidermal electronics (Fig. 9.2c) [69, 70]. With proper engineering design, the devices fabricated using this technique can be strained up to 200 % [71].

Devices fabricated using metal films or wires deform plastically with prolonged use leading to permanent deformation that reduces the device efficiency or totally disconnects the circuit [72–75]. Other conductors, such as liquid metals and ionic liquids, have been recently used as alternatives that do not suffer from this limitation.

9.2.2.2 Liquid Metal

There is a growing interest in using liquid metals in flexible electronics as alternatives to conventional metals. The most famous liquid metals are eutectic alloys such as eutectic gallium-indium (eGaIn) and eutectic gallium-indium-tin (Galinstan). They have high metallic conductivity ($\approx 3.4 \times 10^6 \, \Omega^{-1} \cdot m^{-1}$ for EGaIn and $\approx 3.1 \times 10^6 \, \Omega^{-1} \cdot m^{-1}$ for Galinstan [76]) and form a thin gallium-oxide skin that allows them to form stable nonspherical structures despite their high surface tension [77, 78]. In contrast to mercury, gallium-indium alloys are non-toxic and therefore have potential applications in biocompatible sensors. Gallium-indium alloys can be injected into microchannels due to their low viscosity [16, 25, 26, 32, 37, 79], and their liquid nature allows them to take the shape of the microchannel even at very high strains (up to 700 %) without failure [11].

Researchers have developed fabrication techniques that are more automated than manual injection [80]. For instance, a microtip wet with the liquid metal can transfer patterns on a substrate by direct contact [81]. The tip can be replaced by a syringe needle that continuously extrudes the liquid metal and directly writes onto a surface [28, 78]. Liquid metal can selectively wet parts of a substrate using a predesigned mask [82, 83], or by treating the substrate surfaces such that liquid metal self-assembles into the desired areas [33, 84–86]. Liquid metals are not suitable for inkjet printing due to the high surface tension, its corrosive nature to most other metals and the presence of surface oxide. However, a dispersion of liquid metal nanoparticles in a volatile solvent can be inkjet printed since the dispersion properties are dictated by the carrier solvent rather than the liquid metal [87]. Finally, liquid metal films maybe subtractively patterned by selective laser ablation [88].

The various patterning techniques have enabled the fabrication of different types of liquid metal-based sensors, such as capacitive pressure sensors [35–37], resistive strain sensors [18, 28, 33, 87], resistive pressure sensors [4, 26, 27, 31, 32, 79, 89], curvature sensors [16, 25] and shear sensors [89]. Examples of other liquid metal-based devices are antennas [90–92], soft wires [6, 39], self-healing wires [93], diodes [94], and capacitors [81].

9.2.2.3 Ionic Liquids and Solutions

Ionic liquids and salt solutions can also be used in sensor skins. Ionic liquids are molten salts while salt solutions are salts dissolved in a solvent, typically water. These solutions are capable of reflowing and are typically used within an elastomer substrate, where preformed microchannels are filled with the conductive solutions by injection [6, 29, 39] or vacuum [95].

The most familiar form of a salt solution is a sodium chloride (NaCl) solution, which has been used to make sensors [95] and sensor arrays [96]. Researchers have also demonstrated the use of potassium chloride solution (KCl), sodium hydroxide (NaOH), and hydrogen chloride (HCl) in their sensors, but found that both NaOH and HCl were corrosive to the sensor interface and that NaCl had a better range than KCl [95]. An example of an ionic liquid is 1-ethyl-3-methylimidazolium ethyl sulfate [6, 29, 38]. Ionic liquids and salt solutions have been used in fabricating tactile sensors [29, 95, 96], strain sensors [39], curvature sensors [38], diodes [94] and wearable devices [6].

The major drawback of using salt solutions is that popular substrates, like elastomers, are gas permeable, which means that water will slowly evaporate and leave a salt residue in the substrate. Water evaporation can be slowed down by adding glycerol to the solution [39, 97].

9.2.2.4 Conductive Inks

A conductive ink is a solvent that contains a suspension of conductive particles, such as metallic nanoparticles, organometallic compounds, carbon nanotubes and graphene [98–100]. Volatile organic solvents (ethanol, toluene, etc.) are commonly used and leave behind the conductive particles on the substrate as they evaporate. Additives are often used to keep particles suspended, increase adhesion onto the surface, or reduce surface tension. In some cases, a means of coalescing or sintering is necessary to bridge gaps between nanoparticles and ensure conductivity [87, 101, 102].

The flow properties of the ink are dictated by the properties of the carrier solvent, therefore conductive inks have a lower viscosity relative to many of the previously described liquid conductors. Hence, conductive inks are compatible with inkjet printing [87, 103–105], screen-printing (polymer thick film) [99, 106] and direct-writing [101, 107, 108]. Researchers invented conductive silver inks that can be directly written on different surfaces using rollerball pens [109, 110]. As previously mentioned, liquid metal dispersion inks have also been developed, which bridge the gap between liquid metal conductors and conductive inks using liquid metal nanoparticles suspended in a carried solvent [87].

9.2.2.5 Conductive Polymer Composites

Conductive polymer composites generally consist of a polymer mixed with a conductive material that is packed tightly enough in the polymer to maintain conductivity. Example conductive materials include silver nanoparticles [111–113], graphite [112, 114], graphene [115, 116], carbon black [114, 117], carbon nanotubes [114, 116, 118, 119] and liquid metals [120]. Different polymers can be used, such as polydimethylsiloxane [114, 116, 119, 120], polyisoprene [117], polyvinylidene fluoride [112], rubber fibers [113], and a number of other polymers [115]. Conductive polymers have been used to create tactile sensors [114, 117–119] and strain sensors [47, 117, 118, 120].

It is possible to pattern the conductive polymer using different techniques such as extrusion [119], screen printing [111], spray deposition [118] and hot-rolling [112]. The flow properties of the polymer composite affect the patterning process; therefore, it is common to add thinners to the composite in order to reduce its viscosity. Examples of thinners are reverse micelle solution, which also controls the hardness of the final conductive polymer [119], and cyclohexane [114]. Researchers have developed a novel composite material by embedding liquid metal nanoparticles in elastomer. Initially, the composite is not conductive due to the absence of a conductive path between the nanoparticles; however, applying local pressure on the composite breaks the boundaries between the nanoparticles and creates a conductive path within the composite [120, 121].

9.3 Structures and Designs

It is important to choose the proper materials to fabricate the sensor to ensure stability of the structure and compatibility with the host under representative operating conditions. Therefore, the designer should be aware of different structures and design approaches in order to fabricate a properly functioning device. This section highlights the common features and systems of sensor skins.

9.3.1 Features

There are several features of sensor skins that affect how sensor elements function. These features include microchannels containing liquid conductors (which we discussed in Sect. 9.2.2) and interfaces within the sensor skins. Examples of the latter include interfaces between two different conductors, between substrates, between the sensor skin and external electronics, and, in the case of wearables, between the sensor skin and the human.

9.3.1.1 Microchannels

Microchannels are defined as flow passages with dimensions on the order of tens to hundreds of microns [24, 122]. In sensor skins, these microchannels can be used for sensing or as communication pathways. Chossat et al. demonstrated both of these uses in a wearable glove, where microchannels filled with an ionic liquid serve as the sensing component and microchannels filled with liquid metal serve as a communication pathway [6].

Elastomeric substrates deform under the influence of pressure or strain, therefore changing the dimensions of the embedded microchannels filled with the conductive liquid. This is important for resistive sensors, where the resistance of the channel is guided by $R = \rho L/A$, where R is the resistance, ρ is the resistivity, L is the length and A is the cross-sectional area of the sensor. This is the operational concept behind the variety of liquid-embedded sensors and devices described in Sects. 9.2.2.2 and 9.2.2.3.

Replica molding, subtractive and additive manufacturing are the common techniques used to manufacture microchannels in elastomeric matrices. Molds can be made via 3D printing [26, 27, 89], photolithography [123], patterning films [16] or laser engraving [39]. Microchannels can also be made by subtractively removing material via laser ablation [18] or by adding material via direct printing [28, 119]. The choice of the mold fabrication technique depends on the required resolution of the mold, available time and equipment. For instance, fabricating molds using photolithography produces small features with high resolution, but it is a time-consuming process [124, 125]. 3D printing is a fully automated process; however, the feature sizes of the mold are limited by the nozzle size of the printer. Laser ablation is also capable of achieving small feature sizes, but has less control over channel geometry than 3D printing [126].

9.3.1.2 Interfaces

At physical interfaces within devices, the change in stiffness from a highly deformable substrate to a rigid component or interconnect is a common cause of device failure. As the device is flexed or stretched, rigid parts are unable to follow the change in conformation of more flexible parts. For example, in Fig. 9.1a, while the silicone elastomer substrate is capable of withstanding strains up to 150 %, the electrical interface between the liquid metal and the copper wire limits the usable strain to 50 %. Beyond this limit, the copper wires lose contact with the liquid metal and fail to deliver strain data out of the sensor. This often also results in permanent failure of the device, as the wires contacting the liquid metal in this device will pull out of the channels, thus breaking the path from the sensing element to the rest of the system.

Researchers have addressed this deficiency in interfacing with liquid metal-based sensors using a variety of methods. For instance, copper wires can be replaced by ionic liquids to detect strain signals, which are transferred to external control circuit using liquid metal wires [6, 29]. Other approaches include using stretchable interconnects [127, 128], stretchable wires [11] and stretchable metal films [129, 130]. Uniaxially conductive polymer composites serve as a signal transmitter between the liquid metal circuit and the skin [131].

In addition to interfaces within the device, the interface between the human and the device greatly affects the device performance efficiency. Though elastomers are biocompatible and useful for wearable electronics, in reality, elastomer devices are difficult to secure onto the skin. Researchers have used skin adhesives to adhere electronic devices to the skin [61, 132]. With regard to garments, sensors have been held in place around joints via straps to create sensory suits [5]. Researchers ensured that the interface between the sensor and the strap was robust during motion by creating a stiffness gradient that transitions from the relatively stiff strap to the much softer sensor.

9.3.2   Systems

Moving beyond our discussion of the individual components and fabrication techniques for soft sensor skins, we can start looking at examples and applications. In this section, we focus on sensor skins for robotics and wearables as they fit with the scope of this chapter.

9.3.2.1   Sensor Skins for Robotics

Most traditional robotic systems have very fine tuned position control and can operate very quickly and efficiently. However, they generally lack any knowledge of their environment, which poses a potential safety risk for robots working alongside humans. Applying sensor skins to robots would provide environmental information and increase their awareness with their surroundings [133]. Tactile sensors tend to either use capacitive [30, 35–37, 111, 118, 134, 135] or resistive [4, 26, 31, 32, 38, 79, 117, 136] means of measuring pressure or normal force on the surface. Work has also been done to sense shear forces as well as normal forces on a surface [89]. While a lot of these works are on single sensors, these sensors can be arrayed into a sensor skin.

Soft robots need proprioceptive feedback through soft sensors that are mechanically compatible with the bulk of their structure. This proprioceptive feedback can come from strain sensors [6, 18, 26, 28, 34, 39, 47, 87, 97, 117, 120] or curvature sensors [16, 25, 137]. Yuen et al. demonstrated a robotic fabric skin which included a strain sensor that could differentiate between bending and compressing motion [138]. Resistive strain and pressure sensors have also been combined with existing pneumatic actuators to provide data about current state [139, 140]. A modular capacitive sensor skin has been developed to provide tactile information to existing robots [141].

9.3.2.2 Wearables

Many wearable sensor skin applications are designed for proprioception on conformal interfaces. Proprioceptive devices are used to estimate the state or pose of part of the human body. This can be applied on a smaller scale to measure the state of various joints on the fingers [16, 25] and across the entire hand and wrist [6, 142]. On a larger scale, exosuits have been developed to determine the pose of the lower body [5, 8, 143]. These devices are all composed primarily of elastomers with embedded liquid conductors injected into molded microchannels within the elastomer [144, 145]. Alternate designs rely upon direct adhesion of the strain sensor to the skin [61, 146]. These devices leverage thin film mechanics to measure strain due to skin stretch during joint flexion. The principles used to gather proprioceptive information can also be used to develop user interface devices, such as a wearable keypad [32, 119, 147].

9.4 Frontier and Outlook

The previous sections covered much of the published research in the field of stretchable skin sensors. Here, we discuss a few examples of the ongoing research and look at future opportunities.

Current research focuses on using new materials and novel fabrication strategies to develop sensors that can do multiple sensing tasks, have higher sensitivity and better mechanical properties, in addition to having a long lifetime. There is a need to develop methods to integrate these devices in a larger soft-bodied system, or to design the sensor as a built-in part of it. Such improvements will have a tremendous influence on the future sensor skins applications. An example of an integrated system is the exosuit that we discussed in Sect. 9.3.2.2. In the current state, exosuits require large power supplies or power cables, which is a significant drawback that needs to be addressed in order to make these devices practical.

Biocompatible soft sensors have significant potential in surgical robots. Integrating soft sensors in surgical tools will not only allow sensing of the force exerted on the surface but also determine the type of tissue onto which the force is being applied. Furthermore, the developing field of soft robotics holds the promise of creating new soft surgical tools that are mechanically compatible with soft tissue.

9.5 Conclusion

This chapter gives an overview of the current research on sensor skins. Sensor skins are sensor-embedded substrates that have the ability to flex, bend or stretch. Sensor skins are used to estimate large-deformation motions and changes in system states. There is a growing interest in applying sensor skins to human skin and tissues, since most of the materials used are biocompatible.

Sensor skins are generally made of two components: a conductive material which is the sensing and/or signal transmitting element, and a stretchable encasing substrate. The wide variety of materials that have been used in skin sensors provide a diverse foundation for researchers to develop new devices, fabrication techniques and designs. Within the current state-of-the-art, we are able to control the shape and dimensions of the devices and therefore the resulting sensing and mechanical properties. However, it is important to ensure both compatibility and stability of the sensor skin with the target host of the device in order to meet performance goals. Much of the current research in this field is aimed at integrating multiple sensing elements together into complex sensor skins. This poses new challenges in signal processing and networks that do not exist at the single element level. Major challenges that researchers are working to overcome include a lack of highly scalable manufacturing techniques for soft materials and integration of miniaturized electronics. Together, solving these challenges will significantly improve the utility of sensor skins outside of a laboratory environment.

References

1. C.M.A. Ashruf, Thin flexible pressure sensors. Sens. Rev. **22**(4), 322–327 (2002)
2. C. Pang, C. Lee, K.Y. Suh, Recent advances in flexible sensors for wearable and implantable devices. J. Appl. Polym. Sci. **130**(3), 1429–1441 (2013)
3. S. Khan, L. Lorenzelli, R.S. Dahiya, Technologies for printing sensors and electronics over large flexible substrates: a review. IEEE Sens. J. **15**(6), 3164–3185 (2015)
4. Patrick J. Codd, Arabagi Veaceslav, Andrew H. Gosline, Pierre E. Dupont, Novel pressure-sensing skin for detecting impending tissue damage during neuroendoscopy. J. Neurosurg.: Pediatr. **13**(1), 114–121 (2013)
5. A.T. Asbeck, S.M.M. De Rossi, K.G. Holt, C.J. Walsh, A biologically inspired soft exosuit for walking assistance. Int. J. Robot. Res. 0278364914562476 (2015)
6. J.-B. Chossat, Y. Tao, V. Duchaine, Y.L. Park, Wearable soft artificial skin for hand motion detection with embedded microfluidic strain sensing, in *2015 IEEE International Conference on Robotics and Automation (ICRA)*, pp. 2568–2573, May 2015
7. K.C. Galloway, P. Polygerinos, C.J. Walsh, R.J. Wood, Mechanically programmable bend radius for fiber-reinforced soft actuators, in *2013 16th International Conference on Advanced Robotics (ICAR)*, pp. 1–6, Nov 2013
8. M. Wehner, B. Quinlivan, P.M. Aubin, E. Martinez-Villalpando, M. Baumann, L. Stirling, K. Holt, R. Wood, C. Walsh, A lightweight soft exosuit for gait assistance, in *2013 IEEE International Conference on Robotics and Automation (ICRA)*, pp. 3362–3369, May 2013

9. D.H. Kim, Y.S. Kim, J. Wu, Z. Liu, J. Song, H.S. Kim, Y.Y. Huang, K.C. Hwang, J.A. Rogers, Ultrathin silicon circuits with strain-isolation layers and mesh layouts for high-performance electronics on fabric, vinyl, leather, and paper. Adv. Mater. **21**(36), 3703–3707 (2009)
10. J.C. McDonald, G.M. Whitesides, Poly(dimethylsiloxane) as a material for fabricating microfluidic devices. Acc. Chem. Res. **35**(7), 491–499 (2002)
11. S. Zhu, J.-H. So, R.L. Mays, S. Desai, W.R. Barnes, B. Pourdeyhimi, M.D. Dickey, Ultrastretchable fibers with metallic conductivity using a liquid metal alloy core. Adv. Fun. Mat. **32**(18), 2308–2314 (2013)
12. L. Mullins, Effect of stretching on the properties of rubber. Rubber Chem. Technol. **21**(2), 281–300 (1948)
13. W.N. Findley, F.A. Davis, *Creep and Relaxation of Nonlinear Viscoelastic Materials*. Courier Corporation (2013)
14. N.G. McCrum, C.P. Buckley, C.B. Bucknall, *Principles of Polymer Engineering*. Oxford University Press (1997)
15. A. Bratov, J. Muñoz, C. Dominguez, J. Bartroli, Photocurable polymers applied as encapsulating materials for ISFET production. Sens. Actuators, B: Chem. **25**(13), 823–825 (1995)
16. R.K. Kramer, C. Majidi, R. Sahai, R.J. Wood. Soft curvature sensors for joint angle proprioception, in *2011 IEEE/RSJ International Conference on Intelligent Robots and Systems (IROS)*, pp. 1919–1926, 2011
17. R.F. Shepherd, F. Ilievski, W. Choi, S.A. Morin, A.A. Stokes, A.D. Mazzeo, X. Chen, M. Wang, G.M. Whitesides, Multigait soft robot. Proc. Natl. Acad. Sci. **108**(51), 20400–20403 (2011)
18. J.C. Case, E.L. White, R.K. Kramer, Soft material characterization for robotic applications. Soft Robot. **2**(2), 80–87 (2015)
19. M.A. Eddings, M.A. Johnson, B.K. Gale, Determining the optimal PDMSPDMS bonding technique for microfluidic devices. J. Micromech. Microeng. **18**(6), 067001 (2008)
20. D.H. Kim, Z. Liu, Y.S. Kim, J. Wu, J. Song, H.S. Kim, Y. Huang, K.C. Hwang, Y. Zhang, J. A. Rogers, Optimized structural designs for stretchable silicon integrated circuits. Small **5**(24), 2841–2847 (2009)
21. D.Y. Khang, H. Jiang, Y. Huang, J.A. Rogers, A stretchable form of single-crystal silicon for high-performance electronics on rubber substrates. Science **311**(5758), 208–212 (2006)
22. D.H. Kim, J.A. Rogers, Stretchable electronics: materials strategies and devices. Adv. Mater. **20**(24), 4887–4892 (2008)
23. J.A. Fan, W.H. Yeo, Y. Su, Y. Hattori, W. Lee, S.Y. Jung, Y. Zhang, Z. Liu, H. Cheng, L. Falgout, M. Bajema, T. Coleman, D. Gregoire, R.J. Larsen, Y. Huang, J.A. Rogers, Fractal design concepts for stretchable electronics. Nat. Commun. **5** (2014)
24. G.M. Whitesides, The origins and the future of microfluidics. Nature **442**(7101), 368–373 (2006)
25. C. Majidi, R. Kramer, R.J. Wood, A non-differential elastomer curvature sensor for softer-than-skin electronics. Smart Mater. Struct. **20**(10), 105017 (2011)
26. Y.L. Park, B.R. Chen, R.J. Wood, Design and fabrication of soft artificial skin using embedded microchannels and liquid conductors. IEEE Sens. J. **12**(8), 2711–2718 (2012)
27. A. Anderson, Y. Menguc, R.J. Wood, D. Newman, Development of the polipo pressure sensing system for dynamic space-suited motion. IEEE Sens. J. **15**(11), 6229–6237 (2015)
28. J.W. Boley, E.L. White, G.T.-C. Chiu, R.K. Kramer, Direct writing of gallium-indium alloy for stretchable electronics. Adv. Funct. Mater. **24**(23), 3501–3507 (2014)
29. J.B. Chossat, H.S. Shin, Y.L. Park, V. Duchaine, Soft tactile skin using an embedded ionic liquid and tomographic imaging. J. Mech. Rob. **7**(2), 021008 (2015)
30. A.P. Gerratt, H.O. Michaud, S.P. Lacour, Elastomeric electronic skin for prosthetic tactile sensation. Adv. Funct. Mater. **25**(15), 2287–2295 (2015)

31. F.L. Hammond, R.K. Kramer, Q. Wan, R.D. Howe, R.J. Wood, Soft tactile sensor arrays for micromanipulation, in *2012 IEEE/RSJ International Conference on Intelligent Robots and Systems (IROS)*, pp. 25–32, Oct 2012

32. R.K. Kramer, C.Majidi, R.J. Wood, Wearable tactile keypad with stretchable artificial skin, in *2011 IEEE International Conference on Robotics and Automation (ICRA)*, pp. 1103–1107 (2011)

33. R. Matsuzaki, K. Tabayashi, Highly stretchable, global, and distributed local strain sensing line using GaInSn electrodes for wearable electronics. Adv. Funct. Mater. **25**(25), 3806–3813 (2015)

34. J.T.B. Overvelde, Y. Mengüç, P. Polygerinos, Y. Wang, Z. Wang, C.J. Walsh, R.J. Wood, K. Bertoldi, Mechanical and electrical numerical analysis of soft liquid-embedded deformation sensors analysis. Extreme Mech. Lett. **1**, 42–46 (2014)

35. J. Choi, S. Kim, J. Lee, B. Choi, Improved capacitive pressure sensors based on liquid alloy and silicone elastomer. IEEE Sens. J. **15**(8), 4180–4181 (2015)

36. S. Baek, D.J. Won, J.G. Kim, J. Kim, Development and analysis of a capacitive touch sensor using a liquid metal droplet. J. Micromech. Microeng. **25**(9), 095015 (2015)

37. D. Ruben, P. Wong, J.D. Posner, V.J. Santos, Flexible microfluidic normal force sensor skin for tactile feedback. Sens. Actuators, A **179**, 62–69 (2012)

38. K. Noda, E. Iwase, K. Matsumoto, I. Shimoyama, Stretchable liquid tactile sensor for robot-joints, in *2010 IEEE International Conference on Robotics and Automation (ICRA)*, pp. 4212–4217, May 2010

39. J.-B. Chossat, Y.-L. Park, R.J. Wood, V. Duchaine, A soft strain sensor based on ionic and metal liquids. IEEE Sens. J. **13**(9), 3405–3414 (2013)

40. C.R. Merritt, H.T. Nagle, E. Grant, Textile-based capacitive sensors for respiration monitoring. IEEE Sens. J. **9**(1), 71–78 (2009)

41. M. Stoppa, A. Chiolerio, Wearable electronics and smart textiles: a critical review. Sensors **14**(7), 11957–11992 (2014)

42. C. Mattmann, F. Clemens, G. Tröster, Sensor for measuring strain in textile. Sensors **8**(6), 3719–3732 (2008)

43. L.M. Castano, A.B. Flatau, Smart fabric sensors and e-textile technologies: a review. Smart Mater. Struct. **23**(5), 053001 (2014)

44. L. Capineri, Resistive sensors with smart textiles for wearable technology: from fabrication processes to integration with electronics. Procedia Eng. **87**, 724–727 (2014)

45. R. Xu, K.I. Jang, Y. Ma, H.N. Jung, Y. Yang, M. Cho, Y. Zhang, Y. Huang, J.A. Rogers, Fabric-based stretchable electronics with mechanically optimized designs and prestrained composite substrates. Extreme Mech. Lett. (2014)

46. L. Hu, M. Pasta, F.L. Mantia, L.F. Cui, S. Jeong, H.D. Deshazer, J.W. Choi, S.M. Han, Y. Cui, Stretchable, porous, and conductive energy textiles. Nano Lett. **10**(2), 708–714 (2010)

47. C. Cochrane, V. Koncar, M. Lewandowski, C. Dufour, Design and development of a flexible strain sensor for textile structures based on a conductive polymer composite. Sensors **7**(4), 473–492 (2007)

48. R.L. Crabb, F.C. Treble, Thin silicon solar cells for large flexible arrays. Nature **213**(5082), 1223–1224 (1967)

49. K.A. Ray, Flexible solar cell arrays for increased space power. IEEE Trans. Aerosp. Electron. Syst. **AES-3**(1), 107–115 (1967)

50. K. Jain, M. Klosner, M. Zemel, S. Raghunandan, Flexible electronics and displays: high-resolution, roll-to-roll, projection lithography and photoablation processing technologies for high-throughput production. Proc. IEEE **93**(8), 1500–1510 (2005)

51. D.H. Kim, J. Song, W.M. Choi, H.S. Kim, R.H. Kim, Z. Liu, Z. Liu, Y.Y. Huang, K.C. Hwang, Y.W. Zhang, J.A. Rogers, Materials and noncoplanar mesh designs for integrated circuits with linear elastic responses to extreme mechanical deformations. PNAS **105**(48), 18675–18680 (2008)

52. G.H. Gelinck, H.E.A. Huitema, E. van Veenendaal, E. Cantatore, L. Schrijnemakers, J.B.P. H. van der Putten, T.C.T. Geuns, M. Beenhakkers, J.B. Giesbers, B.H. Huisman, E.J. Meijer, E.M. Benito, F.J. Touwslager, A.W. Marsman, B.J. E. van Rens, D.M. de Leeuw, Flexible active-matrix displays and shift registers based on solution-processed organic transistors. Nat. Mater. **3**(2), 106–110 (2004)
53. J.A. Rogers, Z. Bao, K. Baldwin, A. Dodabalapur, B. Crone, V.R. Raju, V. Kuck, H. Katz, K. Amundson, J. Ewing, P. Drzaic, Paper-like electronic displays: large-area rubber-stamped plastic sheets of electronics and microencapsulated electrophoretic inks. Proc. Nat. Acad. Sci. U.S.A. **98**(9), 4835–4840 (2001) (ArticleType: research-article/Full publication date: Apr. 24, 2001/Copyright 2001 National Academy of Sciences)
54. C. Wang, G.G. Wallace, Flexible electrodes and electrolytes for energy storage. Electrochimica Acta (2015)
55. S.D. Perera, B. Patel, N. Nijem, K. Roodenko, O. Seitz, J.P. Ferraris, Y.J. Chabal, K. J. Balkus, Vanadium oxide nanowire carbon nanotube binder-free flexible electrodes for supercapacitors. Adv. Energy Mater. **1**(5), 936–945 (2011)
56. S.I. Park, Y. Xiong, R.H. Kim, P. Elvikis, M. Meitl, D.H. Kim, J. Wu, J. Yoon, C.J. Yu, Z. Liu, Y. Huang, K.C. Hwang, P. Ferreira, X. Li, K. Choquette, J.A. Rogers, Printed assemblies of inorganic light-emitting diodes for deformable and semitransparent displays. Science **325**(5943), 977–981 (2009)
57. P. Salonen, M. Keskilammi, J. Rantanen, L. Sydanheimo, A novel Bluetooth antenna on flexible substrate for smart clothing, in *2001 IEEE International Conference on Systems, Man, and Cybernetics*, vol. 2, pp. 789–794, 2001
58. C. Cibin, P. Leuchtmann, M. Gimersky, R. Vahldieck, S. Moscibroda, A flexible wearable antenna, in *IEEE Antennas and Propagation Society International Symposium, 2004*, vol. 4, pp. 3589–3592, June 2004
59. J.C.G. Matthews, G. Pettitt, Development of flexible, wearable antennas, in *3rd European Conference on Antennas and Propagation, 2009. EuCAP 2009*, pp. 273–277, March 2009
60. A.J. Baca, J.H. Ahn, Y. Sun, M.A. Meitl, E. Menard, H.S. Kim, W.M. Choi, D.H. Kim, Y. Huang, J.A. Rogers, Semiconductor wires and ribbons for high-performance flexible electronics. Angew. Chem. Int. Ed. **47**(30), 5524–5542 (2008)
61. J.A. Rogers, T. Someya, Y. Huang, Materials and mechanics for stretchable electronics. Science **327**(5973), 1603–1607 (2010)
62. H.C. Ko, G. Shin, S. Wang, M.P. Stoykovich, J.W. Lee, D.H. Kim, J.S. Ha, Y. Huang, K.C. Hwang, J.A. Rogers, Curvilinear electronics formed using silicon membrane circuits and elastomeric transfer elements. Small **5**(23), 2703–2709 (2009)
63. D.H. Kim, J. Xiao, J. Song, Y. Huang, J.A. Rogers, Stretchable, curvilinear electronics based on inorganic materials. Adv. Mater. **22**(19), 2108–2124 (2010)
64. D.H. Kim, N. Lu, Y. Huang, J.A. Rogers, Materials for stretchable electronics in bioinspired and biointegrated devices. MRS Bull. **37**(03), 226–235 (2012)
65. P.J. Hung, K. Jeong, G.L. Liu, L.P. Lee, Microfabricated suspensions for electrical connections on the tunable elastomer membrane. Appl. Phys. Lett. **85**(24), 6051–6053 (2004)
66. S.P. Lacour, J. Jones, S. Wagner, T. Li, Z. Suo. Stretchable interconnects for elastic electronic surfaces. Proc. IEEE **93**(8), 1459–1467 (2005)
67. D.H. Kim, J.H. Ahn, W.M. Choi, H.S. Kim, T.H. Kim, J. Song, Y.Y. Huang, Z. Liu, C. Lu, J.A. Rogers, Stretchable and foldable silicon integrated circuits. Science **320**(5875), 507–511 (2008)
68. S. Xu, Y. Zhang, J. Cho, J. Lee, X. Huang, L. Jia, J.A. Fan, Y. Su, J. Su, H. Zhang, H. Cheng, B. Lu, C. Yu, C. Chuang, T. Kim, T. Song, K. Shigeta, S. Kang, C. Dagdeviren, I. Petrov, P.V. Braun, Y. Huang, U. Paik, J.A. Rogers, Stretchable batteries with self-similar serpentine interconnects and integrated wireless recharging systems. Nat. Commun. **4**, 1543 (2013)

69. D.H. Kim, N. Lu, R. Ma, Y.S. Kim, R.H. Kim, S. Wang, J. Wu, S.M. Won, H. Tao, A. Islam, K.J. Yu, T. Kim, R. Chowdhury, M. Ying, L. Xu, M. Li, H.-J. Chung, H. Keum, M. McCormick, P. Liu, Y.-W. Zhang, F.G. Omenetto, Y. Huang, T. Coleman, J.A. Rogers. Epidermal electronics. Science **333**(6044), 838–843 (2011)
70. J. Kim, A. Banks, H. Cheng, Z. Xie, S. Xu, K.I. Jang, J.W. Lee, Z. Liu, P. Gutruf, X. Huang, P. Wei, F. Liu, K. Li, M. Dalal, R. Ghaffari, X. Feng, Y. Huang, S. Gupta, U. Paik, J.A. Rogers, Epidermal electronics with advanced capabilities in near-field communication. Small **11**(8), 906–912 (2015)
71. X. Hu, P. Krull, de B. Graff, K. Dowling, J.A. Rogers, W.J. Arora, Stretchable inorganic-semiconductor electronic systems. Adv. Mater. **23**(26), 2933–2936 (2011)
72. D.S. Gray, J. Tien, C.S. Chen, High-conductivity elastomeric electronics. Adv. Mater. **16**(5), 393–397 (2004)
73. Y.Y. Hsu, B. Dimcic, M. Gonzalez, F. Bossuyt, J. Vanfleteren, de I. Wolf, Reliability assessment of stretchable interconnects, in *2010 5th International Microsystems Packaging Assembly and Circuits Technology Conference (IMPACT)*, pp. 1–4, Oct 2010
74. F. Bossuyt, J. Guenther, T. Lher, M. Seckel, T. Sterken, J. de Vries, Cyclic endurance reliability of stretchable electronic substrates. Microelectron. Reliab. **51**(3), 628–635 (2011)
75. S.P. Lacour, D. Chan, S. Wagner, T. Li, Z. Suo, Mechanisms of reversible stretchability of thin metal films on elastomeric substrates. Appl. Phys. Lett. **88**(20), 204103–204103-3 (2006)
76. N.B. Morley, J. Burris, L.C. Cadwallader, M.D. Nornberg, GaInSn usage in the research laboratory. Rev. Sci. Instrum. **79**(5), 056107 (2008)
77. M.D. Dickey, R.C. Chiechi, R.J. Larsen, E.A. Weiss, D.A. Weitz, G.M. Whitesides, Eutectic gallium-indium (EGaIn): a liquid metal alloy for the formation of stable structures in microchannels at room temperature. Adv. Funct. Mater. **18**(7), 1097–1104 (2008)
78. C. Ladd, J.H. So, J. Muth, M.D. Dickey, 3d Printing of free standing liquid metal microstructures. Adv. Mater. **25**(36), 5081–5085 (2013)
79. Y.L. Park, C. Majidi, R. Kramer, P. Brard, R.J. Wood, Hyperelastic pressure sensing with a liquid-embedded elastomer. J. Micromech. Microeng. **20**(12), 125029 (2010)
80. M.D. Dickey, Emerging applications of liquid metals featuring surface oxides. ACS Appl. Mater. Interfaces (2014)
81. A. Tabatabai, A. Fassler, C. Usiak, C. Majidi, Liquid-phase gallium indium alloy electronics with microcontact printing. Langmuir **29**(20), 6194–6200 (2013)
82. J. Wissman, T. Lu, C. Majidi, Soft-matter electronics with stencil lithography, in *2013 IEEE Sensors*, pp. 1–4, 2013
83. Q. Zhang, Y. Gao, J. Liu, Atomized spraying of liquid metal droplets on desired substrate surfaces as a generalized way for ubiquitous printed electronics. Appl. Phys. A 1–7 (2013)
84. D. Kim, D.W. Lee, W. Choi, Jeong-Bong Lee, A super-lyophobic 3-D PDMS channel as a novel microfluidic platform to manipulate oxidized galinstan. J. Microelectromech. Syst. **22**(6), 1267–1275 (2013)
85. R.K. Kramer, J. William Boley, H.A. Stone, J.C. Weaver, R.J. Wood, Effect of microtextured surface topography on the wetting behavior of eutectic gallium indium alloys. *Langmuir* **30**(2), 533–539 (2014)
86. G. Li, X. Wu, D.W. Lee, Selectively plated stretchable liquid metal wires for transparent electronics. Sens. Actuators B: Chem. **221**, 1114–1119 (2015)
87. J.W. Boley, E.L. White, R.K. Kramer, Mechanically sintered gallium indium nanoparticles. Adv. Mater. **27**(14), 2355–2360 (2015)
88. T. Lu, L. Finkenauer, J. Wissman, C. Majidi, Rapid prototyping for soft-matter electronics. Adv. Funct. Mater. (2014)
89. D.M. Vogt, Y.L. Park, R.J. Wood, Design and characterization of a soft multi-axis force sensor using embedded microfluidic channels. IEEE Sens. J. **13**(10), 4056–4064 (2013)
90. J.H. So, J. Thelen, A. Qusba, G.J. Hayes, G. Lazzi, M.D. Dickey, Reversibly deformable and mechanically tunable fluidic antennas. Adv. Funct. Mater. **19**(22), 3632–3637 (2009)

91. M. Kubo, X. Li, C. Kim, M. Hashimoto, B.J. Wiley, D. Ham, G.M .Whitesides, Stretchable microfluidic radiofrequency antennas. Adv. Mater. **22**(25), 2749–2752 (2010)
92. Z. Wu, Microfluidic stretchable radio frequency devices, in *Proceedings of the IEEE*, 2015
93. E. Palleau, S. Reece, S.C. Desai, M.E. Smith, M.D. Dickey, Self-healing stretchable wires for reconfigurable circuit wiring and 3d microfluidics. Adv. Mater. **25**(11), 1589–1592 (2013)
94. J.H. So, H.J. Koo, M.D. Dickey, O.D. Velev, Ionic current rectification in soft-matter diodes with liquid-metal electrodes. Adv. Funct. Mater. **22**(3), 625–631 (2012)
95. W.-Y. Tseng, J.S. Fisher, J.L. Prieto, K. Rinaldi, G. Alapati, A.P. Lee, A slow-adapting microfluidic-based tactile sensor. J. Micromech. Microeng. **19**(8), 085002 (2009)
96. N. Wettels, V.J. Santos, R.S. Johansson, G.E. Loeb, Biomimetic tactile sensor array. Adv. Robot. **22**(8), 829–849 (2008)
97. Y.N. Cheung, Y. Zhu, C.H. Cheng, C. Chao, W.W.F. Leung, A novel fluidic strain sensor for large strain measurement. Sens. Actuators, A **147**(2), 401–408 (2008)
98. G. Cummins, M.P.Y. Desmulliez, Inkjet printing of conductive materials: a review. Circuit World **38**(4), 193–213 (2012)
99. Y. Zhang, P. Zhu, G. Li, T. Zhao, X. Fu, R. Sun, F. Zhou, C.P. Wong, Facile preparation of monodisperse, impurity-free, and antioxidation copper nanoparticles on a large scale for application in conductive ink. ACS Appl. Mater. Interfaces **6**(1), 560–567 (2014)
100. S. Merilampi, T. Laine-Ma, P. Ruuskanen, The characterization of electrically conductive silver ink patterns on flexible substrates. Microelectron. Reliab. **49**(7), 782–790 (2009)
101. S. Hong, J. Yeo, G. Kim, D. Kim, H. Lee, J. Kwon, H. Lee, P. Lee, S.H. Ko, Nonvacuum, maskless fabrication of a flexible metal grid transparent conductor by low-temperature selective laser sintering of nanoparticle ink. ACS Nano **7**(6), 5024–5031 (2013)
102. M. Grouchko, A. Kamyshny, C.F. Mihailescu, D.F. Anghel, S. Magdassi, Conductive inks with a built-in mechanism that enables sintering at room temperature. ACS Nano **5**(4), 3354–3359 (2011)
103. A. Kamyshny, M. Ben-Moshe, S. Aviezer, S. Magdassi, Ink-jet printing of metallic nanoparticles and microemulsions. Macromol. Rapid Commun. **26**(4), 281–288 (2005)
104. F. Loffredo, A. De Girolamo Del Mauro, G. Burrasca, V. La Ferrara, L. Quercia, E. Massera, G. Di Francia, D. Della Sala, Ink-jet printing technique in polymer/carbon black sensing device fabrication. Sens. Actuators B: Chem. **143**(1), 421–429 (2009)
105. S.M. Bidoki, D.M. Lewis, M. Clark, A. Vakorov, P.A. Millner, D. McGorman, Ink-jet fabrication of electronic components. J. Micromech. Microeng. **17**(5), 967 (2007)
106. T.H. Kang, C. Merritt, B. Karaguzel, J. Wilson, P.D. Franzon, B. Pourdeyhimi, E. Grant, T. Nagle, Sensors on textile substrates for home-based healthcare monitoring, in *Proceedings of the 1st Transdisciplinary Conference on Distributed Diagnosis and Home Healthcare (D2H206)*, pp. 5–7, 2006
107. Y.L. Tai, Z.G. Yang, Fabrication of paper-based conductive patterns for flexible electronics by direct-writing. J. Mater. Chem. **21**(16), 5938 (2011)
108. H.T. Wang, O.A. Nafday, J.R. Haaheim, E. Tevaarwerk, N.A. Amro, R.G. Sanedrin, C.Y. Chang, F. Ren, S.J. Pearton, Toward conductive traces: dip pen nanolithography of silver nanoparticle-based inks. Appl. Phys. Lett. **93**(14), 143105 (2008)
109. A. Russo, B.Y. Ahn, J.J. Adams, E.B. Duoss, J.T. Bernhard, J.A. Lewis, Pen-on-paper flexible electronics. Adv. Mater. **23**(30), 3426–3430 (2011)
110. L.Y. Xu, G.Y. Yang, H.Y. Jing, J. Wei, Y.D. Han, Aggraphene hybrid conductive ink for writing electronics. Nanotechnology **25**(5), 055201 (2014)
111. S. Khan, L. Lorenzelli, R.S. Dahiya, Screen printed flexible pressure sensors skin, in *2014 25th Annual SEMI on Advanced Semiconductor Manufacturing Conference (ASMC)*, pp. 219–224, May 2014
112. K.Y. Chun, Y. Oh, J. Rho, J.H. Ahn, Y.J. Kim, H.R. Choi, S. Baik, Highly conductive, printable and stretchable composite films of carbon nanotubes and silver. Nat. Nanotechnol. **5**(12), 853–857 (2010)

113. M. Park, J. Im, M. Shin, Y. Min, J. Park, H. Cho, S. Park, M.B. Shim, S. Jeon, D.Y. Chung, J. Bae, J. Park, U. Jeong, K. Kim, Highly stretchable electric circuits from a composite material of silver nanoparticles and elastomeric fibres. Nat. Nanotechnol. 7(12), 803–809 (2012)
114. Y.J. Yang, M.Y. Cheng, W.Y. Chang, L.C. Tsao, S.A. Yang, W.P. Shih, F.Y. Chang, S.H. Chang, K.C. Fan, An integrated flexible temperature and tactile sensing array using PI-copper films. Sens. Actuators, A 143(1), 143–153 (2008)
115. R. Verdejo, M.M. Bernal, L.J. Romasanta, M.A. Lopez-Manchado, Graphene filled polymer nanocomposites. J. Mater. Chem. 21(10), 3301–3310 (2011)
116. M. Chen, T. Tao, L. Zhang, W. Gao, C. Li, Highly conductive and stretchable polymer composites based on graphene/MWCNT network. Chem. Commun. 49(16), 1612 (2013)
117. M. Knite, V. Teteris, A. Kiploka, J. Kaupuzs, Polyisoprene-carbon black nanocomposites as tensile strain and pressure sensor materials. Sens. Actuators, A 110(13), 142–149 (2004)
118. D.J. Lipomi, M. Vosgueritchian, B.C.K. Tee, S.L. Hellstrom, J.A. Lee, C.H. Fox, Z. Bao, Skin-like pressure and strain sensors based on transparent elastic films of carbon nanotubes. Nat. Nano 6(12), 788–792 (2011)
119. S. Jung, J.H. Kim, J. Kim, S. Choi, J. Lee, I. Park, T. Hyeon, D.H. Kim, Reverse-micelle-induced porous pressure-sensitive rubber for wearable human machine interfaces. Adv. Mater. 26(28), 4825–4830 (2014)
120. A. Fassler, C. Majidi, Liquid-phase metal inclusions for a conductive polymer composite. Adv. Mater. 27(11), 1928–1932 (2015)
121. Y. Lin, C. Cooper, M. Wang, J.J. Adams, J. Genzer, M.D. Dickey, Handwritten, soft circuit boards and antennas using liquid metal nanoparticles. Small (2015)
122. S.G. Kandlikar, W.J. Grande, Evolution of microchannel flow passages thermohydraulic performance and fabrication technology. Heat Transfer Eng. 24(1), 3–17 (2003)
123. Y. Xia, G.M. Whitesides, Soft lithography. Annu. Rev. Mater. Sci. 28(1), 153–184 (1998)
124. L. Geppert, Semiconductor lithography for the next millennium. IEEE Spectr. 33(4), 33–38 (1996)
125. S. Okazaki, Resolution limits of optical lithography. J. Vac. Sci. Technol., B 9(6), 2829–2833 (1991)
126. E.A. Waddell, Laser ablation as a fabrication technique for microfluidic devices, in Microfluidic Techniques, ed. by S.D. Minteer, Number 321 in Methods In Molecular Biology (Humana Press, Totowa, 2006), pp. 27–38. doi:10.1385/1-59259-997-4:27
127. H.J. Kim, T. Maleki, P. Wei, B. Ziaie, A biaxial stretchable interconnect with liquid-alloy-covered joints on elastomeric substrate. J. Microelectromech. Syst. 18(1), 138–146 (2009)
128. H.J. Kim, C. Son, B. Ziaie, A multiaxial stretchable interconnect using liquid-alloy-filled elastomeric microchannels. Appl. Phys. Lett. 92(1), 011904–011904-3 (2008)
129. T. Li, Z. Huang, Z. Suo, S.P. Lacour, S. Wagner, Stretchability of thin metal films on elastomer substrates. Appl. Phys. Lett. 85(16), 3435–3437 (2004)
130. Y. Arafat, I. Dutta, R. Panat, Super-stretchable metallic interconnects on polymer with a linear strain of up to 100 %. Appl. Phys. Lett. 107(8), 081906 (2015)
131. T. Lu, J. Wissman, F.N.U. Ruthika, C. Majidi, Soft anisotropic conductors as electric vias for Ga-based liquid metal circuits. ACS Appl. Mater. Interfaces (2015)
132. Y.L. Zheng, X.R. Ding, C.C.Y. Poon, B.P.L. Lo, H. Zhang, X.L. Zhou, G.Z. Yang, N. Zhao, Y.T. Zhang, Unobtrusive sensing and wearable devices for health informatics. IEEE Trans. Biomed. Eng. 61(5), 1538–1554 (2014)
133. N. Lu, D.H. Kim, Flexible and stretchable electronics paving the way for soft robotics. Soft Rob. 1(1), 53–62 (2014)
134. H.K. Lee, S.I. Chang, E. Yoon, A flexible polymer tactile sensor: fabrication and modular expandability for large area deployment. J. Microelectromech. Syst. 15(6), 1681–1686 (2006)

135. I.M. Koo, K. Jung, J.C. Koo, J.D. Nam, Y.K. Lee, H.R. Choi, Development of soft-actuator-based wearable tactile display. IEEE Trans. Rob. **24**(3), 549–558 (2008)
136. J. Engel, J. Chen, C. Liu, Development of polyimide flexible tactile sensor skin. J. Micromech. Microeng. **13**(3), 359 (2003)
137. J.K. Paik, R.K. Kramer, R.J. Wood, Stretchable circuits and sensors for robotic origami, in *2011 IEEE/RSJ International Conference on Intelligent Robots and Systems (IROS)*, pp. 414–420, 2011
138. M. Yuen, A. Cherian, J.C. Case, J. Seipel, R.K. Kramer, Conformable actuation and sensing with robotic fabric, in *2014 IEEE/RSJ International Conference on Intelligent Robots and Systems (IROS 2014)*, pp. 580–586. IEEE, 2014
139. Y.L. Park, B.R. Chen, C. Majidi, R.J. Wood, R. Nagpal, E. Goldfield, Active modular elastomer sleeve for soft wearable assistance robots, in *2012 IEEE/RSJ International Conference on Intelligent Robots and Systems (IROS)*, pp. 1595–1602, 2012
140. Y.L. Park, R.J. Wood, Smart pneumatic artificial muscle actuator with embedded microfluidic sensing, in *2013 IEEE Sensors*, pp. 1–4, 2013
141. G. Berselli (ed.), *Smart Actuation and Sensing Systems—Recent Advances and Future Challenges* (InTech, Rijeka, 2012)
142. P. Polygerinos, K.C. Galloway, E. Savage, M. Herman, K. O'Donnell, C.J. Walsh, Soft robotic glove for hand rehabilitation and task specific training, in *2015 IEEE International Conference on Robotics and Automation (ICRA)*, pp. 2913–2919, May 2015
143. A.T. Asbeck, K. Schmidt, I. Galiana, D. Wagner, C.J. Walsh, Multi-joint soft exosuit for gait assistance, in *2015 IEEE International Conference on Robotics and Automation (ICRA)*, pp. 6197–6204, May 2015
144. A. Asbeck, S. De Rossi, I. Galiana, Y. Ding, C. Walsh, Stronger, smarter, softer: next-generation wearable robots. IEEE Robot. Autom. Mag. **21**(4), 22–33 (2014)
145. Y. Menguc, Y.L. Park, E. Martinez-Villalpando, P. Aubin, M. Zisook, L. Stirling, R. J. Wood, C.J. Walsh, Soft wearable motion sensing suit for lower limb biomechanics measurements, in *2013 IEEE International Conference on Robotics and Automation (ICRA)*, pp. 5309–5316, May 2013
146. N. Lu, C. Lu, S. Yang, J. Rogers, Highly sensitive skin-mountable strain gauges based entirely on elastomers. Adv. Funct. Mater. **22**(19), 4044–4050 (2012)
147. F. Gemperle, N. Ota, D. Siewiorek. Design of a wearable tactile display, in *Proceedings of the Fifth International Symposium on Wearable Computers, 2001*, pp. 5–12, 2001

Chapter 10
Multifunctional Epidermal Sensor Systems with Ultrathin Encapsulation Packaging for Health Monitoring

Milan Raj, Shyamal Patel, Chi Hwan Lee, Yinji Ma, Anthony Banks, Ryan McGinnis, Bryan McGrane, Briana Morey, Jeffrey B. Model, Paolo DePetrillo, Nirav Sheth, Clifford Liu, Ellora Sen-Gupta, Lauren Klinker, Brian Murphy, John A. Wright, A.J. Aranyosi, Moussa Mansour, Ray E. Dorsey, Marvin Slepian, Yonggang Huang, John A. Rogers and Roozbeh Ghaffari

Abstract Wearable sensors have the potential to enable longitudinal, objective health monitoring in patients with chronic diseases, including cardiac rhythm disorders, neurological and movement disorders, diabetes, and pain. However, conventional wearable devices are typically comprised of rigid, packaged electronics, which may compromise overall signal fidelity and wearer comfort during activities of daily living and sleep. In this chapter, we present recent advances in the development of thin and stretchable epidermal systems for biometric data measurements.

M. Raj · S. Patel · R. McGinnis · B. McGrane · B. Morey · J.B. Model · P. DePetrillo · N. Sheth · C. Liu · E. Sen-Gupta · L. Klinker · B. Murphy · J.A. Wright · A.J. Aranyosi · R. Ghaffari (✉)
MC10 Inc., Lexington, MA, USA
e-mail: rghaffari@mc10inc.com

C.H. Lee · A. Banks · J.A. Rogers
Department of Materials Science and Engineering, Beckman Institute
for Advanced Science and Technology, University of Illinois Urbana-Champaign, Urbana, IL, USA

Y. Ma · Y. Huang
Department of Mechanical Engineering and Department of Civil and Environmental Engineering, Northwestern University, Evanston, IL, USA

M. Mansour
Massachusetts General Hospital, Harvard Medical School, Boston, MA, USA

R.E. Dorsey
Department of Neurology, University of Rochester Medical Center, Rochester, NY, USA

M. Slepian
Sarver Heart Center, University of Arizona, Tucson, AZ, USA

193

These non-invasive epidermal systems are fully integrated with multiple sensors, an analog front end module, a radio for wireless communication, onboard flash memory, a rechargeable battery all encapsulated in a soft, stretchable and water-resistant silicone, and with an air permeable adhesive layer that interfaces with the human skin. The encapsulated system intimately couples with the skin at multiple locations on the body. We present results showing the potential of this technology to quantitatively assess bio-kinematics and electrophysiological signals. Finally, we provide perspectives on remaining challenges and opportunities to achieve clinical validation and commercial adoption of these technologies.

Keywords Stretchable electronics · Epidermal electronics · Movement disorders · Cardiac disease · Sleep monitoring · Bare die · Electromyography · Electrocardiography · Biomechanics

10.1 Introduction

Conventional electronics-enabled sensors and body worn devices have enabled long-term continuous monitoring of cardiac, neural, muscular, and motor symptoms of patients in the clinic and in the home environment [1–3]. Although conventionally packaged sensors are able to capture continuous data, the size, weight, geometry, and mechanical properties of established systems typically limit seamless integration with the human body. This mechanical mismatch can cause skin frustration and lead to poor signal quality, particularly over extended periods of use. There is thus a need for establishing a new class of soft, conformal electronics, and sensors that better interface with the human body to minimize discomfort and optimize skin coupling during prolonged use.

Advances in the miniaturization and mechanics of inorganic electronics have enabled soft, flexible, and stretchable sensing/actuating systems that conformally integrate with biological tissue [4–6]. These remarkable properties of soft and stretchable bioelectronics have driven the rapid development of a new class of epidermal sensors with many desirable physical attributes, including low elastic modulus, thin profile, and ability to accommodate cyclical strains in a way that is imperceptible to the human body [4–7]. Extremely compliant electronic systems that incorporate thin film semiconductors and integrated arrays of ultrathin biosensors [4–15] represent powerful tools that address many of the constraints observed in existing health tracking devices. Unlike rigid wearable devices that are typically strapped on to the wrist, ankles or torso, conformal skin-coupled systems are mechanically optimized for integration with the human skin, enabling enhanced wearability and discreet tracking of limb specific fine motor movements and cardiac biopotentials [16].

Assessment of cardiac disease, movement, and neurological diseases has been limited to self-reporting, patient diaries, and infrequent physical, neural, and cognitive examinations by physicians [1–3]. Although patient diaries are widely used in the home, there are compliance issues and concerns about bias and the qualitative nature of the collected data. Furthermore, detailed clinical examinations typically occur with variable frequency and provide a very brief window of insight, which may not be sufficient to observe the full spectrum of symptoms and fluctuations (e.g., freezing of gait, sleep disturbances) that can occur on a daily basis.

In this chapter, we present a review of epidermal sensor systems that integrate multiple sensing modalities, wireless connectivity, and memory for continuous data capture in ultralow modulus elastomers. These systems establish general design rules for emerging wearable systems and can be used to track the progression of diseases in the home and clinical environments with minimal obtrusiveness and discomfort to the patient.

10.2   Epidermal System Architectures

Figure 10.1a shows a schematic drawing of a representative conformal epidermal sensor system consisting of stretchable circuits, electrodes, and an adhesive layer to interface with the skin. The thin, stretchable circuits contain multiple "islands" with serpentine metal routing lines connecting individual islands. The interconnect geometries are optimized for stretching and bending, enabling highly comfortable wear for the user. Figure 10.1b shows a cross-sectional view of the device stack with biocompatible interfacing materials. This electromechanical construct enables tight integration with the soft, curvilinear surfaces of the human skin in ways that are unachievable with conventional rigid and printed circuit board technologies. Optical images of the representative system are shown in Fig. 10.1c and d under conditions of stretching and bending while peeling from the forearm.

These fully-integrated systems contain a rechargeable battery, flash memory, Bluetooth® communication capability, a low-power microcontroller unit, and sensors for movement tracking (accelerometers and gyroscope) and electrophysiological (50–1000 Hz sampling rate) measurements. A low-power microcontroller conditions signals from the 3-axis accelerometer, gyroscope and electrodes, which are connected to a single channel analog front end. The sensor data is high-pass filtered to remove DC offset noise from the high-impedance electrodes. The analog-to-digital output from the sensors is then processed and sampled by the microcontroller, which in turn, transmits data to flash memory and/or broadcasts wirelessly via Bluetooth low energy. These components are powered with a rechargeable battery, which is recharged through physical connection via onboard microconnectors. The operation time of the system extends over a few days for raw data capture. Onboard algorithms can increase memory capacity by up to a factor of 3.

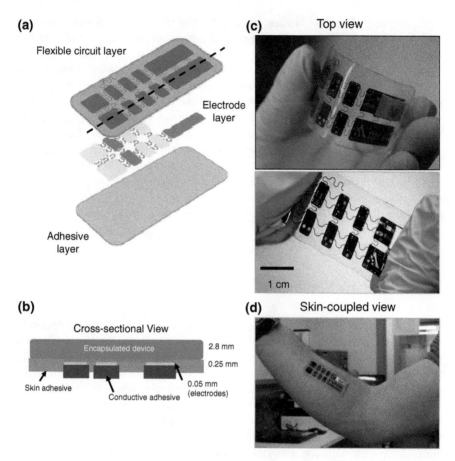

Fig. 10.1 a Schematic drawing of epidermal sensor system consisting of flexible circuit layer, electrodes, and nonconductive skin adhesive layer. **b** *Cross-sectional view* shows placement of conductive hydrogel relative to the nonconductive layer. **c** Optical images of epidermal sensor system in *bended* and *stretched state*. **d** Epidermal sensor system adhered to the forearm of human skin

10.2.1 Sensor Performance

To assess electrode signal quality, these systems are tested on the body during prescribed activity and repeated performance of maximum voluntary contractions (MVCs) in healthy subjects. Subjects are asked to perform a series [5] of MVCs with the flexor digitorum and tibialis anterior muscles. For the flexor digitorum muscle, the experimenter placed their hand across the palm of the subject and while the subject held a sustained contraction by flexing as hard as they could for approximately 2–3 s. For the tibialis anterior muscle, a downward force is applied on the subject's foot while the subject sustains muscle contraction by performing dorsiflexion for approximately 2–3 s. Signal-to-noise-ratio (SNR) is computed by dividing the root mean square (RMS) value of surface electromyography (s-EMG) signal over the RMS value of baseline noise. To segment the EMG signal and baseline noise, the s-EMG envelope is computed to determine a threshold (V_{th}) based on the amplitude of the signal. Baseline noise is defined as the signal under this threshold (V_{th}), whereas s-EMG signal is everything above. The EMG electrodes typically achieves an SNR value greater than 10 at both sensor locations, which is comparable with SNR values reported in the literature using s-EMG signals measured at a single location [17].

10.2.2 Signal Processing

Once the sensor data is filtered and segmented, a set of time and frequency domain signal features are extracted from processed accelerometer and s-EMG data using defined time windows. Signal features are extracted from such windows assigned across the duration of each clinical task. EMG-based time domain signal features [18] include root mean square value, kurtosis, and signal entropy. Frequency domain features [19] extracted from raw s-EMG and s-EMG envelope include dominant frequency, center frequency, bandwidth, and root mean square frequency [19–21].

Accelerometer-based time domain signal features include range, signal entropy, root mean square value, correlation coefficient between pairs of axes (*XY*, *YZ* and *ZX*), and range of autocovariance. Frequency domain signal features include spectral entropy, dominant frequency, and ratio of energy in the dominant frequency to the entire frequency spectrum of the signal.

10.3 Mechanical Characteristics

The soft epidermal systems described in Fig. 10.1 require an ultra-elastic, biocompatible, and protective interface to facilitate interaction with soft biological tissues. To test the mechanical robustness and failure modes of these epidermal electronics systems, we applied finite element analysis (Abaqus) leveraging geometries and material properties of novel-encapsulated techniques to protect the packaged electronics and serpentine interconnections between the islands, providing isolation of stresses in regions of the active electronics. Figure 10.2 shows finite element simulation results of the epidermal systems undergoing mechanical bending. The overall stress within the islands is reduced to < 0.1 % while the system is under 15–20 % mechanical strain. A typical encapsulation consists of a low durometer material and an adhesive interface layer that offers additional advantages of coupling with the epidermis, allowing the system to be wrapped around most curvilinear surfaces and applied anywhere on the body. Although these

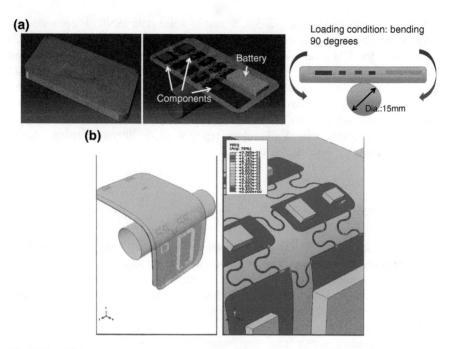

Fig. 10.2 **a** Finite element analysis (FEA) results of the *stress/strain* relationship for an epidermal sensor system and encapsulation layers. The loading condition was defined as *bending* about a 15 mm diameter *cylinder*. **b** FEA results of *bending* 90° at the circuit, interconnect and chip interface

systems contain high-modulus internal circuit components, they nevertheless, can laminate softly and noninvasively onto human skin by virtue of system level soft mechanical properties.

10.3.1 Microfluidic Enclosure Configurations

Microfluidic suspensions of interconnected electronics sandwiched between elastomeric enclosures enable even softer mechanics at the system level [5]. The layout in this novel configuration contains a matrix of microfluidic spaces with electronics embedded in between. The constituent devices and stretchable interconnections are attached to the top and bottom surfaces of the encapsulating silicone as shown in Fig. 10.3a. The bonded electronics are filled with a dielectric fluid that fills the voids (Fig. 10.3b). The liquid silicone has several important properties to facilitate high efficiency RF energy transfer, low reactivity, and chemical stability for long-term operation without corrosion or chemical degradation. The suspended electronics and interconnects can deform with minimal stress concentration in response to external deformations of the system. Moreover, the interconnects can

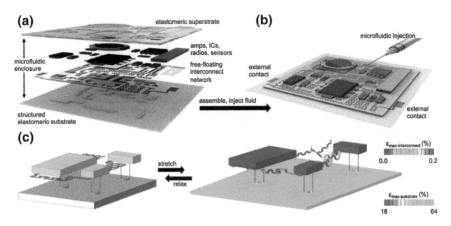

Fig. 10.3 **a** Schematic illustration of the stack up of the microfluidic-enclosed epidermal system. **b** Illustration of the epidermal system encapsulated and injected with *liquid silicone* using a syringe. **c** 3-dimensional FEA results for the epidermal system, highlighting mechanics during *stretch* and *relaxation states*. (Reproduced with Permission from Ref. [5], Copyright Science)

move freely in and out of plane, enabling stretchability relative to more constrained encapsulation strategies [5]. These mechanical attributes are apparent in the three-dimensional finite element analysis results (Fig. 10.3c). The modeling results show the maximum principal strains that are located in the stretchable interconnect regions (~ 0.2 % for biaxial stretch of 50 % of the system).

These microfluidic-enclosed epidermal systems are laminated on the chest with electrodes positioned in the fourth intercostal spaces. This location allows for collection of electrocardiography data with high signal-to-noise-ratios and clearly identifiable Q, R, and S waveforms. The same hardware system was tested on the forehead to effectively capture electroencephalography (EEG) signal from the brain. The low interfacial stresses associated with the low-modulus encapsulations and ultralow modulus-enclosed fluid enable intimate coupling with the skin through van der Waals forces alone [5].

10.3.2 Soft/Core Shell Configurations

In addition to fluid encapsulated epidermal electronics, similar classes of electronics have been encapsulated in an ultralow modulus silicone material that serves as the core structure surrounded by a more stiff shell material. Figure 10.4a shows this core/shell structure compared to a standard low durometer silicone material. The device in the core/shell configuration intimately couples with the epidermis without delaminating. In contrast, the device in standard encapsulation tends to delaminate

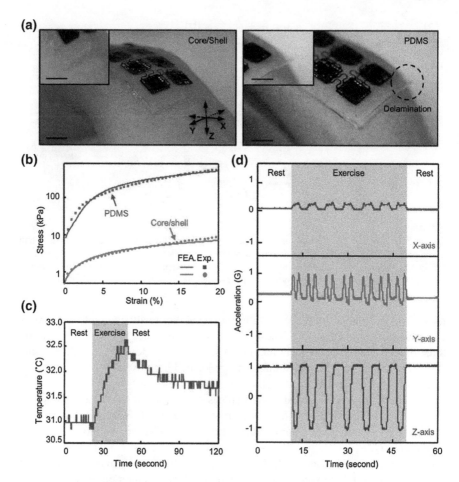

Fig. 10.4 a Optical images of soft *core/i* encapsulation strategy and a standard package encapsulation for epidermal sensor system (*scale bar*, 1 cm). **b** Experimental and FEA results of *stress/strain* responses for these two encapsulated systems. **c** Real time monitoring of temperature changes characterized during *exercise* and *rest*. **d** Real time monitoring of activity during *exercise* and *rest* captured with an accelerometer sensor onboard in the epidermal sensor system. (Reproduced with Permission from Ref. [6], Copyright Advanced Functional Materials)

around its edges. This difference in lamination is governed by differences in the tensile modulus of the system, which can be improved by up to 70 times, as confirmed by experimental and theoretical measurements (Fig. 10.4b). This representative epidermal system allows for wireless transmission of acceleration, electrophysiological, and temperature data streams during daily activity and rigorous exercise (Fig. 10.4c, d).

These systems have differentiated mechanical properties as evidenced by their extremely low tensile modulus. Figure 10.5a, b shows stress/strain curves of the core/shell package obtained from mechanical experimentation and finite element

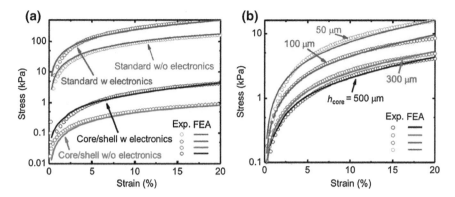

Fig. 10.5 a Experimental and FEA analysis results for *core/shell* and standard silicone encapsulation approaches with and without electronics present in the system. **b** Experimental and FEA analysis results for *core/shell* silicone with different *core thicknesses* (h_{core}) from 50–500 μm. (Reproduced with Permission from Ref. [6], Copyright Advanced Functional Materials)

analysis. The results indicate that stresses are controlled by h_{core}. Increasing h_{core} effectively reduces stresses on the skin. In contrast, h_{shell} has very little effect on stress for the range of values tested in this study. These configurations and associated stresses suggest that the systems impose minimal mechanical constraints on the motion of the human skin.

The core/shell encapsulations and embedded electronics can be textured or painted using a broad spectrum of colors to modulate the physical appearance and esthetics of the device. Figure 10.6a shows an example of a commercially available device laminated on the inner wrist. The design consists of round and layered corners to minimize edge stresses and delamination (Fig. 10.6b). The texture can also be designed to mimic the microscale surface of human skin, as shown in the scanning electronic microscope (SEM) image in Fig. 10.6c.

10.3.3 Ultrathin Adhesive Substrate Configurations

In addition to surface mounted components, even thinner constructs consisting of unpackaged electronics components have achieved tattoo-like form factors. These systems typically consist of bare dies, antennae for power and data transfer, and a few passive components. Figure 10.7a–f shows schematic illustrations and images of epidermal near field communication (NFC) devices with coil designs. Thin layers of polyimide (from above and below) encapsulate the coils to physically isolate the copper metal layer within the neutral mechanical plane. The antenna and NFC bare die were connected using a modified flip chip technique [7]. For integration on the skin, a thin low-modulus acrylic adhesive with a thickness of 25 μm serves as the substrate. This design allows conformal contact with the skin and allows for communication with a range of up to 3 cm under various states of deformations.

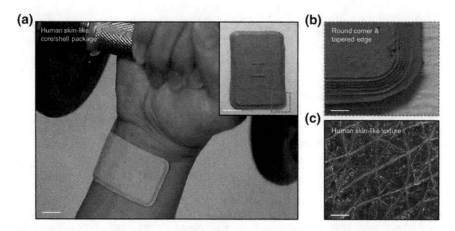

Fig. 10.6 **a** Optical image of *core/shell* encapsulated epidermal sensor system. The system is attached to the wrist during dumbbell lifting. Inset shows an image of *core/shell* structure with the logo of the University of Illinois in the *middle* of the structure. **b** *Magnified view* of the *core/shell* encapsulation near the *rounded corner* with a tapered edge. **c** SEM image of the human skin-like textured surface of the encapsulation (*scale bar*, 0.5 mm). (Reproduced with Permission from Ref. [6], Copyright Advanced Functional Materials)

The ability to encapsulate and laminate these NFC-based systems with adhesives provides new routes for epidermal systems using bare die and thin adhesive-based layers serving as the substrate.

10.4 Implications for Biomedicine

This chapter reviews flexible/stretchable electronics systems that are fully integrated with multifunctional sensors, antennas, wireless connectivity, microprocessors, near field communication energy harvesting modules, and memory. The ability to laminate these systems on the skin for continuous health monitoring introduces new possibilities for health monitoring in the home and clinical settings [22–24]. While there are many devices that exploit conventional electronics materials [16], the systems described in this chapter have the potential to attach on the skin in ways that are imperceptible to the subject, and thereby improve comfort and compliance for patients [25]. Furthermore, these systems are able to capture and transmit high signal fidelity biometric data for continuous monitoring of symptoms across a broad range of disorders, including movement disorders, cardiac rhythm disorders, and frailty [16, 22–29].

Objective measures captured during the performance of activities of daily living such as gait [28] and sleep [30] highlight the potential of this epidermal sensor technology for monitoring subjects outside the clinic. Sleep disturbances captured by measuring postural transitions, as well as nighttime ambulation can provide a

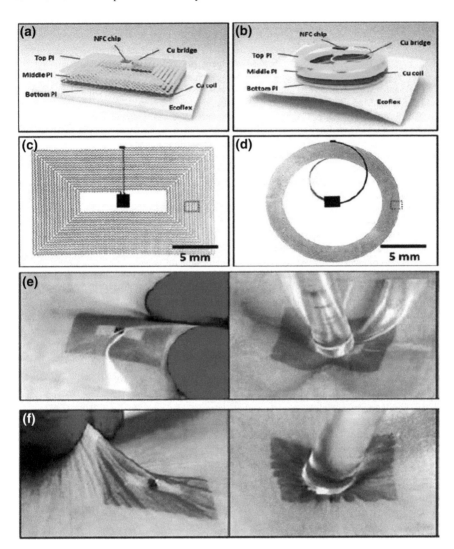

Fig. 10.7 Schematic illustration and optical microscope images of skin-mounted NFC devices with different **a** *rectangular* and **b** *circular coil* geometries. **c, d** Optical images of the functional NFC-enabled epidermal sensor systems prior to encapsulation. **e** Optical images of NFC system on the human skin in *bending* and *stretch states* induced by pinching with fingers and plunging with glass rod, respectively. **f** Optical images of NFC system undergoing extreme mechanical deformations during pinching and glass rod plunging (Reproduced with Permission from Ref. [7], Copyright Small)

unique window into nonmotor characteristics, which is not feasible to assess during clinical examinations. The wearability and usability of these emerging epidermal systems are now being investigated in clinical studies to identify potential advantages and limitations of this technology compared to gold standards of care.

References

1. S. Patel, Z. Park, P. Bonato, L. Chan, M. Rodgers, A review of wearable sensors and systems with application in rehabilitation. J. Neuroeng. Rehabil. **9**, 21–38 (2012)
2. P. Bonato, Wearable sensors and systems. From enabling technology to clinical applications. IEEE Eng. Med. Biol. Mag. **29**, 25–36 (2010)
3. X.F. Teng, Y.T. Zhang, C.C.Y. Poon, P. Bonato, Wearable medical systems for p-health. IEEE Rev. Biomed. Eng. **1**, 62–74 (2008)
4. D.H. Kim et al., Epidermal electronics. Science **333**, 838–843 (2011)
5. S. Xu et al., Soft microfluidic assemblies of sensors, circuits, and radios for the skin. Science **344**, 70–74 (2014)
6. C.H. Lee et al., Soft core/shell packages for stretchable electronics. Adv. Funct. Mater. **25**, 3698–3704 (2015)
7. J. Kim et al., Epidermal electronics with advanced capabilities in near field communication. Small **11**, 906–912 (2014)
8. J.A. Rogers, T. Someya, Y.G. Huang, Materials and mechanics for stretchable electronics. Science **327**, 1603–1607 (2010)
9. M. Kaltenbrunner et al., An ultra-lightweight design for imperceptible plastic electronics. Nature **499**, 458–463 (2013)
10. S.P. Lacour, J. Jones, Z. Suo, S. Wagner, Design and performance of thin metal film interconnects for skin-like electronic circuits. IEEE Electron Device Lett. **25**, 179–181 (2004)
11. T. Someya, T. Sekitani, S. Iba, Y. Kato, H. Kawaguchi, T. Sakurai, A large-area, flexible pressure sensor matrix with organic field-effect transistors for artificial skin applications. Proc. Natl. Acad. Sci. USA. **101**, 9966–9970 (2004)
12. S.J. Benight, C. Wang, J.B.H. Tok, Z.A. Bao, Stretchable and self-healing polymers and devices for electronic skin. Prog. Polym. Sci. **38**, 1961–1977 (2013)
13. S. Wagner, S. Bauer, Materials for stretchable electronics. MRS Bull. **37**, 207–217 (2012)
14. D.Y. Khang, H.Q. Jiang, Y. Huang, J.A. Rogers, A stretchable form of single-crystal silicon for high-performance electronics on rubber substrates. Science **311**, 208–212 (2006)
15. D.H. Kim, J.Z. Song, W.M. Choi, H.S. Kim, R.H. Kim, Z.J. Liu et al., Materials and noncoplanar mesh designs for integrated circuits with linear elastic responses to extreme mechanical deformations. PNAS **105**, 18675–18680 (2008)
16. S.S. Lobodzinski, ECG patch monitors for assessment of cardiac rhythm abnormalities. Prog. Cardiovasc. Dis. **56**, 224–229 (2013)
17. E.A. Clancy, N. Hogan, Multiple site electromyograph amplitude estimation. IEEE Trans. Biomed. Eng. **42**, 203–211 (1995)
18. S.H. Roy, B.T. Cole, L.D. Gilmore, C.J. De Luca, C.A. Thomas, M.M. Saint-Hilaire, S.H. Nawab, High-resolution tracking of motor disorders in Parkinson's disease during unconstrained activity. Mov. Disord. **28**, 1080–1087 (2013)
19. A.I. Meigal, S. Rissanen, M.P. Tarvainen, P.A. Karjalainen, I.A. Iudina-Vassel, O. Airaksinen, M. Kankaanpää, Novel parameters of surface EMG in patients with Parkinson's disease and healthy young and old controls. J. Electromyogr. Kinesiol. **19**, e206–e213 (2009)
20. R. Moddemeijer, On estimation of entropy and mutual information of continuous distributions. Sig. Process. **16**, 233–248 (1989)
21. J.W. Sammon, A nonlinear mapping for data structure analysis. IEEE Trans. Comput. **5**, 401–409 (1969)
22. M. Morris, E.F. Huxham, J. McGinley, K. Dodd, R. Iansek, The biomechanics and motor control of gait in Parkinson disease. Clin. Biomech. **16**, 459–470 (2001)
23. E. Tandberg, J.P. Larsen, K. Karlsen, A community-based study of sleep disorders in patients with Parkinson's disease. Mov. Disord. **13**, 895–899 (1998)
24. A. Luke, K.C. Maki, N. Barhey, R. Cooper, D. McGee, Simultaneous monitoring of heart rate and motion to assess energy expenditure. Med. Sci. Sports Exerc. **29**, 144–148 (1997)

25. D. Son et al., Multifunctional wearable devices for diagnosis and therapy of movement disorders. Nat. Nanotechnol. **9**, 397–404 (2014)
26. S. Parvaneh et al., Regulation of cardiac autonomic nervous system control across frailty statuses: a systematic review. Gerontology **62**(1), 3–15 (2015)
27. M. Schwenk et al., Frailty and technology: a systematic review of gait analysis in those with frailty. Gerontology **60**, 79–89 (2014)
28. B. Najafi, T. Khan, J. Wrobel, Laboratory in a box: wearable sensors and its advantages for gait analysis, in *Conference Proceedings IEEE Engineering in Medicine Biology Society* (2011), pp. 6507–6510
29. M. Achey et al., The past, present, and future of telemedicine for Parkinson's disease. Mov. Disord. **29**, 871–883 (2014)
30. J.M. Kelly, R.E. Strecker, M.T. Bianchi, Recent developments in home sleep-monitoring devices. ISRN Neurol. **2012**, 1–10 (2012)

Chapter 11
Laser-Enabled Fabrication Technologies for Low-Cost Flexible/Conformal Cutaneous Wound Interfaces

Manuel Ochoa, Rahim Rahimi and Babak Ziaie

Abstract Laser-enabled fabrication methods, in particular laser surface modification of low-cost materials such as paper, is an attractivetechnology for fabrication of flexible sensors and microsystems. Such devices are uniquely suited for cutaneous wound interfaces in which one has to sense multiple parameters and deliver drugs using a disposable low-cost platform. In this chapter, we discuss our recent efforts towards using laboratory scale CO_2 lasers to modify commercial hydrophobic papers (e.g., parchment paper, wax paper, palette paper, etc.) and thermoset polymers (e.g., polyimide) by controlled surface ablation. Such treatment imparts unique physical and chemical properties (hydrophilicity, extreme porosity, carbonization, etc.) to the material and allows for selective surface functionalization. Using this method, we fabricated a variety of sensors (pH, oxygen, silver, strain) and chemical delivery (oxygen) modules on low-cost commercial substrates for chronic wound management.

Keywords Laser processing · Carbonization · Pyrolization · Paper · Polyimide · Hydrophobic · Wound · Physical/chemical sensors · Drug delivery · Flexible electronics

11.1 Introduction

Cutaneous wounds pose major health and financial burdens for millions of people in the US and even more so throughout the world. Chronic (non-healing) wounds, in particular, affect 6 million Americans and cost the health care system $20 billion annually [1]. Such wounds are those which do not follow the standard cascade of biological processes (i.e., inflammation, proliferation, and maturation) by which common acute wounds heal [2]. Instead, their healing is hampered by conditions

M. Ochoa · R. Rahimi · B. Ziaie (✉)
School of Electrical and Computer Engineering, Purdue University,
West Lafayette, IN, USA
e-mail: bziaie@purdue.edu

© Springer International Publishing Switzerland 2016   207
J.A. Rogers et al. (eds.), *Stretchable Bioelectronics for Medical Devices and Systems*, Microsystems and Nanosystems,
DOI 10.1007/978-3-319-28694-5_11

such as local hypoxia, irregular vascular structure, external mechanical pressure, and bacterial infections [2–8]. Traditionally, these wounds are treated using specialized wound dressings that help to maintain an optimum level of moisture in the wound bed while enabling gas exchange (for oxygenation) [9–12]. This approach is often complemented by other therapeutic attention-intensive procedures including continual dressing replacement [13, 14], wound debridement [15, 16], vacuum therapy [16, 17], and drug administration [18, 19]. Due to the complex nature of these wounds, however, such methods are often insufficient to promote healing; thus, their successful treatment depends on ongoing research efforts to develop novel devices and technologies for chronic wound management.

Rather than relegating treatment to passive wound dressings, it is more beneficial to use active dressings (i.e., 'smart' dressings) comprised of arrays of sensors and drug delivery modules that can address the therapeutic requirements of the wound in a localized, responsive manner. This can be accomplished by the incorporation of bio/chemical sensors (e.g., pH, oxygen, inflammatory signals) and drug delivery capabilities (e.g., via nanomedicine [20] or microfluidics [21]) into the dressings [22, 23] by either (i) developing new wound dressing platforms or (ii) embedding flexible sensors into existing commercial ones. These smart systems can help optimize the healing process, decrease the healing time, and prevent infections. They can evaluate the local wound environment, release wound healing agents as needed, detect the optimal replacement time, and alert the patient/caregiver of any unusual phenomena (preferably through a wireless link). The fabrication of such systems requires a deviation from traditional MEMS and transducer fabrication methods in order to create devices with mechanical and electrical properties which are optimized for the unique environment of wounds (i.e., soft, deformable, wet, and warm). Additionally, the techniques should be economical, adaptable for moderate-volume production, and preferably customizable (i.e., for a precision medicine approach). As a result, researchers have embraced the use of commercially available materials and rapid prototyping equipment for the development of low-cost, conformable, disposable devices for wound healing and other wearable applications. Processes such as inkjet printing, screen printing, micro-gravure coating, and laser machining are particularly suited for these applications due to their scalability and ease of implementation [24, 25].

Among these rapid fabrication technologies, laser machining offers a unique set of capabilities directly beneficial for the development of flexible/stretchable low-cost systems for cutaneous wound interfaces. Laser machining provides the ability to cut materials, etch (ablate) them, alter their surface morphology, and induce surface chemical changes, all of which increase the functionality of the material. For example, laser can be used to tune the hydrophilicity of materials such as paper, or (using different parameters) it can be used to pyrolize polymers to create active carbon materials. Furthermore, the availability of commercial, reliable, and precise laser systems allows them to become part of large-scale (e.g., roll-to-roll) production lines.

This chapter highlights the utility of laser processing for creating flexible device platforms for cutaneous wound interfaces. Section 11.2 briefly discusses the unique properties of an exposed wound environment which dictate the physical characteristics of a typical dressing. Section 11.3 reviews laser technology with an emphasis on common types of laser systems for flexible device manufacturing. Finally, Sects. 11.4 and 11.5 present two laser-enabled flexible platforms utilized to create smart wound dressings: laser-patterned hydrophobic paper, and laser-carbonized polymeric substrates.

11.2 Cutaneous Wound Interfaces

Cutaneous wounds present a unique environment in which dermal monitoring devices must operate Fig. 11.1. They consist of an exposed subcutaneous tissue (dermis; and in deep wounds hypodermis, fat tissue, and possibly muscle) that is wet/moist, warm, and loaded with cellular and bio/chemical components (red and white blood cells, plasma, bacteria in case of infection, and inflammatory/regenerative biomolecules) [2, 15]. Furthermore, the wound bed contains delicate regenerating tissue which may be sensitive to chemical, thermal, or mechanical stimuli. As a result, any device designed for such cutaneous interfaces must conform to a set of requirements mimicking the properties of typical wound dressings. Despite the large variation in structural and physical properties, current wound dressings have certain core requirements for their functional efficacy. First is flexibility, i.e., such dressings must be sufficiently flexible to conform to the wound and not limit the patient's mobility [26]. Second is gas permeability, which is essential for maintaining an adequate oxygen supply; alternatively, some dressings may supply oxygen at required levels [27]. Third is moisture control [28]; the dressing should keep the wound bed moist but absorb excess exudate. Finally, the material in contact with the wound bed should be sufficiently soft to avoid causing mechanical insults and interfering with the epithelization process (a minimal adhesion is often preferred to reduce the mechanical load).

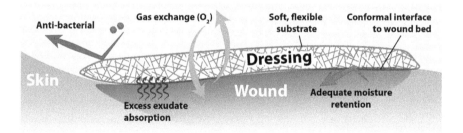

Fig. 11.1 Wound dressing requirements for proper healing. The dressing should conform to the wound bed, prevent bacterial infections, retain adequate moisture, remove excess exudate, and promote oxygenation

In addition to the above standard requirements, any sensor-equipped dressing must provide enhanced capabilities such as wound status reporting and (semi) automated micro-environment control. To achieve this goal, wound sensors require physical and operational specifications that differ from those of typical industrial and process control sensors. For example, although sensitivity and specificity remain critical for wound sensors, other parameters such as working life and response time are less important due to the disposable nature of wound dressings and the relatively slow rate of biological wound healing processes. In accordance with typical wound dressing replacement schedules, most embedded sensors require a working life of at most a few days to a week, after which they must be discarded [29, 30]. Hence, sensor deterioration and biofouling pose a reduced threat to wound sensor design (unlike long-term implantable sensors). Additionally, response times in the range of tens of seconds or even a few minutes are often acceptable depending on the sensing parameter. The level of oxygen, for example, or pH in a wound is not expected to drastically change within such time frames [31–33]. Similarly, one does not need micrometer-scale spatial resolution.

As a result of such less stringent requirements, wound sensor fabrication can deviate from traditional MEMS or microfabrication techniques. Instead, they can take advantage of emerging rapid prototyping technologies that allow the processing of more unusual materials while simultaneously enabling the personalization of wound dressing geometries and sensor distribution. Laser machining is particularly attractive for the fabrication of such devices since it can be easily adapted to commercial manufacturing setups for processing sensors on a variety of flexible substrates including polymers (e.g., polyimide, Parylene) or non-woven fiber mesh (e.g., paper, fabrics).

11.3   Lasers for Flexible Device/System Fabrication

Laser processing provides an attractive method for fabricating inexpensive microsystems by delivering a confined controlled energy onto the surface of the material. The absorbed energy can be used to alter the surface chemistry, etch the substrate, or change the chemical structure of the material, all on a wide range of soft and hard materials from paper to metal alloys. The specific effect on the material is dictated by the laser processing parameters, which are controllable to high specificity in modern systems. Laser systems do heat up the substrate at the focal point and are not compatible with heat-sensitive materials such as polyvinylidene fluoride (PVDF); however, they are very attractive for the rapid machining of many other emerging microsystem materials such as polymers and paper. This section briefly discusses the principles of two commercial lasers commonly used for creating flexible medical devices and their interactions with various materials used in flexible electronics. For a detailed description, the reader may consult any standard textbook on laser material processing (e.g., [34, 35]).

11.3.1 Laser Systems Used in Flexible Material Processing

Commercial laser engraving systems offer adequate resolution, control, and processing speed for fabricating wound healing devices. These systems consist of a laser module connected to a machining enclosure that contains a working stage and a software-controlled lens. The substrate is placed on the stage and the lens module guides the laser beam on the surface of the substrate to cut or ablate regions as defined in a CAD drawing. Most commercial systems use either a 10.6 μm CO_2 laser (typical powers of 30–150 W) suitable for cutting polymers and wood or a 1.06 μm fiber laser (typical powers of about 40 W) that can mark metals and cut thin foils [36–38]. These systems have a linear scanning speed of a few meters per second and the output power and laser spot size/focus can be adjusted by software.

11.3.1.1 Solid State Lasers: Fiber Laser

The active gain medium for fiber lasers consists of an optical fiber doped with rare-earth elements (e.g., Nd, Yb, Er). Commercial fiber laser systems typically operate at the wavelength of 1.06 μm, which is sufficiently small for creating device features with micrometer resolution. Additionally, the wavelength is more easily absorbed by metallic materials allowing for the processing of metal films. The output beam of the fiber laser can be operated in continued, pulsed, or Q-switched modes. A typical commercial fiber system is the PLSMW from Universal Laser Systems, Inc., which offers pulsed frequencies of up to hundreds of kHz, powers of up to 45 W, and scanning speeds of up to 4 mm/ms with optics allowing for laser spot size as small as 6 μm.

11.3.1.2 Gas Lasers

In gas lasers, the active medium consists of gas molecules offering advantages such as lasing media homogeneity, ease of cooling and replenishment, and low cost. The wavelength of the emitted light depends on the primary gas used; for instance, xenon chloride (XeCl) produces 308 nm, xenon fluoride (XeF) produces 351 nm, argon fluoride (ArF) produces 191 nm, argon produces 488 nm, and carbon dioxide (CO_2) produces 10.6 μm. Of these, CO_2 is the most common in industrial engraving systems due to its ability to cut a broad range of materials. During its operation, the CO_2 molecules are excited by vibrational excitation of nitrogen (intermixed with the CO_2) using high voltage electrical discharge. The excited nitrogen molecules correspond to highly unstable (001) vibrational levels of CO_2. The transition between (001) and the ground level of (100) results in a 10.6 μm laser radiation. The properties of CO_2 laser are mainly determined by the gas flow in the discharge tube and can be operated in both pulsed as well as continuous wave (CW) mode.

An example of a commercially available CO_2 laser system is the Universal Laser System, Inc., Professional Series, with a maximum power of 120 W, a maximum speed 4 mm/ms, wavelength of 10.6 μm, and continuous laser processing mode with optics allowing for laser spot size as small as 30 μm.

11.3.2 Laser-Material Interaction

CO_2 and fiber laser systems are routinely used for modifying materials in various ways, such as cutting, marking, welding, and chemical alteration [39, 40]. The specific result is determined by the interactions caused by the thermo-physical properties of the material and the electromagnetic radiation of the laser. During these interactions, a portion of the light is reflected, another transmitted, and the rest absorbed. The absorbed energy causes thermal effects (e.g., local heating, melting, vaporizing, or pyrolyzing). When laser machining the materials, the absorbance of the material is of utmost importance [41]. In general, metals have a lower absorbance at larger wavelengths. For instance, copper has a low absorbance of 0.05 at 10.6 μm (CO_2 laser) and higher absorbance of 0.3 at 1.6 μm (fiber laser). Hence, fiber lasers are often used as the primary laser pro tool for metallic materials. The resulting effect of laser absorbance can be classified into two categories: physical and chemical, as described below.

11.3.2.1 Physical Material Alterations via Laser

Physical changes to the material include removal of material (i.e., for through-hole cutting or surface ablation) and texturing (e.g., surface roughness). Both processes result from thermal effects and are a function of the laser fluence [34, 35]. The optical energy delivered by the laser per surface area of the material is known as the fluence. Ablation occurs at energy densities greater than the material's fluence threshold, which is between 1 and 10 J/cm^2 for metals and between 0.1 and 1 J/cm^2 for organic materials [42]. In thermoplastics, this process locally melts the material causing some to evaporate [40], Fig. 11.2a. The amount and rate of material removal can be precisely controlled by the laser scanning speed and the intensity. High intensity (or lower scanning speed) result is deep material removal which is commonly used for cutting materials, whereas lower intensity (or higher scanning speed) can be used for surface texturing. During the latter process, some of the molten material can redeposit on the surface creating local surface roughness. The power can be controlled to minimize ablation while still permitting redeposition to occur. Such laser surface texturing has been widely used for applications such as improving the adhesion and increasing the griping performance of steel sheets [39, 43]. In more recent developments, laser surface texturing has been used to create micro- and nano-scale super-hydrophobic roughness on different materials for self-cleaning applications [44]. In thermoset polymers, laser treatment above the

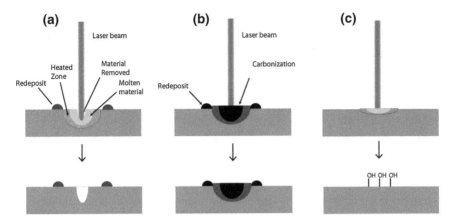

Fig. 11.2 Interactions of laser with materials. **a** Physical interaction with thermoplastics causes material removal via melting and evaporation. Molten material can redeposit on the surface. High power results in through-hole features, but lower power allows controlled ablation/texturing. **b** Physical interaction with thermosets decomposes the material (pyrolysis) with redeposited debris. High power results in through-hole features, but lower power allows controlled ablation/texturing. **c** Very low power allows minimal material damage but alters the surface chemistry via interactions with atmospheric gasses

fluence threshold results in material decomposition (e.g., pyrolysis, or carbonization in organic materials), Fig. 11.2b. This process also results in surface roughness due to non-uniform surface pyrolization as well as redeposition of pyrolized material. The physical surface texturing/ablative effects can be used to create porous and conductive regions on polymeric or paper substrates which form the basis for fabricating a variety of passive and active components used in flexible smart wound dressings described in the subsequent subsections of this chapter.

11.3.2.2 Laser Activated Chemical Processes

In addition to physical surface modification, the laser-beam interaction with material surfaces may also induce chemical modifications via a photo-thermal effect [45]. The high temperature generated by the laser can decompose the material and cause interactions with ambient gasses (e.g., oxygen, nitrogen) which can form additional functional groups Fig. 11.2c. Oftentimes, laser processing in such conditions results in the formation of hydrophilic functional groups allowing laser to be used for controlling surface wettability [46, 47]. This technique is particularly useful when working with natively hydrophobic substrates such as parchment paper, wax paper, or freezer paper; processing techniques for these are described in the following section.

11.4   Laser Treatment of Paper Substrates

11.4.1   Paper—Its Use for Device Fabrication

Paper is a classic material whose invention dates back to ancient China. Besides its traditional application as a writing medium, paper can be repurposed using modern technologies to create novel devices with complex functionality to use it as flexible sensors for wound monitoring. Paper offers many unique properties including biocompatibility, low cost, and ubiquity [48]. Moreover, its cellulose mesh composition invites customization of many physical parameters including thickness, fiber size, porosity, and hydrophilicity. Cellulose is a natural, hydrophilic fiber with strong hydrogen bonding between polymer chains which render it insoluble in water and most organic solvents. These properties have enabled the manufacturing of many different types of commercial paper including filter paper [49], wax paper [50], and parchment paper [47], among others. These three, in particular, have found niche applications in the field of disposable sensor development due to their favorable, tunable physical properties. The first of these is known to have excellent wicking properties. The latter two are naturally hydrophobic but their surface properties can be altered by plasma treatment or by laser ablation [51]. Many researchers have used these properties for the fabrication of paper-based systems for various biomedical applications including controlled drug delivery, tissue engineering, sutures, biodegradable vascular grafts [52, 53], low-cost microfluidics [54], and various flexible sensors [49, 55–60].

11.4.2   Laser Treatment of Hydrophobic Paper

Patterning paper to create hydrophilic-hydrophobic structures for microfluidic platforms used in medical diagnostic tests was first demonstrated by Müller and Clegg [61]. They created a narrow channel in paper via hot stamping of paraffin which led to faster diffusion in chromatography experiments with a significant reduction in the required analyte volume. In 2007, the Whitesides group reinvented the idea of using hydrophobic barriers to create a network of microchannels in paper. In their original work [49], they used SU-8-soaked chromatography paper which was subsequently patterned by lithography to create microfluidic channels. Since then, many researchers have used SU-8-based lithography, wax printing, screen printing, inkjet printing, and xerography for paper surface patterning. However, these methods suffer from one or more of the following shortcomings: (1) limited resolution (millimeter scale in wax printing), (2) multi-step processing sequence (wax remelting is done after printing in order to ensure its complete penetration across the paper thickness), and (3) low mechanical strength of the substrate when submerged in aqueous media for long periods of time. An alternative technique, developed by our group, is laser surface treatment of commercially available hydrophobic papers (e.g., parchment paper, wax paper, and palette paper) [47]. In contrast to the methods described by the Whitesides group which start with a hydrophilic plain paper and are predominantly additive, our approach is subtractive and can selectively convert hydrophobic areas to hydrophilic ones in a single-step process. Since the hydrophobic agent is already present throughout the thickness of the paper, our method does not require heat treatment after patterning to create islands of hydrophilic patterns, as is the case in the wax printing technique. This approach brings several major improvements and allows for the fabrication of more complicated platforms not feasible with other methods. These include: (1) higher resolution, (2) single-step patterning, (3) simultaneous surface processing and micromachining, and (4) greater robustness.

Laser patterning of paper can be achieved using a broad variety of commercially available paper substrates including parchment paper (Reynolds Parchment Paper, 50 μm thick), wax paper (Reynolds Cut-Rite Wax Paper, 30 μm thick), and palette paper (Canson Palette Paper, 70 μm thick). Parchment consists of a compressed cellulose fiber sheet encapsulated in a thin coating of silicone to achieve a hydrophobic, heat-resistant surface. Wax and palette paper use wax and plastic (polyethylene), respectively, as the coating material. Among these, parchment paper is particularly well-suited for creating biomedical microdevices due to its robustness and its silicone coating which allows seamless integration with other silicone (e.g., PDMS) devices. To pattern hydrophilic designs onto the paper substrates, the paper is placed flat on the cutting surface of a commercial laser engraver system (Universal Laser Systems, Inc., maximum power 60 W, maximum speed 4 mm/ms, wavelength 10.6 μm, CW mode), while a computer is used to control the laser parameters to transfer the pattern from a CAD design. The two primary parameters are laser power and beam scanning speed. Increasing the power or lowering the speed imparts higher energy onto the material resulting in deeper ablation (useful for creating though-paper defects). To avoid completely cutting through the paper (instead of selective surface modification), the laser power and scanning speed must be carefully selected. For the parchment paper selected in our experiments (50 μm-thick silicone coated), the laser source was controlled at 15 % of its maximum power and was operated at a maximum scanning speed. Even for other types of paper, namely wax and palette paper, successful surface treatment is achieved with these parameters.

Laser processing of paper creates physical and chemical surface changes which are observable via SEM and XPS analyses. Figure 11.3 shows SEM images of treated and untreated samples of parchment paper. The non-treated areas show cellulose fibers clearly protected by the surface coating, creating a natively hydrophobic surface Fig. 11.3a–c. In contrast, once the surface is laser treated (Fig. 11.3d–f), the top silicone layer is modified increasing the surface roughness via the creation of micro- and nano structures caused by thermal decomposition of the material. Nanofibers are also visible in laser-treated wax paper and laser-treated freezer paper; on such thermoplastic-coatings, the fibers can be attributed to melting and redeposition of the polymer. Such increase in surface roughness promotes a

Fig. 11.3 a–f *Top-view* SEM images of laser-treated parchment paper under different magnifications [47]. **a–c** Areas without laser treatment, and **d–f** laser-treated areas. **g, h** Paper-based devices for wound monitoring and healing. **g** Oxygen sensor [54]. **h** Array of pH sensors [63]. **i** Selective oxygenation patch driven by a finger-actuated pump

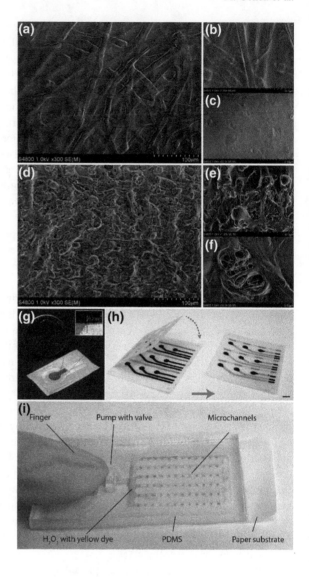

hydrophilic behavior. In addition to these physical changes, XPS analysis reveals the formation of hydrophilic functional groups on the surface of the paper on the laser-treated regions. In particular, the C1s spectra of laser-treated regions reveal additional oxidation peaks (C–OH at 286.8 eV and O=C–OH at 288.6 eV) compared to the spectra of untreated regions in the treated regions. Similarly, the Si2p spectrum of the treated region shows the presence of the SiO_2 peak at 103.7 eV, which is not observed in the case of untreated regions. These laser-induced chemical groups in combination with the increased surface roughness are capable of changing the water contact angle of parchment paper from its native 115° down to 20°, thus achieving very hydrophilic surfaces.

We have used this technique to fabricate paper-based devices such as sensors, batteries, and drug delivery platforms [47, 50, 51, 62]. As related to smart dressings for chronic wound management, we have developed oxygen and pH sensors on a parchment paper substrate. The oxygen sensor consists of a silver cathode and a zinc anode patterned on a parchment paper substrate by screen printing [54], Fig. 11.3g. Hydrophilic traces are first defined on hydrophobic paper using laser modification prior to screen printing. An electrolyte and a PDMS membrane seal the two electrodes in a single chamber. The paper behaves as the gas permeable membrane allowing oxygen to be reduced at the cathode. For oxygen concentrations between 0 and 45 %, this sensor produces an electric current that changes linearly from 2 to 129 μA (sensitivity of 2.6 μA/%O_2) with a time response of 17 s. The dynamic range and response time are acceptable for assessing hypoxia in chronic wounds and the structure can be readily embedded into traditional wound dressings.

Since the pH of the wound bed is an indicator of the healing progress (and infection status), we also developed a flexible, disposable array of pH sensors on paper [63], Fig. 11.3h. The sensors were fabricated on palette paper to allow practical packaging via thermal self-lamination. Each individual sensor of the array consists of two electrodes, one being an Ag/AgCl reference electrode and the other a carbon electrode coated with a conductive proton-selective polymer, polyaniline (PANI). The sensors are screen printed on an acrylic-coated paper (i.e., palette paper) using a laser-patterned tape as a mask. This is achieved by first applying a layer of Scotch MagicTape^{TM} onto the palette paper and then defining electrode patterns by directly writing on the mask (magic tape) using laser ablation. The laser changes the surface energy of the magic tape resulting in evaporation of the cellulose acetate in the tape. This process produces a shadow mask directly on the paper substrate without the requirement for transfer and alignment. Laser machining is further used to create an insulating layer that is sealed over the sensors by lamination technology. Characterization of the pH sensors reveals a linear ($r^2 = 0.9734$) relationship between the output voltage and pH in the 4–10 pH range with an average sensitivity of −50 mV/pH. The sensors feature a rise and fall time of 12 and 36 s, respectively, for a pH swing of 8–6–8.

Another laser-assisted paper-based device is a platform for selective oxygenation of the wound bed [62]. The platform consists of a PDMS microfluidic network bonded to a parchment paper substrate. Generation of oxygen occurs by flowing H_2O_2 through the channels and chemically decomposing it via a catalyst embedded in laser-defined regions of the parchment paper; these regions are hydrophilic due to an increased surface roughness as well as an increase in hydrophilic functional groups. PDMS is bonded to parchment paper using partially-cured PDMS followed by a brief air plasma treatment, resulting in a strong bond. Using a peroxide flow rate of 250 μL/min, oxygen generation in the catalyst spots raises the oxygen level on the opposite side of the parchment paper from atmospheric levels (21 %) to 25.6 %, with a long-term (30 h) generation rate of 0.1 μL O_2/min/mm^2. This rate is comparable to clinically proven levels for adequate healing. A prototype of the oxygen delivery patch driven by a finger-actuated pump is shown in Fig. 11.3i.

In addition to sensors, it is also possible to create electrical components such as batteries. Such flexible disposable batteries are attractive for their ability to directly integrate with other paper electronics [64]. In one example from our group, paper-based electrochemical cells were fabricated on wax paper using CO_2 laser surface treatment and micromachining. A four-cell zinc–copper battery shows a steady open-circuit voltage of ~ 3 V and can provide 0.25 mA for at least 30 min when connected to a 10 kΩ load. Higher voltages and current values can be obtained by adjusting the number and size of electrochemical cells in the battery without changing the fabrication process.

11.5 Laser Carbonization of Thermoset Polymers

11.5.1 Pyrolized Carbon for Sensing and Flexible Electronic Applications

Nanoscale carbon (carbon nanotubes and graphene) has unique chemical and electrical properties which have garnered significant attention over the past two decades [65]. In addition to their use in nanoelectronics, carbon nanoparticles printed onto flexible substrates or loaded onto various polymeric binders have been used to fabricate flexible and stretchable systems [66]. Examples of these include carbon-based pastes for screen printing conductive films [67], carbon inks for ink-jet printing [68], carbon-PLA filaments for extrusion-based 3D printers [68], and carbon-loaded elastomers [69]. These manufacturing materials are suitable for creating medical microsystems which can interface conformally with the human body and living tissue using rapid prototyping technologies Fig. 11.4.

Using carbon-based composites, researchers have been able to create a variety of electrical and mechanical sensors and actuators which are applicable for monitoring

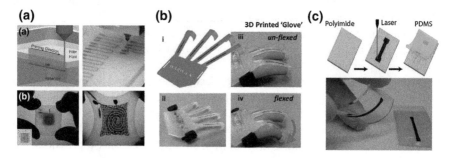

Fig. 11.4 a Carbon grease-based conductive liquid directly 3D printed into uncured elastomer via a dispenser printer and optical images for various three-layer strain and pressure sensors [88]. **b** Printed wearable glove motion sensors fabricated by commercial conductive carbon filaments and 3D printing technology [68]. **c** Stretchable carbon nanocomposite with using laser pyrolization and transferring onto PDMS elastomer [74]

or treating cutaneous wounds. For example, carbon-based inks have been used to define electrically conductive traces on ceramic and polymeric substrates using screen printing techniques [70, 71]. The natural biocompatibility and chemical stability of carbon allow these traces to be used as electrical conduits on wearable devices/systems including smart tattoos [72]. Meanwhile, materials comprising carbon nanoparticles embedded in stretchable binders (e.g. carbon-loaded PDMS or carbon-loaded PLA) have been used to create soft, elastomeric arrays strain/pressure sensors [73, 74].

Despite numerous reported carbon-based devices and sensors, few have been commercialized. This is primarily due to challenges with scaling the production of carbon-based materials. For example, many of the nano/micro particles used in these systems are not economical to mass-produce for practical use in medical devices since they must typically be made into (possibly non-biocompatible) inks/pastes with general applicability (rather than specifically for medical applications) [75]. It would be more economical to create the carbon composites directly on the substrate without the use of additional binder materials (e.g., for inks) which may interfere with the biocompatible aspects of the material. One approach is to fabricate devices using carbonizable material. Many thermoset organic polymers can be pyrolized by raising their temperature to above 1000 °C [76]. The result is pure carbon which is electrically conductive. Researchers have used this idea to create conductive carbon traces by pyrolyzing photoresists using a high temperature furnace [77]. Such materials have been used to fabricate supercapacitors [78], batteries [79], electrodes and biosensors for biomedical applications [80]. While the process is economical and straightforward, it does not allow precise patterning of the material in a straightforward process. Moreover, the various components of the device must be resistant to high temperatures, lest they too be thermally decomposed.

11.5.2 Laser Carbonization

An alternative approach is to selectively carbonize a thermoset polymeric substrate via laser ablation. In this technique, laser energy is imparted onto a substrate to locally heat it to a sufficiently high temperature to induce carbonization. The use of such precise energy control allows selective carbonization limited to only the trajectory of the laser beam. Lasers (mostly excimer with wavelengths ranging from 193 to 351 nm) have been used to carbonize a wide range of polymeric materials [81]. Carbon films generated using UV-laser irradiation in air are of glassy nature and have also been used to fabricate conductive traces and electrodes on polyimide substrates [82, 83].

A practical material for laser-carbonizing is polyimide [84, 85], an organic thermoset which can be casted as a liquid and cured into a film, or alternatively, procured in thin sheets with or without metallic backing. Upon pyrolization, the polyimide is converted into a porous nano-scale carbonized layer with a very high surface area (analogous to "activated carbon" which is known to be very absorbent of gasses and small molecules) [86, 87] that can be used for creating functional materials/devices. Unlike bulk pyrolization methods (e.g., furnaces), the laser-based technique offers unprecedented control over carbon nanoparticle deposition and patterning. In addition, the degree of carbonization (and the density of carbon traces) can be controlled by the laser parameters.

Recent work in our group has been focused on developing stretchable biomedical devices via laser-induced pyrolysis and their incorporation into stretchable platforms [73]. Using a commercial CO_2 laser engraver, we were able to create partially-aligned graphene and carbon nanotube (CNT) conductive traces on polyimide substrates (Fig. 11.5a). The traces were then transferred and embedded into an elastomer (e.g., PDMS or Ecoflex®) to fabricate stretchable and conductive composites. The surface morphology of the carbon traces, before and after embedment in the PDMS, was qualitatively investigated by scanning electron microscopy (SEM). The top view of carbon traces clearly reveals highly porous carbon micro- and nanoparticles arranged in a parallel pattern. This phenomenon is related to the method by which the laser beam is scanned across the sample during the fabrication. Since the practical spot size (diameter) of the laser beam in our system is 30 µm, ablation of areas larger than 30 µm requires multiple sweeps of the laser beam over the targeted area, thus generating carbon particles in a parallel orientation to the direction of laser motion. Higher magnified pictures of the carbon particles in the pyrolized lines are shown in Fig. 11.5b. Partially oriented carbon flakes and high aspect ratio filaments (some of them as small as ∼70 nm wide with lengths of up to ∼2 µm) can be seen on the carbon traces. A cross-sectional view of the carbon patterns shows that the entire thickness of the pyrolized carbon is comprised of highly porous nanomaterials (Fig. 11.5c). This enables the PDMS to penetrate deep into the carbon patterns resulting in a uniform transfer of carbon nanoparticles to the elastomeric matrix. The thickness of the carbonized regions embedded in the PDMS is ∼30 µm (Fig. 11.5d), which is close to their original thickness on the polyimide (before transfer). The PDMS-carbon composites showed high stretchability (up to 100 % strain) and sensitive (gauge factor of up to 20,000)

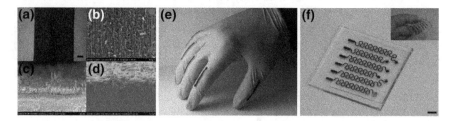

Fig. 11.5 Surface and film architecture details [73]. **a** Photograph of the carbonized polyimide. *Scale bar* 250 µm. **b** SEM image of the aligned particles in the traces with the *arrow* showing the direction of laser ablation. **c** Cross-sectional image of the carbon traces showing the porosity of the carbonized material. **d** Cross-sectional SEM of stretchable carbon traces embedded in PDMS. **e** Glove equipped with strain sensors at the joints. **f** Stretchable array of pH sensors

with a high degree of sharp anisotropy. It was found that the high sensitivity and unidirectional strain sensitivity of the conductive composite stems from the partially aligned graphene and CNT particles which are created in the direction of the laser scanning.

Conductivity in the stretchable composite is achieved by the large number of electrical pathways that are formed by interconnection of the carbon particles in the network. The increase in the electrical resistance of the composite due to applied strain is a result of increase in the spacing between the carbon particles and reduce the number of electrical pathways in the composite. The conductivity can be tuned by varying the laser processing parameters. Our experiments reveal that there is an optimum combination of laser power (4.5–8.25 W) and speed (0.5–1.9 m/s) needed for producing high-quality, high-conductivity traces. Our data show a linear relationship with speed and power (a maximum conductivity of 0.02 Mhos/square with 6.75 W and 1.3 m/s) with the optimal energy density being 62 000 J/m^2. The sensors have sufficiently fast response for many practical applications such as biological kinetics (e.g., wound healing or muscle repair) and human motion.

By forming the stretchable composite into the meander shaped patterns, the strain sensitivity of the conductive composite can significantly be reduced, which makes it suitable for highly stretchable electrode applications Fig. 11.5e. The carbon traces can be further functionalized with other materials such as polyaniline to create sensors with high surface area and specific binding to chemical/biological analytes. For example, our group recently developed a stretchable pH sensor for wound monitoring applications, Fig. 11.5f. The process combines selective laser pyrolization/carbonization and embedment into an elastomeric matrix (meander interconnects show small sensitivity to strains of up to 50 %). The stretchable conductive composites are functionalized with a pH-sensitive polymer (polyaniline) to create stretchable pH sensors with a sensitivity response of −51 mV/pH over a physiologically relevant range of pH 4–10. The mechanical flexibility and stretchability of the sensor along with its chemical performance are sufficient for low-cost disposable wound dressings with smart sensing functionality.

11.6 Conclusion

Laser processing of materials offers a practical and economical technique for physical and chemical alteration of low-cost flexible material surfaces that can subsequently be used to create sensors and drug delivery devices for use in the next-generation smart wound dressings. This technology allows processing of flexible/stretchable materials ranging from polymers to metallic films allowing for rapid fabrication of devices using commercial substrates. In particular, a CO_2 laser system can be used to selectively tune the surface energy (i.e., wettability) of commercial hydrophobic papers (e.g., wax paper, parchment paper) to create paper-based low-cost microfluidic platforms and sensors for wound monitoring and healing (e.g., oxygen detection and generation). Additionally, the system can be used for pyrolyzing materials such as organic thermosets to create patterns of carbon nanomaterials which exhibit exceptional electrical conductivity and chemical properties. By incorporating these nanomaterials on stretchable substrates, one can fabricate highly-sensitive strain sensors for monitoring the effect of strain/stress on wounds. These examples demonstrate the possibilities offered by the combination of modern commercial laser systems with off-the-shelf flexible substrates for the development of flexible low-cost devices that are particularly suited for cutaneous wound interfaces.

References

1. L.K. Branski, G.G. Gauglitz, D.N. Herndon, M.G. Jeschke, A review of gene and stem cell therapy in cutaneous wound healing. Burns 35(2), 171–180 (2009)
2. A. Stojadinovic, J.W. Carlson, G.S. Schultz, T.A. Davis, E.A. Elster, Topical advances in wound care. Gynecol. Oncol. 111(2 Suppl), S70–S80 (2008)
3. D.M. Castilla, Z.-J. Liu, O.C. Velazquez, Oxygen: implications for wound healing. Adv. Wound Care 1(6), 225–230 (2012)
4. N.S. Greaves, S.A. Iqbal, M. Baguneid, A. Bayat, The role of skin substitutes in the management of chronic cutaneous wounds. Wound Repair Regen. 21(2), 194–210 (2013)
5. C.K. Sen, Wound healing essentials: let there be oxygen. Wound Repair Regen. 17(1), 1–18 (2010)
6. V. Falanga, Classifications for wound bed preparation and stimulation of chronic wounds. Wound Repair Regen. 8(5), 347–352 (2000)
7. R.O.Y.W. Tarnuzzer, G.S. Schultz, Biochemical analysis of acute and chronic wound environments. Wound Repair Regen. 4, 321–325 (1996)
8. A.A. Tandara, T.A. Mustoe, Oxygen in wound healing—more than a nutrient. World J. Surg. 28(3), 294–300 (2004)
9. K. Bumpus, M.A. Maier, The ABC's of wound care. Curr. Cardiol. Rep. 15(4), 346 (2013)
10. J.G. Powers, L.M. Morton, T.J. Phillips, Dressings for chronic wounds. Dermatol. Ther. 26(3), 197–206 (2013)
11. D. Queen, H. Orsted, H. Sanada, G. Sussman, A dressing history. Int. Wound J. 1(1), 59–77 (2004)
12. D. Okan, K. Woo, E.A. Ayello, G. Sibbald, The role of moisture balance in wound healing. Adv. Skin Wound Care 20(1), 39–53 (2007)
13. J.L. Richard, J.M. Rochet, N. Sales-aussias, Dressings for acute and chronic wounds. Arch. Dermatol. 143(10), 1297–1304 (2007)
14. N.F. Watson, W. Hodgkin, Wound dressings. Surgery 23(2), 52–55 (2005)
15. G.S. Schultz, R.G. Sibbald, V. Falanga, E.A. Ayello, C. Dowsett, K. Harding, M. Romanelli, M.C. Stacey, L. Teot, W. Vanscheidt, Wound bed preparation: a systematic approach to wound management. Wound Repair Regen. 11, 1–28 (2003)
16. C.L. Hess, M.A. Howard, C.E. Attinger, A review of mechanical adjuncts in wound healing: hydrotherapy, ultrasound, negative pressure therapy, hyperbaric oxygen, and electrostimulation. Ann. Plast. Surg. 51(2), 210–218 (2003)
17. A. Sood, M.S. Granick, N.L. Tomaselli, Wound dressings and comparative effectiveness data. Adv. Wound Care 00(00), 130716103126002 (2013)
18. Z. Aziz, S.F. Abu, N.J. Chong, A systematic review of silver-containing dressings and topical silver agents (used with dressings) for burn wounds. Burns 38(3), 307–318 (2012)

19. S.M. O'Meara, N.A. Cullum, M. Majid, T.A. Sheldon, Systematic review of antimicrobial agents used for chronic wounds. Br. J. Surg. **88**(1), 4–21 (2001)
20. P. Boisseau, B. Loubaton, Nanomedicine, nanotechnology in medicine. C. R. Phys. **12**(7), 620–636 (2011)
21. M. Ochoa, C. Mousoulis, B. Ziaie, Polymeric microdevices for transdermal and subcutaneous drug delivery. Adv. Drug Deliv. Rev. **64**(14), 1603–1616 (2012)
22. N. Mehmood, A. Hariz, R. Fitridge, N.H. Voelcker, Applications of modern sensors and wireless technology in effective wound management. J. Biomed. Mater. Res. B. Appl. Biomater. 1–11, 2013
23. T.R. Dargaville, B.L. Farrugia, J.A. Broadbent, S. Pace, Z. Upton, N.H. Voelcker, Sensors and imaging for wound healing: a review. Biosens. Bioelectron. **41**, 30–42 (2013)
24. B. Derby, Inkjet printing of functional and structural materials: fluid property requirements, feature stability, and resolution. Annu. Rev. Mater. Res. **40**(1), 395–414 (2010)
25. S. Hengsbach, A.D. Lantada, Rapid prototyping of multi-scale biomedical microdevices by combining additive manufacturing technologies. Biomed. Microdevices **16**(4), 617–627 (2014)
26. D. Queen, J.H. Evans, J.D. Gaylor, J.M. Courtney, W.H. Reid, An in vitro assessment of wound dressing conformability. Biomaterials **8**(5), 372–376 (1987)
27. T.E. Wright, W.G. Payne, F. Ko, D. Ladizinsky, N. Bowlby, R. Neeley, B. Mannari, M.C. Robson, The effects of an oxygen-generating dressing on tissue infection and wound healing. J. Appl. Res. **3**(4), 363–370 (2003)
28. D.W. Brett, A review of moisture-control dressings in wound care. J. Wound Ostomy Cont. Nurs. **33**(6 Suppl), S3–S8 (2006)
29. S.-F. Lo, M. Hayter, C.-J. Chang, W.-Y. Hu, L.-L. Lee, A systematic review of silver-releasing dressings in the management of infected chronic wounds. J. Clin. Nurs. **17** (15), 1973–1985 (2008)
30. G. Chaby, P. Senet, M. Vaneau, P. Martel, J.-C. Guillaume, S. Meaume, L. Téot, C. Debure, A. Dompmartin, H. Bachelet, H. Carsin, V. Matz, J.L. Richard, J.M. Rochet, N. Sales-Aussias, A. Zagnoli, C. Denis, B. Guillot, O. Chosidow, Dressings for acute and chronic wounds: a systematic review. Arch. Dermatol. **143**(10), 1297–1304 (2007)
31. S.L. Percival, S. McCarty, J.A. Hunt, E.J. Woods, The effects of pH on wound healing, biofilms, and antimicrobial efficacy. Wound Repair Regen. **22**(2), 174–186 (2014)
32. G. Gethin, The significance of surface pH in chronic wounds. Wounds UK **3**(3), 52–55 (2007)
33. S. Schreml, R.M. Szeimies, L. Prantl, S. Karrer, M. Landthaler, P. Babilas, Oxygen in acute and chronic wound healing. Br. J. Dermatol. **163**(2), 257–268 (2010)
34. E. Kannatey-Asibu, *Principles of Laser Materials Processing* (Wiley, New York, 2009)
35. W. Steen, I. Mazumder, K.G. Watkins, *Laser Material Processing*, 4th edn. (Springer, New York, 2010)
36. S. Mueller, B. Kruck, P. Baudisch, LaserOrigami: laser-cutting 3D objects, in *Proceedings of the SIGCHI Conference on Human Factors in Computing Systems*, 2013, pp. 2585–2592
37. A. Toossi, M. Daneshmand, D. Sameoto, A low-cost rapid prototyping method for metal electrode fabrication using a CO_2 laser cutter. J. Micromech. Microeng. **23**(4), 047001 (2013)
38. J. Yuan, J. Chen, C. He, Research of micro removing copper foil of FCCL assisted with laser, in *2011 IEEE International Conference on Mechatronics and Automation*, 2011, pp. 749–754
39. D. Waugh, J.B. Griffiths, J.R. Lawrence, C. Dowding, *Lasers in Surface Engineering, Surface Engineering Series* (ASM International, Materials Park, 1998)
40. N.B. Dahotre, S.P. Harimkar, *Laser Fabrication and Machining of Materials* (Springer, Heidelberg, 2007)
41. W. Steen, *Laser Materials Processing* (Springer, London, 1991)
42. D. Bäuerle, *Laser Processing and Chemistry* (Springer, Heidelberg, 2000)
43. C. Hallgren, H. Reimers, D. Chakarov, J. Gold, A. Wennerberg, An in vivo study of bone response to implants topographically modified by laser micromachining. Biomaterials **24**(5), 701–710 (2003)

44. M. Psarski, J. Marczak, J. Grobelny, G. Celichowski, Superhydrophobic surface by replication of laser micromachined pattern in epoxy/alumina nanoparticle composite. J. Nanomater. **2014**, 41 (2014)
45. H. Klank, J.P. Kutter, O. Geschke, CO_2-laser micromachining and back-end processing for rapid production of PMMA-based microfluidic systems. Lab Chip **2**(4), 242 (2002)
46. J. Lawrence, L. Li, *Laser Modification of the Wettability Characteristics of Engineering Materials* (Professional Engineering, London, 2001)
47. G. Chitnis, Z. Ding, C. Chang, C.A. Savran, B. Ziaie, Laser-treated hydrophobic paper: an inexpensive microfluidic platform. Lab Chip **11**(6), 1161–1165 (2011)
48. A.K. Yetisen, M.S. Akram, C.R. Lowe, Paper-based microfluidic point-of-care diagnostic devices. Lab Chip **13**(12), 2210–2251 (2013)
49. A.W. Martinez, S.T. Phillips, M.J. Butte, G.M. Whitesides, Patterned paper as a platform for inexpensive, low-volume, portable bioassays. Angew. Chem. Int. Ed. Engl. **46**(8), 1318–1320 (2007)
50. G. Chitnis, T. Tan, B. Ziaie, Laser-assisted fabrication of batteries on wax paper. J. Micromech. Microeng. **23**(11), 114016 (2013)
51. G. Chitnis, B. Ziaie, Waterproof active paper via laser surface micropatterning of magnetic nanoparticles. ACS Appl. Mater. Interfaces **4**(9), 4435–4439 (2012)
52. L.S. Nair, C.T. Laurencin, Polymers as biomaterials for tissue engineering and controlled drug delivery. Adv. Biochem. Eng. Biotechnol. **102**, 47–90 (2006)
53. P.B. Maurus, C.C. Kaeding, Bioabsorbable implant material review. Oper. Tech. Sports Med. **12**(3), 158–160 (2004)
54. R. Rahimi, G. Chitnis, P. Mostafalu, M. Ochoa, S. Sonkusale, B. Ziaie, A low-cost oxygen sensor on paper for monitoring wound oxygenation, in *The 7th International Conference on Microtechnologies in Medicine and Biology*, 2013
55. A.W. Martinez, S.T. Phillips, Z. Nie, C.-M. Cheng, E. Carrilho, B.J. Wiley, G.M. Whitesides, Programmable diagnostic devices made from paper and tape. Lab Chip **10**(19), 2499–2504 (2010)
56. C. Rivet, H. Lee, A. Hirsch, S. Hamilton, H. Lu, Microfluidics for medical diagnostics and biosensors. Chem. Eng. Sci. **66**(7), 1490–1507 (2011)
57. D. Nilsson, An all-organic sensor–transistor based on a novel electrochemical transducer concept printed electrochemical sensors on paper. Sens. Actuators B Chem. **86**, 193–197 (2002)
58. P. Spicar-Mihalic, B. Toley, J. Houghtaling, T. Liang, P. Yager, E. Fu, CO_2 laser cutting and ablative etching for the fabrication of paper-based devices. J. Micromech. Microeng. **23**(6), 067003 (2013)
59. W. Karlos, D.P. De Jesus, A. Fracassi, C. Lucio, W.K.T. Coltro, D.P. de Jesus, J.A.F. da Silva, C.L. do Lago, E. Carrilho, Toner and paper-based fabrication techniques for microfluidic applications. Electrophoresis **31**(15), 2487–2498 (2010)
60. A. Russo, B.Y. Ahn, J.J. Adams, E.B. Duoss, J.T. Bernhard, J.A. Lewis, Pen-on-paper flexible electronics. Adv. Mater. **23**(30), 3426–3430 (2011)
61. R.H. Müller, D.L. Clegg, Automatic paper chromatography. Anal. Chem. **21**(9), 1123–1125 (1949)
62. M. Ochoa, R. Rahimi, T.L. Huang, N. Alemdar, A. Khademhosseini, M.R. Dokmeci, B. Ziaie, A paper-based oxygen generating platform with spatially defined catalytic regions. Sens. Actuators B Chem. **198**, 472–478 (2014)
63. R. Rahimi, M. Ochoa, X. Zhao, M.R. Dokmeci, A. Khademhosseini, B. Ziaie, A flexible ph sensor array on paper using laser pattern definition and self-aligned laminated encapsulation, in *Hilton Head 2014: A Solid-State Sensors, Actuators and Microsystems Workshop*, 2014
64. T.H. Nguyen, A. Fraiwan, S. Choi, Paper-based batteries: a review. Biosens. Bioelectron. **54**, 640–649 (2014)

65. A. Qureshi, W.P. Kang, J.L. Davidson, Y. Gurbuz, Review on carbon-derived, solid-state, micro and nano sensors for electrochemical sensing applications. Diam. Relat. Mater. **18**(12), 1401–1420 (2009)
66. S. Park, M. Vosguerichian, Z. Bao, A review of fabrication and applications of carbon nanotube film-based flexible electronics. Nanoscale **5**(5), 1727–1752 (2013)
67. W.-J. Guan, Y. Li, Y.-Q. Chen, X.-B. Zhang, G.-Q. Hu, Glucose biosensor based on multi-wall carbon nanotubes and screen printed carbon electrodes. Biosens. Bioelectron. **21** (3), 508–512 (2005)
68. S.J. Leigh, R.J. Bradley, C.P. Purssell, D.R. Billson, D.A. Hutchins, A simple, low-cost conductive composite material for 3D printing of electronic sensors. PLoS ONE **7**(11), e49365 (2012)
69. S. Pyo, J.-I. Lee, M.-O. Kim, T. Chung, Y. Oh, S.-C. Lim, J. Park, J. Kim, Development of a flexible three-axis tactile sensor based on screen-printed carbon nanotube-polymer composite. J. Micromech. Microeng. **24**(7), 075012 (2014)
70. J. Wang, B. Tian, V.B. Nascimento, L. Angnes, Performance of screen-printed carbon electrodes fabricated from different carbon inks. Electrochim. Acta **43**, 3459–3465 (1988)
71. K. Grennan, A.J. Killard, M.R. Smyth, Physical characterizations of a screen-printed electrode for use in an amperometric biosensor system. Electroanalysis **13**(8–9), 745–750 (2001)
72. A.J. Bandodkar, V.W.S. Hung, W. Jia, G. Valdés-Ramírez, J.R. Windmiller, A.G. Martinez, J. Ramírez, G. Chan, K. Kerman, J. Wang, Tattoo-based potentiometric ion-selective sensors for epidermal pH monitoring. Analyst **138**(1), 123–128 (2013)
73. R. Rahimi, M. Ochoa, W. Yu, B. Ziaie, Highly stretchable and sensitive unidirectional strain sensor via laser carbonization. ACS Appl. Mater. Interfaces, 150220090153008, 2015
74. M.-Y. Cheng, C.-M. Tsao, Y.-Z. Lai, Y.-J. Yang, The development of a highly twistable tactile sensing array with stretchable helical electrodes. Sens. Actuators A Phys. **166**(2), 226–233 (2011)
75. J. Zhong, Carbon nanofibers and their composites: a review of synthesizing, properties and applications. Materials (Basel) **7**(5), 3919–3945 (2014)
76. O.C. Jeong, S. Konishi, Three-dimensionally combined carbonized polymer sensor and heater. Sens. Actuators A Phys. **143**(1), 97–105 (2008)
77. N.E. Hebert, B. Snyder, R.L. Mccreery, W.G. Kuhr, S.A. Brazill, Performance of pyrolyzed photoresist carbon films in a microchip capillary electrophoresis device with sinusoidal voltammetric detection. Anal. Chem. **75**(16), 4265–4271 (2003)
78. M.S. Kim, B. Hsia, C. Carraro, R. Maboudian, Flexible micro-supercapacitors with high energy density from simple transfer of photoresist-derived porous carbon electrodes. Carbon N. Y. **74**, 163–169 (2014)
79. H.-S. Min, B.Y. Park, L. Taherabadi, C. Wang, Y. Yeh, R. Zaouk, M.J. Madou, B. Dunn, Fabrication and properties of a carbon/polypyrrole three-dimensional microbattery. J. Power Sources **178**(2), 795–800 (2008)
80. K.C. Morton, C.A. Morris, M.A. Derylo, R. Thakar, L.A. Baker, Carbon electrode fabrication from pyrolyzed parylene C. Anal. Chem. **83**(13), 5447–5452 (2011)
81. T. Lippert, E. Ortelli, J. Panitz, F. Raimondi, J. Wambach, J. Wei, A. Wokaun, Imaging-XPS/ Raman investigation on the carbonization of polyimide after irradiation at 308 nm. Appl. Phys. A **69**, 651–654 (1999)
82. G. Shafeev, P. Hoffmann, Light-enhanced electroless Cu deposition on laser-treated polyimide surface. Appl. Surf. Sci. **138–139**, 455–460 (1999)
83. J.M. Ingram, M. Greb, J.A. Nicholson, A.W. Fountain, Polymeric humidity sensor based on laser carbonized polyimide substrate. Sens. Actuators B Chem. **96**(1–2), 283–289 (2003)
84. M. Inagaki, S. Harada, T. Sato, T. Nakajima, Y. Horino, K. Morita, Carbonization of polyimide film 'Kapton'. Carbon N. Y. **27**(2), 253–257 (1989)
85. F. Raimondi, S. Abolhassani, R. Brütsch, F. Geiger, T. Lippert, J. Wambach, J. Wei, A. Wokaun, Quantification of polyimide carbonization after laser ablation. J. Appl. Phys. **88**(6), 3659 (2000)

86. M. Park, L.N. Cella, W. Chen, N.V. Myung, A. Mulchandani, Carbon nanotubes-based chemiresistive immunosensor for small molecules: detection of nitroaromatic explosives. Biosens. Bioelectron. **26**(4), 1297–1301 (2010)
87. J. Li, Y. Lu, Q.L. Ye, J. Han, M. Meyyappan, Carbon nanotube based chemical sensors for gas and vapor detection. Chem. Phys. Lett. **313**(2), 91 (1999)
88. J.T. Muth, D.M. Vogt, R.L. Truby, Y. Mengüç, D.B. Kolesky, R.J. Wood, J.A. Lewis, Embedded 3D printing of strain sensors within highly stretchable elastomers. Adv. Mater. **26**, 6307–6312 (2014)

Chapter 12
Nanomaterials-Based Skin-Like Electronics for the Unconscious and Continuous Monitoring of Body Status

J.H. Lee, H.S. Kim, J.H. Kim, I.Y. Kim and S.-H. Lee

Abstract Long-term continuous monitoring of body condition from the skin has been one of the critical issues in the ubiquitous healthcare. For this purpose, skin-like stretchable and flexible electrodes have been highly required and diverse electrodes have been developed. However, these electrodes have limits such as lower electrical property, biocompatibility, and discomfort to patients. To address these challenges, nanomaterial-based electronic devices have been developed. In this chapter, current status of nanomaterial-based skin-like electronics with mechanical properties comparable to those of skin is reviewed, and their applications in biomedical fields are described. The types of clinically significant biosignals that can be measured from skin using soft electrodes are briefly summarized. The requirements of electrode for long-term, continuous, and unconscious measurement of these biosignals are also briefly described. Among several nanomaterials for soft electronics, carbon nanotube (CNT), graphene, and metallic nanowire are mainly commented and diverse flexible and stretchable electrodes using nanomaterials and their fabrication methods were described. For the biomedical applications, safety for the human use is a critical requirement, and their biocompatibility, future research directions, and possible additional applications in various fields are assessed.

Keywords Skin-like electronics · Soft material · Conductive nanomaterial · Biopotential

J.H. Lee · H.S. Kim · J.H. Kim · S.-H. Lee
KU-KIST Graduate School of Converging Science and Technology,
Korea University, Seoul 136-701, Republic of Korea

I.Y. Kim
Department of Biomedical Engineering, Hanyang University,
Seoul 133-791, Republic of Korea

S.-H. Lee (✉)
Department of Biomedical Engineering, College of Health Science,
Korea University, Seoul 136-701, Republic of Korea
e-mail: dbiomed@korea.ac.kr

© Springer International Publishing Switzerland 2016
J.A. Rogers et al. (eds.), *Stretchable Bioelectronics for Medical Devices and Systems*, Microsystems and Nanosystems,
DOI 10.1007/978-3-319-28694-5_12

12.1 Introduction

The brain and nervous system control the human body, coordinating its voluntary and involuntary actions. The central nervous system and peripheral nervous system send and receive biosignals from different parts of the body and they play a pivotal role in transmitting electrochemical waves that form neural circuits. Notably, malfunctions of the nervous system may cause serious diseases. Thus, continuous monitoring of biopotentials generated from the nervous system is critically important in healthcare and can help clinicians to diagnosis and treat diseases such as epilepsy, heart disease such as arrhythmia and Parkinson's. Therefore, progress in related technologies will also contribute to enhancing the quality of patients' lives through improved diagnosis and treatment of disease and advanced prosthetic applications. In addition to biosignals, other electrical, mechanical, and optical information such as impedance, strain, pressure, and imaging are important in health care and prosthetics applications. Although most of these parameters were measured at hospital previously, recent progress in ubiquitous healthcare technology allows them to be continuously monitored regardless of time and location for earlier and accurate diagnosis of disease. For this purpose, development of continuous and long-term measurement methods of these parameters without causing any inconvenience to the patients has been one of the hottest issues. To date, diverse measurement devices have been developed; however, recording these biosignals continuously over weeks without causing biocompatible issues and discomfort to patients remains a considerable challenge [1]. To address this challenge, diverse skin-attachable soft electrodes have been developed. They offer a facile and safe recording solution that compares favorably to invasive or conventional wet electrode-based strategies. During recent decades, the technology for these electrodes has rapidly progressed, and a number of diverse electrodes usable for patients have been developed. Some of them are soft enough to be conformally attached to the skin and enable the continuous measurement of biosignals for extended periods of time without critical effect of motion and sweat [2]. Recent progress of micro- and nanotechnology allows the marvelous development of this skin-like electrode minimizing patients' incontinence and enhancing the quality of their daily life. Roger's group demonstrated striking epidermal electronic systems (EES) by integrating multiple components such as transistors, light-emitting diodes, and wireless power coils for the multifunctional biosignal detection system [3]. Such skin-attachable devices are expected to contribute much in the diverse fields including medicine, robotics, game, and smart devices. To date, diverse technologies to fabricate skin-like electronics have been developed and introduced at several review papers [4, 5]. However, their fabrication technologies are based on photolithography and the process is complicate and expensive. As an alternative approach, nanomaterial-based skin-like electronics attract much attention and are extensively used for the stretchable and flexible electronics. Carbon nanotubes (CNTs) and graphene, the two latest emerging carbon allotropes, demonstrated the supreme electrical, thermal, optical, and mechanical properties and attracted great attention to 1D and 2D nanomaterials. Metallic nanowires and nanoparticles have also attracted a lot of attention for potential applications as transparent and flexible electrodes. The fabrication of nanomaterial-based electronics is simple and cost-effective and will be used as prerequisite materials of skin-like electronics. Despite the extensive potential, nanomaterial-based skin-like electrodes applicable to the biomedical areas are not popular in comparing with metal-based electrodes because of difficulty in precise patterning, low conductivity, and increasing resistance with applied strain. The biocompatibility issue of carbon material is still debatable, which prevents the expansion of nanomaterials' applications.

In this chapter, the current status of nanomaterial-based skin-like electronics with mechanical properties comparable to those of skin, and survey of their applications is reviewed in biomedical fields. The types of clinically significant biosignals that can be measured from skin using soft electrodes are briefly summarized and also described about the requirements of electrodes for long-term, continuous, and unconscious measurement of these biosignals. Although several nanomaterials were developed and used for soft electronics, carbon nanotube (CNT), graphene, and metallic nanowire are representative nanomaterials and will be mainly described. For the biomedical applications, safety for the human use is critical requirement, and their biocompatibility will be described. Several methods for fabrication of skin-like electrodes with nanomaterials and polymers will be briefly discussed in this chapter. Finally, future research directions and possible additional diverse applications are assessed.

12.2 Measurable Biosignals from the Skin Using Skin-Like Electrode Only

Various biosignals are generated from a living body and they reflect the status of health, activity, and organ functions (Fig. 12.1). By analyzing diverse biosignals, a quantity of information about the subject's physical and biological state can be obtained. Biosignals recorded from the skin are classified according to their physical characteristics as electrical, mechanical, optical, acoustic, and thermal biosignals. In this section, electrical, mechanical (motion and pressure), and epidermal impedance, which are critical information for health monitoring and other biomedical applications, are discussed.

12.2.1 Electrical Biosignals

Electrical biosignals are generated from the signals transmitted through the nervous system and commonly used for monitoring patient's health and the early detection

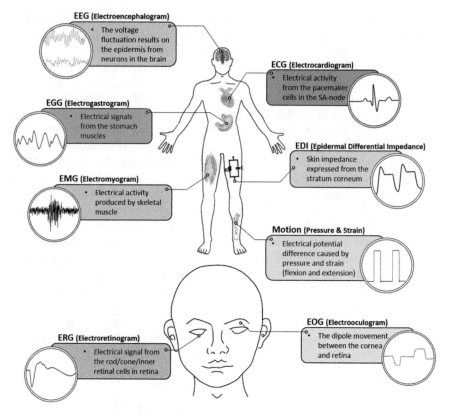

Fig. 12.1 Electrical biosignals from the biosignals measured from the skin

and diagnosis of disease. Table 12.1 summarizes the representative electrical biosignals s and their details are as follows:

- The electrocardiogram (ECG) is a technique for measuring the electrical activity over the heart using two or more electrodes. The heart electrical conduction sequence starts from the depolarization of pacemaker cells in the sinoatrial (SA) node and proceeds to the atrium and, through the atrioventricular (AV) node to the bundle of His and Purkinje fibers, ultimately spreading into the ventricles. In this sequence, the ECG has a specific waveform that can be divided into P-, Q-, R-, S-, and T-waves. The P-wave represents atrial depolarization; the QRS-wave represents ventricular depolarization; and the T-wave represents heart repolarization. Heart diseases, including arrhythmia and other conduction disturbances, can be diagnosed by ECG waveform analysis.
- The electroencephalogram (EEG) is used to measure the activity and function of the brain. This technique typically uses noninvasive electrodes attached to the scalp to measure the voltage fluctuations resulting from ionic current of neurons in the brain. EEG signals can be divided according to band frequencies into

Table 12.1 Electrical biosignals measurable from the skin

Biosignal	Measuring site	Frequency (Hz)	Amplitude range (peak)	Signal source	Application
ECG	Chest	0.01–250	<5 mV	Heart muscle movement	Arrhythmia, abnormal conduction, and cardiac monitoring
EEG	Head	0.01–150	2–100 µV	Brain neuron activation	Brain activity, function research, and brain computer interface (BCI)
EGG	Stomach	0.01–0.25	200–500 µV	Gastric muscle movement	Abnormal stomach function
EMG	Arm and leg	0.01–10,000	50 µV–5 mV	Muscle movement	Diagnostic muscle activity, exercise cure, and human machine interface (HMI)
EOG	Around eye	0.01–50	10 µV–5 mV	Eye dipole field	Eye movement

delta band (<4 Hz), theta band (4–7 Hz), alpha band (8–15 Hz), beta band (16–31 Hz), gamma band (>32 Hz), and mu band (8–12 Hz). In application, brain function and activity are assessed by performing a band-frequency analysis of the EEG. Besides diagnosis, brain–computer interface, human emotion recognition, and neuromarketing are recent hot issues of EEG applications [6, 7]. Especially, neuromarketing, which studies consumers' sensorimotor, cognitive, and affective response to marketing stimuli, is one of hot issues in marketing research and many big companies use the neuromarketing technology to measure consumer thoughts on their advertisements or products.

- The electrogastrogram (EGG) is the recording of electrical activity from the stomach contractions using three or more electrodes placed around the stomach. By measuring EGG before and after a meal, stomach disease and malfunction can be diagnosed. An EGEG (electrogastroenterogram) similarly records the electric signals from the intestines and stomach in a similar way. Although EGG is not measurable from the skin, it is important information to monitor the activity of digestive system.

- The electromyogram (EMG), generally surface EMG, records the electrical activity produced by skeletal muscles, activated electrically or neurologically, using electrodes placed on the skin. Surface-EMG use is restricted only to superficial muscles, and is influenced by the depth of the subcutaneous tissue. EMG is used for studying symptoms including muscle weakness, muscle pain, paralysis, and involuntary muscle twitching. Recently, EMG signal is broadly used for the control of automatic prosthetics devices including limb prostheses and games [8, 9].

- The electrooculogram (EOG) measures the dipole movement between the cornea and retina. Pairs of electrodes are placed around eye (above and below/left and right) to record eye movement and perform ophthalmological diagnosis. The EOG signal enables the operation of computer mouse [10] or other devices such as wheelchair [11] just by movement of eye and blinking of eye, which has vast potential applications such as game and prosthetics.
- The electroretinogram (ERG) is the measurement of electrical signals from rod cells, cone cells, and inner retinal cells in the retina. In this application, electrodes are placed on the cornea and the skin near the eye. ERG is used for the diagnosis of various retinal diseases, including cone dystrophy and night blindness and for the artificial vision system.

12.2.2 Motions and Pressure

Motions are reflections of body condition and behavior, and continuous monitoring of body movements (e.g., leg, arm, and volume change in the breast) is one of important issues in breath monitoring, prosthetics, sports medicine, and wearable devices. Pressure is another important physical parameter for the diagnosis of disease and body functions. The representative human body motions and pressures (or vibration) are as follows:

- Breast volume: Change in breast volume with time indicates an underlying respiratory condition. Respiratory rate during normal activity, exercise, and sleep is frequently measured to diagnose sleep apnea, lung function, and lung injury. Changes in breast volume are closely related to lung function and are a primary determinant of lung injury [12].
- Joint angle: Most body motions occurring at a joint (e.g., finger and knee) can provide significant information in the context of prosthetics, physical therapy, and sports medicine. In a remedial context, the degree of angle with joint flexion and extension is used for diagnosis of muscular disease progression.
- Pulse of blood flow: The pulse generated by the blood flow is generally measured using plethysmogram and the pulse waveform is related to the change in blood flow to the skin. Each stroke volume produces a measurable change in the plethysmogram waveform according to the flow delivered to that specific segment of skin.
- Vocal folds vibration: To generate sound, vocal folds composing of mucous membrane across the larynx regulate the flow of air being expelled from the lungs during phonation. By measuring the pressure from the throat, the performance of vocal folds could be evaluated.

12.2.3 Epidermal Impedance (Skin Impedance)

Epidermal impedance changes in response to the environment (e.g., hydration and temperature), time, climate (seasons), volume changes (e.g., breast changes), and composition of the body (e.g., fat and water) and reflects the hydration status of the stratum corneum [13]. Impedance of the skin is measured by stimulating the skin with a small alternating current (AC), and observing changes caused by sweat on the skin. Representative parameters inferred from skin impedance are as follows:

- Hydration of skin: In dermatology, monitoring of skin's hydration is a crucial technique for diagnosing diseases and determining therapeutic methods. Cosmetology and dermatologic treatments are assessed by measuring hydration. Methods for determining skin hydration include measurement of thermal conductivity, reflective index, and electric impedance, the latter of which is known to be a highly reliable and compelling method because it is cost effective and very simple to use [14, 15].
- Body fat percentage (BFP): BFP, defined as the total mass of fat divided by total body mass, is an important parameter for measuring the degree of obesity, a representative lifestyle disease that requires monitoring in daily life. BFP has previously been determined by underwater weighing and plethysmography, among other methods. However, bioelectrical impedance analysis has become the most popular method and is broadly used, even in fitness centers and the home.
- Emotion monitoring: The change of impedance by sweat when a subject is emotionally upset or nervous provides information regarding the subject's emotional state to be inferred. Continuous measurement of impedance is one of useful methods to monitor the emotion and will be applied extensively in diverse area.

12.3 Technologies to Fabricate Skin-Like Electronics Using Nanomaterials

12.3.1 Requirement for the Skin-Like Electronics

For the comfort, daily life of patients wearing skin-like electronics without trouble and inconvenience, the long-term, continuous, and unconscious measurement of biosignal is one of crucial requirements and total size of whole system should be minimized as possible. For the long-term and continuous wearing on the skin, the electrodes should be flexible and stretchable because our body is curved and stretchable, and mechanical mismatch between skin and electrode may prevent the smooth motion of body. Such mechanical property enables the minimization of artifact generated by environmental interference and motion. The conductivity of skin-like electronics should be high for the artifact-free signal measurement, and be maintained constant as possible while it is stretched or twisted. The material should be amenable to be fabricated to any shape and be inert to sweat. Evaporation of sweat is another critical requirement. The skin-like electronics should be biocompatible for long-term and continuous signal measurement without trouble to the skin. Finally, the electrical connection of skin-like electronics to the measuring and power system should be easy and simple.

12.3.2 Materials for Skin-Like Electronics

For the biomedical applications of skin-like electronics, choice of materials satisfying aforementioned requirements is critical and the materials [16]. Material for soft electronics can be classified into following two categories: (1) substrate materials for skin-like electronics, and (2) conductive nanomaterials.

12.3.2.1 Substrate Materials for Skin-Like Electronics

Various research groups have used soft and elastic materials because they enable conformal contact of electrodes over large skin areas and can be easily fabricated using soft- and photolithographic methods. Moreover, soft materials have other properties that are suitable for use as skin-like materials, such as elasticity, gas permeability, optical transparency, durability, and biocompatibility. In this section, the following elastic and soft materials are described: polydimethylsiloxane (PDMS), polyurethane (PU), polyimide (PI) and parylene (PL) (Fig. 12.2). The properties of these materials are summarized in Table 12.2.

(1) Polydimethylsiloxane

PDMS is a popular elastomer for the use in micro/nano-fabrication. PDMS is highly permeable to gas, capable of forming any structure and is biocompatible; it also has good thermal and chemical stability. PDMS has been extensively used as a flexible and stretchable substrate for skin-like electronics [17]. Using oxygen plasma radiation, PDMS surfaces can be transformed from hydrophobic to hydrophilic. Because of its high softness, PDMS can maintain excellent conformal contact with human skin during natural movements. The PDMS membrane readily allows gas diffusion, enabling easy evaporation of sweat. PDMS exhibits thermal stability up to ~ 186 °C in air, and is optically transparent down to ~ 300 nm. Because of these excellent properties, PDMS has been broadly used in diverse applications, including lab-on-a-chip, skin-like electronics and implantable electrodes, and systems for screening drugs and toxic materials. Microscale PDMS structures can be easily fabricated by a soft lithography process that includes mold-based replication, contact printing, and stamping. These fabrication methods enable to simply used PDMS as substrate of skin-like electronics.

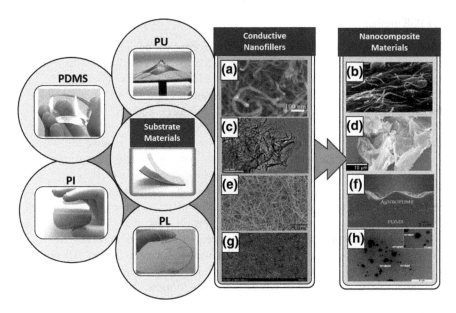

Fig. 12.2 Schematic of materials for skin-like electronics. Nanocomposite materials (**b, d, f, h**) are made by mixing conductive nanofillers (CNT **a**, graphene **c**, metal nanowires **e**, metal nanoparticles **g**) into soft substrate materials. **a** FESEM image of raw MWCNTs. **b** CNT/PDMS. **c** TEM images of chemically modified graphene. **d** SEM of agglomerated graphene oxide nanosheets. **e** SEM image of the AgNW film on Si substrate. **f** Cross-sectional SEM image of the AgNW/PDMS layer. **g** Scanning electron microscopy image of the packing of 14 nm gold nanoparticles forming the wires. **h** Spherical gold nanoparticles are dispersed in PDMS

Table 12.2 Properties of substrate materials for skin-like electronics

Properties	PDMS	PU	PI	PL-C
(A) Surface (interfacial free energy)	Low ($\sim$21.6 dyn/cm^2)	Low	Low	20 dyn/cm
(B) Hydrophobicity (contact angle)	Hydrophobic (90°–120°)	Hydrophobic (71.1°–94.2°)	Hydrophobic (82°)	Hydrophobic (80°)
(c) Permeability	High	High	Low	Low
(d) Thermal stability	Stable (up to $\sim$186 °C in air)	260–360 °C	Does not melt, decomposes at 520 °C	290 °C
(e) Color	Transparent (down to $\sim$300 nm)	Colored	Yellow/orange	Gray/white
(f) Mechanical property (E-modulus)	1.8 MPa	0.025 GPa	2.5 GPa	2.8 GPa
(g) References	[17, 93–95]	[18, 19]	[20, 21]	[22]

(2) Polyurethane

PU is the most commonly used polymer in biomedical devices. PU is synthesized by carbamate (urethane) links between alcohols and reactive hydroxyl (–OH) groups (polyols) and an isocyanate group [18]. The first synthesis of PU was achieved in the 1930s. Since then, diverse applications of PU have been developed, including as rigid foams, adhesives, resins, polymers, and coatings. PU has been broadly used in biomedical fields as skin dressings and tapes, vascular grafts, and patches. Surface modification with certain ionic functional groups, such as poly sodium vinyl sulfonate and propyl sulfonate reduces platelet deposition on PU surfaces. Such modified PU enables its applications to be extended to skin-like electronics as well as implantable devices requiring good blood compatibility, such as artificial hearts and valves. The permeability of PU is an important feature for biomedical applications such as controlled membranes, dialysis membranes and wound dressings, among others, and its permeability can be modified by controlling the type of extender chains and molecular weights of polyols [19]. These useful biocompatible and mechanical properties of PU provide skin-like electronics as well as invasive application in biomedical areas to useful substrate

(3) Polyimide

PI, a polymer of imide monomers, is widely used in biomedical applications as a flexible substance and passivation layer. PI is an excellent insulating material; photosensitive PI, in particular, protects electrodes against diverse sources of damage, including chemical (corrosion, absorption, and ion transport) and physical (splitting and cracking) insults [20]. It is also a hydrophobic material, creating a surface that repels water for easy cleaning, but it can be made hydrophilic by the radiation of oxygen plasma to the PI surface. The permeability of PI depends not only on chemical structure (diamine and dianhydride components), but also on imidization process conditions (chemical vs. thermal imidization), and PI has good chemical and thermal resistance. PIs are generally orange/yellow in color, but some PIs exhibit optical clarity. PIs have good mechanical properties and durability. In spite of these good chemical properties, one major advantage of PI is its suitability for photolithography processes, including stable and durable metal deposition; importantly, its thickness can be controlled to within a few microns [21]. Diverse metal layers can be deposited on a PI substrate using vapor deposition, sputtering, and reactive ion etching processes that are cost-effective and allow specified structure patterning with conventional etching process and high repeatability. These diverse fabrication methods enable PI to be applied on substrate of skin-like electronics.

(4) Parylene

PL is a green polymer that does not require initiating or terminating solvents; a representative PL-based coating and substrate commonly used for biomedical applications. Because it provides excellent biocompatibility, PL-C has been widely used as a substrate material for implantable electronic devices [22]. Various research groups have verified the usefulness of PL-C coatings and substrates for thin, flexible, and stretchable electrodes. PL also shows good chemical resistance and thermal stability, with decomposition beginning at 290 °C. Even though PL has a critical issue like a delamination, well-coated PL substrate has been applied on substrate material that provides good mechanical and chemical properties. As the other application, PL coating is commonly used by deposition techniques such as chemical vapor deposition.

Beside these materials, diverse materials could be used as substrate. Super-stretchable Ecoflex, liquid crystal, polyethylene terephthalate (PET) and polyethylene (PEN) are commercially available polymers and they have been broadly used for flexible and stretchable electronics [23–25]. Among these materials, Ecoflex is broadly used as strain sensor due to its highly stretchable property. One of hot materials adapted for use in biomedical devices is silk [26]. To date, silk, with features of nanofibers, films, micro/nano spheres and porous sponges, has gained prominence as a material for biomedical applications including scaffold and drug delivery. It also has the property of controllable degradability and high permeability is good for evaporation of sweat and use in high-humidity conditions [27], which enables extensive potential exploitation in attachable devices on the skin. Some neural-implantable electrode which is ultrathin enabling conformal contact to highly wrinkled brain surface has been reported using silk as biodegradable support [28].

12.3.2.2 Conductive Nanofillers and Conductive Nanocomposite

Conductive nanocomposite materials have attracted considerable attention for flexible and stretchable electronic applications because they do not require metal deposition, eliminating corrosion, and delamination concerns. The conductive nanocomposite-based fabrication methods overcome limitations of metal or silicon-based systems. Nanocomposite conductive material is generally synthesized by blending conductive nanofillers and stretchable polymer with and without using surfactants. Although diverse nanofillers have been developed [29] (N. Saba), carbon nanotubes (CNTs, Fig. 12.2a) [30], graphene (Fig. 12.2c) [31], metal nanowires (NWs, Fig. 12.2e) [32], metal nanoparticles, and flakes (Fig. 12.2g) [33] are generally used as conductive nanofiller, whereas PDMS, PMMA, polystyrene, chitosan, and PVA are generally used as polymers. CNTs are tube-shaped, 3D nanostructures of carbon allotropes, and are classified as single-walled nanotubes (SWNTs) and multi-walled nanotubes (MWNTs), depending on the number of concentric layers of graphene sheets. For over a decade, they have been well known for their strong mechanical properties and high electrical conductivity. By the rapid progress of purification and functionalization (covalent and noncovalent modification, ligand attachment, etc.) of CNTs, they are being actively applied to advanced biosensors, implant coatings, and composite biomaterials [34–36]. The blending ratio of both materials determines the conductivity and degree of mechanical deformability. High loading of CNTs is required for high conductivity, but it also decreases the softness and stretchability of the composite material. Since the first report of CNTs/polymer nanocomposites, various CNT composite materials and potential applications in diverse fields have been and continue to be developed [37]. Among polymers, PDMS and chitosan have been commonly used, and several CNT/PDMS (Fig. 12.2b) skin-like electrodes have been used to measure ECG, EEG, and EMG signals [38–40]. CNTs/polymer nanocomposites have enormous potential for future flexible and stretchable electrodes and circuits, however, they have some critical limits. One of them is that their conductance is still quite low. They can be used in voltage-driven devices, such as capacitive touch screens and electrodes for biopotential sensing, but their use in current-driven devices like organic LED and stimulating electrodes is limited. To enhance conductivity, researchers have mixed in metal NWs (e.g.,: gold, silver, and copper nanowire) or microscale silver flakes (Fig. 12.2f, h) [32, 41]. Another problem of CNT/PDMS is that their conductivity varies with strain and bending, and the resistance of a composite sheet can be stabilized by an initial pretreatment process consisting of repeated cycles of stretching and release. Nanocomposite material consisting of CNTs and chitosan has recently been used in a variety of applications, including biosensors, biofuel cells, drug delivery systems, and electronic circuit [42]. The use of natural and biocompatible polymers for the synthesis of nanocomposite material may expand their applications to include the development of biodegradable conducting materials and devices.

Graphene, another carbon material, has attracted extensive interest owing to its exceptional electrical, mechanical, thermal, and optical properties. Graphene, a two-dimensional single-atom-thick sheet of graphite, was first discovered in the 1960s and produced and isolated in measurable quantities in 2003 [43] Graphene-polymer composites are of interest to researchers in a variety of fields. In recent years, researchers have successfully fabricated graphene-polymer composites similar to CNT-based polymer composites. It has been shown that the combination of graphene and polymer matrices improves electrical conductivity at a low percolation threshold, and increases mechanical strength, elastic modulus, and thermal conductivity and stability. This reinforcement enables several applications, including high-strength and lightweight structural polymers in automobile and aerospace fields. Graphene and polymer matrices are also used in the electronics industry, where they improve thermal conductivity for thermal management and serve as a flexible electronic circuit, and in the energy storage field, where they are used as highly conductive polymers for flexible super capacitors (Fig. 12.2d) [31, 44].

Besides CNTs and graphene, metal NWs are extensively used for skin-like electronics. One appealing approach for obtaining stretchable and flexible electronics is to coat metallic NWs on the surface of an elastic substrate or to infiltrate them into the elastic polymer. In addition to higher conductivity of NWs, transparency is another advantage of such electrodes compared with those fabricated from CNTs or graphene. Among NWs, silver NWs (AgNWs) are the most extensively used because of their intriguing electrical, thermal, and optical properties [45]. Diverse methods can be used to prepared AgNWs, including hydrothermal, electrochemical, UV irradiation and template techniques, and microwave-assisted processes. The diameters and lengths of AgNWs are generally 30–200 nm and 1–20 μm, respectively. NW length is critical for improving conductivity because greater length results in longer percolation paths and fewer NW junctions. Although AgNWs show excellent performance, they are expensive; thus, cheaper nanofillers are desirable for broad applications. Copper NWs (CuNWs) are an appealing alternative to AgNWs because of their high conductivity (comparable to that of silver), cost-effectiveness, and abundance in the earth. Despite these advantages, CuNWs exhibit poor stability against thermal oxidation and chemical corrosion, which degrades conductivity over time;a crucial problem for broader application. To address these problems, researchers have developed several methods for maintaining their conductivity and transparency, including coating with an aluminum-doped zinc oxide (AZO)/aluminum oxide (Al_2O_3) or nickel passivation layer. Although these coating methods considerably increase the conductivity of nanocomposite materials, stretch dramatically degrades their conductivity, limiting them to low-stretch applications. Nanoparticles (NPs) with diameters ranging from 10 to 100 nm are broadly used in flexible sensors, and details for NP-based flexible sensors are well described in Haick's review paper [46]. Ag flake is frequently used as conductive filler, and Someya et al. developed elastic conductor ink comprising of Ag flakes, a fluorine rubber and a fluorine surfactant to develop flexible electronic system [47]. The homogenous dispersion of nanofiller and polymer has been a great interest, and diverse method such as surface modification, ultrasonication [48, 49], hot pressing [50], jet-milling [51], and shear mixing [49, 52, 53] and chemical agents [48, 50, 51, 54, 55] have been employed.

12.3.3 Fabrication Methods for Flexible and Stretchable Electrodes Using Nanomaterials

(1) Geometry and pre- and post-stretching

Although a metal layer deposited on a stretchable substrate cannot be extended beyond a few percent, stretchability can be greatly enhanced by designing a metal pattern shape and depositing metal on a prestrained substrate. To date, several representative geometries that provide excellent stretchability have been designed. A mesh shape is a favored design for covering and contacting curved surfaces, and recent in vivo studies have demonstrated that mesh architectures are highly suited to curvilinear surfaces (Fig. 12.3a) [1, 5, 28]. Recent efforts to fabricate skin-like electrodes have highlighted serpentine and fractal geometries [56, 57]. Rogers and colleagues intensely investigated serpentine and fractal geometries, showing that a thin electrode can be stretched more than 300 %, depending on its specific structure, without disconnection [58]. Although this method has been popularly used for the metal line, its extensive application is expected in the fabrication of soft electronics. Diverse nanomaterial-based stretchable patterns could be achieved on the stretchable substrate, and typical examples are percolation nanomaterial networks [59, 60], net-shaped structures [59, 61] and spring-like structures [62]. Fibrous structure such as CNTs and metal NWs can easily form networked pattern. Dispersing, electrospinning, or cross-stacking of aligned ribbons [63] are popular methods to spread these materials on the substrate. Within a certain level of tensile strain applied, the nanomaterials will rotate and slide against each other to accommodate the strain.

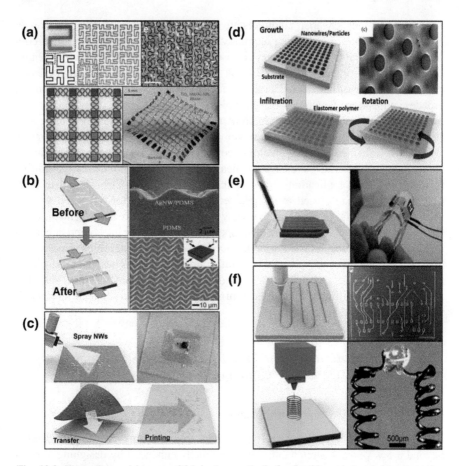

Fig. 12.3 Illustration and images of fabrication methods for flexible and stretchable electrodes. **a** (Design) Implementation of Peano curves for stretchable electronics illustration of three iterations of a two-dimensional Peano curve. Optical image of third order Peano-based wires on skin and a skin-replica (*colorized metal wires*), showing the conformal contact of the wires on the substrate. Scale bar, 2 mm. Wearable electronic patch designed by mesh structure. **b** (Post stretching) 'Wavy' structure made from prestrain and post-strain method. Cross-sectional SEM image of the AgNW/PDMS layer. **c** (Nanomaterial deposition) Air spray coating method for coating the nanowires. Optical image of aligned carbon nanotube polymer hybrid architectures by transferring the cut vinyl patterned mask on a soft substrate. **d** (Infiltration) Schematics illustrating the fabrication steps of the aligned MWNT-PDMS array structures. SEM images showing the MWNT architectures before and after PDMS infiltration in two different scales and shapes: arrays of nanotube pillars (500 μm diameter and 100 μm height) before and after infiltration. Cross section image. **e** (Injection) Elastic conductive patterning method of filling conductive liquid through soft polymer channel. Application of the CNT-chitosan electrode in electrical circuits. Operation of a functional LED in a plate circuit and bending test to determine the flexibility of the CNT-chitosan circuit. **f** (Printing) The solvent-cast 3D printing of nanocomposite microstructures and optical image of two helical sensors mechanically supporting and electrically sourcing a luminous LED bulb

A curvilinear substrate is another well-developed strategy for stretchable electronics [64]. This 'wavy' structure is made prestrain and post-strain method (Fig. 12.3b). For the prestrain process, the elastomeric substrate is mechanically or thermally pre-stretched and metallic or semiconducting layers are transferred or deposited onto the prestrained elastomeric substrate. Finally, prestrained substrate is released and the wavy structures, which can be easily bent, stretched, and compressed, are formed spontaneously. In contrast, post-strain method is to stretch the elastic substrate on which conductive layers are already deposited. At first, a conductive layer is deposited on the elastomeric substrate. After deposition, it is stretched and released forming buckled elastomeric surface. This is a simple and cost-effective method for fabrication of stretchable electrodes. Xu et al. reported a AgNW/PDMS conductor with a wavy structure that becomes stable in the tensile strain range of 0–50 %, exhibiting high conductivity [32]. They stabilized their wavy conductor by repeated stretch and release of AgNW/PDMS after it was cured. These stretchable conductors showed outstanding robustness under repeated mechanical loading. Using this method, diverse applications such as stretchable LED circuit and strain sensor have been demonstrated.

(2) Deposition of nanomaterials on the substrate

Just by coating of CNT, graphene, and metallic NWs on the surface of elastic substrate, the stretchable and flexible electronics could be fabricated, although stretchability is not high compared to other method. Especially, coating of metallic NWs is popularly used for variant applications because they are more conductive and transparent than carbon material and metallic nanoparticles. Transparency of NW is another advantage compared to CNT and graphene, which extends applications to flexible display and optical devices. As NWs, silver NWs (AgNWs) are the most extensively used because of their intriguing electrical, thermal, and optical properties. Several coating methods have been employed to fabricate AgNW-based electrodes from NW solutions, such as air-spray coating (Fig. 12.3c) [65–68], vacuum filtration followed by transfer [59, 69], Meyer rod coating [70], transfer printing, and drop-casting [71]. As aforementioned, NW length is critical to improve the conductivity because of longer percolation paths and reduced NW junctions, and a number of research efforts to manufacture stretchable and transparent conductors with long NWs are currently ongoing [72]. A thermal annealing process, which removes poly-4-vinylphenol (PVP) residues and facilitates nanowelding between NWs, is another critical step in enhancing conductivity [59]

(3) Infiltration of elastomers in nanomaterial layer

Infiltrating elastomers in a pre-connected and aligned thin film of conductive filler on the solid substrate has been reported as an effective method to fabricate stretchable electronics (Fig. 12.3d). The advantage of this method is that the conductive filler contacts directly through the thermal annealing prior to infiltrating the elastomers, which enhances the electrical conductivity compared to the blending method. The shape and dimension of filler's geometry can be adjusted by varying the volume fraction and geometry of the filler. By pouring elastomer pre-polymers onto net-worked filler, crosslinking the polymer, and peeling off the elastomer, a highly conductive and stretchable composite can be produced [6, 7]. As conductive nanofillers, CNTs, NPs, and NW are widely used. Jung et al. fabricated composite structures by impregnating vertically aligned arrays of MWNTs into a transparent PDMS matrix. They used vertically aligned MWNT architectures grown on pre-patterned SiO_2/Si substrates, and encapsulated with PDMS prepolymer solution (Fig. 12.3d) [73]. Li et al. reported new class of highly organized SWCNT network-polymer hybrid structures by incorporating horizontally and vertically aligned SWCNTs in desired locations, orientations, and dimensions [8]. Based on this approach, diverse architectures could be built ranging from two-dimensional suspended SWCNT microlines on microtrenched PDMS substrates. Yang et al. demonstrated SWNTs/PDMS composites [9], and Pei et al. reported a composite of SWNTs and a shape memory polymer, poly(tert-butyl acrylate) [10]. NWs could also be infiltrated, and NWs demonstrated better conductivity and transparency. Zhu et al. fabricated a highly conductive and stretchable Ag nanowires/PDMS conductor, in which the PDMS prepolymer was infiltrated into a thick layer of Ag nanowires and thermally cured [32]. For the enhancement of conductivity when stretched, buckled geometry was used and a high stretchability and conductivity was obtained. The sandwich structure is employed to protect the nanocomposite thin film from physical contact and damage by the complete coverage of the thin film on both sides. Amjadi et al. developed super-stretchable, skin-mountable, and ultrasoft strain sensors based on sandwiched structure with Ecoflex-CNT percolation networks—Ecoflex layer. The sensor demonstrated super-stretchability and high reliability for strains as large as 500 % [74].

(4) Fabrication of elastic conductive pattern using microfluidic channelc

By filling conductive liquid in soft polymer channel, the flexible electronics could be fabricated [29, 60]. The microchannel was fabricated by softlithography process, and the inside surface of the channels was treated with plasma oxidation for the enhancement of wettability. By constructing multilayer structure, this approach enables the fabrication of complex 2D and 3D structures. Ha et al. developed flexible LED arrays by embedded liquid-metal interconnections of eutectic alloy of Ga and In (EGaIn) [75] and stretchable loudspeaker is driven by the dynamic interaction between the liquid metal coil and a permanent Neodymium (Nd) magnet [76]. As filler, nano composite materials could be used instead of specific liquid metal alloy. Compared to liquid metal alloy, control of the injection pressure, the viscosity of nanocomposite material and interfacial tension are critical factors determining the successful patterning and the resolution of the resulting patterns. Injection of viscose liquid into the microfluidic channel is still challenging, and air bubbles are created during injecting process. Contrary to liquid metal, nanocomposite materials dry up as time goes, and it cannot be stretched extremely like liquid metal. As an alternative method, nanofiller could be trapped in the negative relief of PDMS channel forming conductive layer. Hwang et al. have developed CNT-chitosan nanocomposite-based electrical circuit (Fig. 12.3e). The negative relief of PDMS electrical circuit channel was fabricated using conventional soft lithography process. Onto the relief, CNT-chitosan nanocomposite was poured and scrubbed with slide glass, then, nanocomposite was filled into the channels and remnant was completely removed. By covering the channel with thin PDMS layer and connecting commercially available 7-segment light emitting diode (LED) and its driving chip, the metal-free stretchable and flexible electronic circuit was fabricated [42]. Figure 12.3e demonstrates the working 7-segment LED circuit. In a similar method, more complicated electronic digital circuit (adder) was developed using CNT/PDMS nanocomposite materials.

(5) Others

Recent progress of 2D and 3D printing technology including ink-jet, screen, and 3D printing could also be applied in printing electrode pattern using well-dispersed nanocomposite materials as ink (Fig. 12.3f) [77, 78]. Especially, 3D printer enables the construction of complicated 3D electronics. Porous surface of PDMS was used to enhance metal deposition and stretchability. Jeong et al. reported a method for fabricating flexible and stretchable electronics by the metal deposition on the porous PDMS layer [62]. The porous PDMS thin film was fabricated by applying pressurized steam to an uncured PDMS layer and conductivity could be maintained to 80 % strain. The commercial electronic components can be bonded on the highly stretchable substrate just by the soldering. The CNT and graphene layers could be deposited on the porous surface maintaining good conductivity and stretchability.

12.4   Biocompatibility for Biomedical Applications

Medical devices and materials require the strict safety for human use, and safety approval can be obtained from regulatory agencies using standard documentation, such as ISO 10993 ("Biological Evaluation of Medical Devices") [79–81]. The elastic materials that comply with ISO 10993 standards or USP (United States Pharmacopeia) class VI standards are PL-C, silk [20, 82–84], PU (U203-FDA95 polyurethane natural white; Seal Maker), and PDMS (S40, S50, and S70—Liquid Silicone Rubber Part A and B; Dow Corning). Although PI has been used in biomedical applications, and several groups have proven its biocompatibility and low cytotoxicity for use as long-term implanted electrodes, to our knowledge, it has not been certified by any regulatory agencies to our knowledge. For the test of biological safety, in vitro experiments, such as cell-cycle analyses and cytotoxicity tests, as well as in vivo tests, such as those involving direct contact of a material with the animal's skin or tissue, have been used to evaluate and establish material biocompatibility. Similar to polymer, there has been concern about biocompatibility of CNTs and graphene, and several reports have revealed that CNTs could induce genotoxicity via induction of reactive oxygen species in biomedical field [17]. This concern can be partly solved by fully encapsulating CNTs into polymers. From our experience, even though CNTs are encapsulated with polymer, CNT debris sometimes is separated from the surface of electrode. Therefore, complete encapsulation of CNTs into polymer is critically required for the biomedical application. Similar to CNTs, graphene also bring a concern about toxicity behavior when exposed to bio-system [42]. Although considerable research has been devoted to the study of skin electronics, biocompatibility remains a critical issue in developing skin electronics. Thus, if skin-like devices are to be used for medical application, additional studies will be required to satisfy existing regulations. In the near future, these regulations are likely to become stricter than now; ultimately, most products contacting the human body are likely to require regulations comparable to medical grade.

12.5 Applications

12.5.1 Biopotential Sensor

For the stable and continuous monitoring of biopotential from the epidermis, conformal contact, biocompatibility, and skin-like mechanical properties are critical components. In general metal (Au, Ag/AgCl, and Pt)-based electrodes have been broadly used for measuring biopotential. Despite extensive application of these electrodes, measuring signals continuously and unconsciously using them remains difficult because most such electrodes require a gel for conformal contact with the skin. These gels sometimes cause allergic reactions and irritation to the skin and cause electrical noise owing to drying of the gel. To address the limits of gel-type electrodes, researchers have developed diverse dry electrodes; however, long-term and continuous measurement of signals with minimum artifact from motion is still challenging. Recent progress in the development of nanomaterials-based electronics plays a pivotal role in overcoming the problems of conventional dry electrodes. S. Lee et al. developed self-adhesive electrodes using nanocomposite conductive materials [38]. They developed EEG recording patch using CNT and adhesive PDMS composite material, yielding a signal quality comparable to that of commercial electrodes while providing robust conformal contact on the skin. The patch can self-adhere on the skin and is resistant to motion and sweat. Lee et al. reported CNT/PDMS in-ear electrodes developed to measure EEG signals in the form of headphones [39]. The shape of electrode is the same as that of a commercial canal-type earphone. A soft and skin-compatible CNT/PDMS-based conductive cap, which looks like the rubber cap of an earphone, took a role of the metal in conventional electrodes. Several skin-like electrodes to measure EMG signals were reported using Ag flakes/fluorine rubber electrode [74] and AgNW/PDMS dry electrode. To conformal contact between skin and electronics for the long-term and stable monitoring of health, self-adhesion capability is another key method. S. Lee et al. developed a self-adhesive capacitive EEG electrode composed of a CNT and adhesive PDMS nanocomposite material. Due to the softness of adhesive PDMS, EEG signals could be stably measured even though the electrode was placed on the

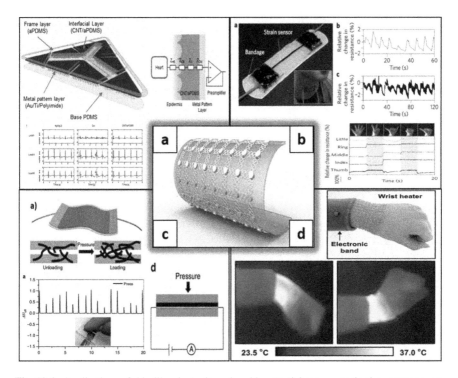

Fig. 12.4 Applications of skin-like electrode such as biopotential sensor, motion/pressure sensors, and stretchable heater. **a** ECG electrode based on CNT/aPDMS and schematic diagram of the impedance from the heart to the preamplifier, which is composed of the impedance from the heart to the epidermis (ZHE), the contact impedance between CNT/aPDMS and the epidermis (ZCE), the intrinsic impedance of CNT/aPDMS (ZC), and the contact impedance between CNT/aPDMS and the metal-patterned layer (ZCM). Motion artifact tests and ECG measurements. **b** Stretchable wearable devices of carbon nanotube strain sensor and results of relative changes in resistance versus time for breathing and data glove configurations, respectively. **c** Pressure sensor based on the AuNWs coated tissue paper and schematic illustration of the sensing mechanism. **d** Stretchable heater based on nanocomposite of AgNWs and IR camera images when the wrist moved downward and upward

hairy skin [85]. Suh et al. developed a self-adhesive patch using unique structural characteristics of gecko foot hair. They fabricated mushroom-shaped pillars made of soft PDMS showing that this structure is less affected by surface contamination and oxidation than conventional acrylic adhesives [86]. Lee et al. demonstrated self-adhesive ECG patch consisting of CNT/adhesive-PDMS and the adhesiveness recovered by washing with alcohol, and they recorded ECG signal while patient is under water bath (Fig. 12.4a) [40]. Jang et al. used ultralow modulus silicone with excellent adhesive property to the skin and high permeability for transepidermal water loss. The electrode is strongly self-adhesive to the skin with using glue [87]. Nevertheless, the continuing death and sloughing of epidermal cells from the skin, an impediment for long-term use, is a problem awaiting a solution.

12.5.2 Stretchable and Flexible Motion and Pressure Sensors

Stretchable and flexible motion and pressure sensors are promising components for a broad range of applications, including wearable devices, consumer electronics, games, robotics, prosthetics, behavioral analysis, and space flight. The most popular application of nanomaterial-based electronics is the motion analysis. If such a motion sensor could be worn for long-term without trouble, its applications would be expanded dramatically. The representative motion sensor is a strain sensor, which is generally used for measuring physical longitudinal deformations of a device. Although metallic strain gauges have historically been used for the biomedical applications, they can maintain only low strain (≤5 %). Various recently developed polymer composite strain sensors offer a maximum measureable strain that is better than that of metallic stain sensors. Strain sensors based on nanomaterials are a focus of development efforts owing to their high stretchability, acceptable conductivity and excellent mechanical properties, and CNTs, graphene, and NWs have been popularly used as conducting nanomaterials. These strain sensors are broadly used in the measurement of human behavior. Yamada et al. developed wearable and stretchable devices fabricated by encapsulating thin films of aligned SWCNT with PDMS. They realized a highly stretchable strain sensor that can withstand strain up to 280 %, with high durability (10,000 cycles at 150 % strain), fast response (delay time, 14 ms), and low creep (3.0 % at 100 % strain) and used them for the detection of knee joint movement, typing using glove, breathing and phonation (Fig. 12.4b) [88]. Lipomi et al. produced conductive, transparent, and stretchable nanotube films by spray-coating of nanotube (length = 2–3 μm) directly onto PDMS [89]. Park et al. devised diverse strain electrodes using CNT/Ecoflex [23], AgNP/PDMS [85], and SiNW/PDMS electrodes [23] in the form of the sandwich structure (i.e., AgNW thin film embedded between two layers of PDMS) and used them for finger motion. Another important motion sensor is a pressure sensor. Flexible and stretchable pressure sensors are of principal importance for rollable touch displays, energy harvesting, biomedical prostheses, and soft robotics. Most of these applications require high sensitivity at low pressure, fast response times, and low-power consumption. Pan et al. introduced conducting, polymer-based, and ultra-sensitive pressure sensors that mimic natural skin. These resistive pressure sensors, which are capable of detecting a pressure of less than 1 Pa and exhibit a short response time, good reproducibility, excellent cycling stability, and temperature-stable sensing, can be used for artificial electronic skin and robotic arm. Gong et al. developed an ultrathin, AuNW-based, wearable, and highly sensitive pressure sensor. This sensor is scalable and can measure pressure patterns by mapping spatial pressure distributions. This flexible pressure sensor is used for wearable, real-time monitoring of blood pulses, detection of small vibrations, and as a tactile sensor and human-like prosthetic skin (Fig. 12.4c) [90]. Hong et al. reported polymer-CNT pressure sensor for the real-time detection of local pressure to monitor hypertrophic scars in burnt skin [88, 91]. Zhao et al. developed a flexible/wearable multifunctional sensor array with simple processes for highly-sensitive contact/pressure/strain detections using PDMS/Ag/Ecoflex/Ag/PDMS sandwich-structured sensor array. Its detection limit is 6 Pa and can be stretched up to 70 % [92]. Pang et al. recorded the physical force of a heartbeat in real time using sensor attached to the artery of a volunteer's wrist. The sensor is based on two interlocked arrays of high-aspect-ratio Pt-coated polymeric nanofibres [86]. Besides these examples, a quantity of strain and pressure sensors were developed, and they are well described at some review papers [93].

12.5.3 Stimulator

Stimulation is one of the most important therapeutic methods for physical therapy, sports, wellness, and pain treatment using surface electrodes and consisted of electrical method and thermal method., Electrical muscle stimulation (EMS) and transcutaneous electrical nerve stimulation (TENS) are representative potential applications of skin-like electrode. In medicine, EMS helps to prevent muscle atrophy, which sometimes occurs after musculoskeletal injuries, such as damage to bones, joints, muscles, ligaments, and tendons. TENS helps to decrease pain signals from the brain, back, cervical muscular, disc syndromes, and other acute/chronic pain signals. The electronics for EMS and TENS require a larger current; as a result, wet electrode is generally used due to high electrical conductivity. Even though conductive nanomaterial-based electronics are widely used, it is still challenging to fabricate stimulating electrode using conductive nanomaterials such as CNTs and graphene because of their lower conductivity than metal. If metal NWs are very stably deposited on the soft substrates, the stimulating electrode could be fabricated; however, fabrication of nanomaterial-based stimulating electrode is still difficult. Nevertheless, long-term wearable skin-like electrodes are necessary for the safe and effective electrotherapy such TENS, neuromuscular electrical stimulation (NMES), and other less common forms of electrotherapy. Highly conductive nanomaterial-based electronics will contribute to the development of stimulating electrode. Contrary to recording electrode, stimulating electrodes require strict regulation for patients' safety due to electric shocking and burns. However, it is challenging to prevent such damage because the degree of the burn (first, second or third) is dependent on the patient's skin condition. Feedback stimulation system will be the clue to prevent the damage. By integrating the sensors and stimulating electrodes on the skin-like electrode, the burning of skin and electrical shock could be detected by the sensor and damages could be prevented. Contrary to electrical stimulation method, thermal stimulation method is widely used by nanomaterial-based fabrications. For example, Chu et al. developed wearable ohmic joule heater using highly conducting CNT/polydimethylsiloxane (PDMS) composite forming a random network of CNTs, and its heating range is from room temperature up to 200 °C [94]. Choi et al. present a soft and stretchable heating element that is lightweight and thin, and is conformally integrated with the human joints and the skin for effective heat transfer and thermotherapy. The heater is composed of highly conductive Ag NW/elastomer nanocomposite (Fig. 12.4d) [95].

Conformal contact is another critical challenge for stable stimulation because their electrode is commonly large for low contact impedance. Therefore, conventional skin electrodes for stimulation are not suitable for long-term and unconsciousness. Soft and flexible electronics enable such long-term wear while maintaining conformal contact because of their skin-like mechanical properties.

12.6 Future Developments

Future skin-like flexible electronics will be more cost-effective, lightweight, soft, portable, biocompatible, and stretchable than currently available devices. From these aspects, nanomaterial-based skin-like electronics has a quantity of potential advantage; however, their applications to healthcare fields are not so widespread because of difficulty in precise patterning, low electrical conductivity and biocompatibility (e.g., carbon material). As strain and pressure sensor, nanomaterial-based sensors are beneficial. For example, to perform sensitive and difficult tasks of robot, such as grabbing fragile objects and performing medical surgery, tactile sensing devices, which have soft, flexible, and highly sensitive tactile sensing electronics, is greatly required. Due to these reasons, the nanomaterial-based strain and pressure sensors will have extensive applications in surgical and prosthetic robots (Fig. 12.5 (right)). It is also anticipated that small devices for sensing various signals (e.g., temperature, environment (air pollution, ultraviolet and radiation), and biochemical compositions) could be integrated into e-skin for the monitoring of diverse body conditions and environmental damages to health (Fig. 12.5 (left)). Recent progress

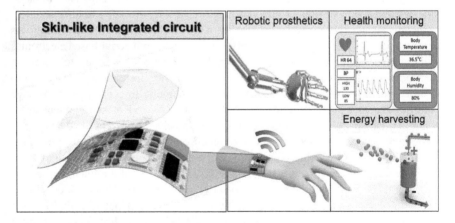

Fig. 12.5 Future perspective of skin-like electrode. Future wearable integrated circuit will be skin-like flexible electronics and multifunctional devices that provide HMI (Human Machine Interface), continuous biosignal monitoring and semipermanent self-energy-supporting from energy harvesting technologies

of technologies to integrate multiple components such as transistors, LED, and electronic elements to the skin-like electronic may enable the production of e-skin to monitor health and environments simultaneously. For the operation of skin-like electronics with multiple components, energy delivery and communication of signal is important. Even though diverse small batteries were developed to date, they are not easily mountable on the human skin because their size and weight are still big. The connection of skin-like electronics to recording system is another critical factor, but it is still challenging. Kim et al. demonstrated the power delivery and radio-frequency communication modules on the soft electronics to establish a multifunctional biosignal detection system. However, for the practical application, reliable and simple connection methods should be developed. Feedback system consisting of sensor and stimulator is hot issue for electrical theragnosis. A stretchable prosthetic skin equipped with ultrathin silicon nanoribbon (SiNR) strain, pressure, and temperature sensor arrays was developed. They are integrated with stretchable humidity sensors and heaters further for the sensation of skin moisture and body temperature regulation. Integrated electrode can transmit electrical signal to the prosthetic skin to stimulate specific nerves. Although it is challenging to integrate multiple components into nanomaterial-based skin-like electronics, the combination of nanomaterials with other technologies may enable the fabrication of electrodes with components. For example, Lee et al. developed the ECG patch by combining the sticky CNT/PDMS layer for self-adhesion and conformal contact to the skin and the PI layer for electrical circuit. This ECG patch could measure ECG signals whose signal quality and reliability are almost comparable to commercial Ag/AgCl electrode even though it does not use gel. Another application of nanomaterial-based electronics is battery and energy harvesting (Fig. 12.5 (right)). While tremendous effort has been concentrated on developing skin-like electrode usable in clinics, it is still challenging because of limit in power delivery. Recent progress in energy storage and harvesting technologies using flexible material could be combined to the skin-like electronics. More information regarding health and environment can be obtained to improve the quality of human life using skin-like electronics.

References

1. J.W. Jeong, G. Shin, S.I. Park, K.J. Yu, L. Xu, J.A. Rogers, Soft materials in neuroengineering for hard problems in neuroscience. Neuron **86**(1), 175–186 (2015)
2. Y. Lee, W.-H. Yeo, Skin-like electronics for a persistent brain-computer interface. J. Nat. Sci. **1**(7), e132 (2015)
3. D.H. Kim, N. Lu, R. Ma, Y.S. Kim, R.H. Kim, S. Wang et al., Epidermal electronics. Science **333**(6044), 838–843 (2011)
4. M.L. Hammock, A. Chortos, B.C.K. Tee, J.B.H. Tok, Z.A. Bao, 25th anniversary article: the evolution of electronic skin (e-skin): a brief history, design considerations, and recent progress. Adv. Mater. **25**(42), 5997–6037 (2013)

5. D.H. Kim, R. Ghaffari, N.S. Lu, J.A. Rogers, Flexible and stretchable electronics for biointegrated devices. Annu. Rev. Biomed. Eng. **14**, 113–128 (2012)

6. Y. Liu, O. Sourina, M.K. Nguyen (eds.), Real-time EEG-based human emotion recognition and visualization, in *2010 International Conference on Cyberworlds (CW)*, IEEE

7. R. Ohme, D. Reykowska, D. Wiener, A. Choromanska, Application of frontal EEG asymmetry to advertising research. J. Econ. Psychol. **31**(5), 785–793 (2010)

8. P. Shenoy, K.J. Miller, B. Crawford, R.P.N. Rao, Online electromyographic control of a robotic prosthesis. IEEE Trans. Bio-Med. Eng. **55**(3), 1128–1135 (2008)

9. H. Converse, T. Ferraro, D. Jean, L. Jones, V. Mendhiratta, E. Naviasky et al., (eds.), An EMG biofeedback device for video game use in forearm physiotherapy, in *SENSORS, 2013 IEEE*, 2013

10. A. Lopez, P. Arevalo, F. Ferrero, M. Valledor, J. Campo (eds.), EOG-based system for mouse control, in *SENSORS, 2014 IEEE*, 2014

11. C.S.L. Tsui, P. Jia, J.Q. Gan, H. Hu, K. Yuan (eds.), EMG-based hands-free wheelchair control with EOG attention shift detection, in *2007 ROBIO 2007 IEEE International Conference on Robotics and Biomimetics*, 2007, IEEE

12. D. Chiumello, E. Carlesso, P. Cadringher, P. Caironi, F. Valenza, F. Polli et al., Lung stress and strain during mechanical ventilation for acute respiratory distress syndrome. Am. J. Respir. Crit. Care Med. **178**(4), 346–355 (2008)

13. T. Yamamoto, Y. Yamamoto, Analysis for the change of skin impedance. Med. Biol. Eng. Compu. **15**(3), 219–227 (1977)

14. X. Huang, H. Cheng, K. Chen, Y. Zhang, Y. Zhang, Y. Liu et al., Epidermal impedance sensing sheets for precision hydration assessment and spatial mapping. IEEE Trans. Bio-Med. Eng. **60**(10), 2848–2857 (2013)

15. X. Huang, W.H. Yeo, Y.H. Liu, J.A. Rogers. epidermal differential impedance sensor for conformal skin hydration monitoring. Biointerphases **7**(1–4) (2012)

16. S. Ramakrishna, J. Mayer, E. Wintermantel, K.W. Leong, Biomedical applications of polymer-composite materials: a review. Compos. Sci. Technol. **61**(9), 1189–1224 (2001)

17. J.Y. Baek, J.H. An, J.M. Choi, K.S. Park, S.H. Lee, Flexible polymeric dry electrodes for the long-term monitoring of ECG. Sensor Actuat. A-Phys. **143**(2), 423–429 (2008)

18. R.J. Zdrahala, I.J. Zdrahala, Biomedical applications of polyurethanes: a review of past promises, present realities, and a vibrant future. J. Biomater. Appl. **14**(1), 67–90 (1999)

19. A. Burke, N. Hasirci, Polyurethanes in biomedical applications. Adv. Exp. Med. Biol. **553**, 83–101 (2004)

20. C. Hassler, T. Boretius, T. Stieglitz, Polymers for Neural Implants. J. Polym. Sci. Polym. Phys. **49**(1), 18–33 (2011)

21. Y. Sun, S.P. Lacour, R.A. Brooks, N. Rushton, J. Fawcett, R.E. Cameron, Assessment of the biocompatibility of photosensitive polyimide for implantable medical device use. J. Biomed. Mater. Res. A **90A**(3), 648–655 (2009)

22. J.-P. Hsu, S. Kammer, E. Jung, L. Rieth, R. Normann, F. Solzbacher (eds.), Characterization of Parylene-C film as an encapsulation material for neural interface devices, in *Conference on Multi-Material Micro Manufacture*, 2007

23. M. Amjadi, Y.J. Yoon, I. Park, Ultra-stretchable and skin-mountable strain sensors using carbon nanotubes-Ecoflex nanocomposites. Nanotechnology **26**(37) (2015)

24. T.J. White, D.J. Broer, Programmable and adaptive mechanics with liquid crystal polymer networks and elastomers. Nat. Mater. **14**(11), 1087–1098 (2015)

25. S. Wagner, S. Bauer, Materials for stretchable electronics. MRS Bull. **37**(3), 207–217 (2012)

26. H. Tao, D.L. Kaplan, F.G. Omenetto, Silk materials—a road to sustainable high technology. Adv. Mater. **24**(21), 2824–2837 (2012)

27. S.W. Hwang, G. Park, H. Cheng, J.K. Song, S.K. Kang, L. Yin et al., 25th anniversary article: materials for high-performance biodegradable semiconductor devices. Adv. Mater. **26**(13), 1992–2000 (2014)

28. D.H. Kim, J. Viventi, J.J. Amsden, J.L. Xiao, L. Vigeland, Y.S. Kim et al., Dissolvable films of silk fibroin for ultrathin conformal bio-integrated electronics. Nat. Mater. 9(6), 511–517 (2010)
29. N. Saba, P.M. Tahir, M. Jawaid, A review on potentiality of nano filler/natural fiber filled polymer hybrid composites. Polymers-Basel. 6(8), 2247–2273 (2014)
30. X.M. Liu, Z.D. Huang, S.W. Oh, B. Zhang, P.C. Ma, M.M.F. Yuen et al., Carbon nanotube (CNT)-based composites as electrode material for rechargeable Li-ion batteries: A review. Compos. Sci. Technol. 72(2), 121–144 (2012)
31. Z.S. Wu, G.M. Zhou, L.C. Yin, W. Ren, F. Li, H.M. Cheng, Graphene/metal oxide composite electrode materials for energy storage. Nano Energy 1(1), 107–131 (2012)
32. F. Xu, Y. Zhu, Highly conductive and stretchable silver nanowire conductors. Adv. Mater. 24 (37), 5117–5122 (2012)
33. C. Farcau, N.M. Sangeetha, H. Moreira, B. Viallet, J. Grisolia, D. Ciuculescu-Pradines et al., High-sensitivity strain gauge based on a single wire of gold nanoparticles fabricated by stop-and-go convective self-assembly. ACS Nano 5(9), 7137–7143 (2011)
34. R.J. Chen, S. Bangsaruntip, K.A. Drouvalakis, N.W.S. Kam, M. Shim, Y.M. Li et al., Noncovalent functionalization of carbon nanotubes for highly specific electronic biosensors. Proc. Natl. Acad. Sci. USA 100(9), 4984–4989 (2003)
35. K. Balani, R. Anderson, T. Laha, M. Andara, J. Tercero, E. Crumpler et al., Plasma-sprayed carbon nanotube reinforced hydroxyapatite coatings and their interaction with human osteoblasts in vitro. Biomaterials 28(4), 618–624 (2007)
36. X.F. Shi, J.L. Hudson, P.P. Spicer, J.M. Tour, R. Krishnamoorti, A.G. Mikos, Injectable nanocomposites of single-walled carbon nanotubes and biodegradable polymers for bone tissue engineering. Biomacromolecules 7(7), 2237–2242 (2006)
37. P.M. Ajayan, O. Stephan, C. Colliex, D. Trauth, Aligned carbon nanotube arrays formed by cutting a polymer resin-nanotube composite. Science 265(5176), 1212–1214 (1994)
38. H.C. Jung, J.H. Moon, D.H. Baek, J.H. Lee, Y.Y. Choi, J.S. Hong et al., CNT/PDMS composite flexible dry electrodes for long-term ECG monitoring. IEEE Trans. Bio-Med. Eng. 59(5), 1472–1479 (2012)
39. J.H. Lee, S.M. Lee, H.J. Byeon, J.S. Hong, K.S. Park, S.H. Lee, CNT/PDMS-based canal-typed ear electrodes for inconspicuous EEG recording. J. Neural Eng. 11(4) (2014)
40. S.M. Lee, H.J. Byeon, J.H. Lee, D.H. Baek, K.H. Lee, J.S. Hong et al., Self-adhesive epidermal carbon nanotube electronics for tether-free long-term continuous recording of biosignals. Sci. Rep.-UK 4 (2014)
41. D. Ryu, K.J. Loh, R. Ireland, M. Karimzada, F. Yaghmaie, A.M. Gusman, In situ reduction of gold nanoparticles in PDMS matrices and applications for large strain sensing. Smart Struct. Syst. 8(5), 471–486 (2011)
42. J.Y. Hwang, H.S. Kim, J.H. Kim, U.S. Shin, S.H. Lee, Carbon nanotube nanocomposites with highly enhanced strength and conductivity for flexible electric circuits. Langmuir 31(28), 7844–7851 (2015)
43. V. Singh, D. Joung, L. Zhai, S. Das, S.I. Khondaker, S. Seal, Graphene based materials: past, present and future. Prog. Mater. Sci. 56(8), 1178–1271 (2011)
44. K. Hu, D.D. Kulkarni, I. Choi, V.V. Tsukruk, Graphene-polymer nanocomposites for structural and functional applications. Prog. Polym. Sci. 39(11), 1934–1972 (2014)
45. D.S. Hecht, L.B. Hu, G. Irvin, Emerging transparent electrodes based on thin films of carbon nanotubes, graphene, and metallic nanostructures. Adv. Mater. 23(13), 1482–1513 (2011)
46. M. Segev-Bar, H. Haick, Flexible sensors based on nanoparticles. ACS Nano 7(10), 8366–8378 (2013)
47. N. Matsuhisa, M. Kaltenbrunner, T. Yokota, H. Jinno, K. Kuribara, T. Sekitani et al., Printable elastic conductors with a high conductivity for electronic textile applications. Nat. Commun. 6 (2015)
48. T.A. Kim, H.S. Kim, S.S. Lee, M. Park, Single-walled carbon nanotube/silicone rubber composites for compliant electrodes. Carbon 50(2), 444–449 (2012)

49. Y.Y. Huang, E.M. Terentjev, Tailoring the electrical properties of carbon nanotube-polymer composites. Adv. Funct. Mater. **20**(23), 4062–4068 (2010)

50. K.Y. Chun, Y. Oh, J. Rho, J.H. Ahn, Y.J. Kim, H.R. Choi et al., Highly conductive, printable and stretchable composite films of carbon nanotubes and silver. Nat. Nanotechnol. **5**(12), 853–857 (2010)

51. T. Sekitani, H. Nakajima, H. Maeda, T. Fukushima, T. Aida, K. Hata et al., Stretchable active-matrix organic light-emitting diode display using printable elastic conductors. Nat. Mater. **8**(6), 494–499 (2009)

52. G.X. Chen, Y.J. Li, H. Shimizu, Ultrahigh-shear processing for the preparation of polymer/carbon nanotube composites. Carbon **45**(12), 2334–2340 (2007)

53. Y.Y. Huang, S.V. Ahir, E.M. Terentjev, Dispersion rheology of carbon nanotubes in a polymer matrix. Phys. Rev. B **73**(12) (2006)

54. T. Sekitani, Y. Noguchi, K. Hata, T. Fukushima, T. Aida, T. Someya, A rubberlike stretchable active matrix using elastic conductors. Science **321**(5895), 1468–1472 (2008)

55. B.K. Price, J.L. Hudson, J.M. Tour, Green chemical functionalization of single-walled carbon nanotubes in ionic liquids. J. Am. Chem. Soc. **127**(42), 14867–14870 (2005)

56. W.H. Yeo, Y.S. Kim, J. Lee, A. Ameen, L.K. Shi, M. Li et al., Multifunctional epidermal electronics printed directly onto the skin. Adv. Mater. **25**(20), 2773–2778 (2013)

57. J.A. Fan, W.H. Yeo, Y.W. Su, Y. Hattori, W. Lee, S.Y. Jung et al., Fractal design concepts for stretchable electronics. Nat. Commun. **5** (2014)

58. S. Xu, Y.H. Zhang, J. Cho, J. Lee, X. Huang, L. Jia et al., Stretchable batteries with self-similar serpentine interconnects and integrated wireless recharging systems. Nat. Commun. **4** (2013)

59. P. Lee, J. Lee, H. Lee, J. Yeo, S. Hong, K.H. Nam et al., Highly stretchable and highly conductive metal electrode by very long metal nanowire percolation network. Adv. Mater. **24** (25), 3326–3332 (2012)

60. S. Han, S. Hong, J. Ham, J. Yeo, J. Lee, B. Kang et al., Fast plasmonic laser nanowelding for a Cu-nanowire percolation network for flexible transparent conductors and stretchable electronics. Adv. Mater. **26**(33), 5808–5814 (2014)

61. Z.P. Chen, W.C. Ren, L.B. Gao, B.L. Liu, S.F. Pei, H.M. Cheng, Three-dimensional flexible and conductive interconnected graphene networks grown by chemical vapour deposition. Nat. Mater. **10**(6), 424–428 (2011)

62. Y. Shang, X. He, Y. Li, L. Zhang, Z. Li, C. Ji et al., Super-stretchable spring-like carbon nanotube ropes. Adv. Mater. **24**(21), 2896–2900 (2012)

63. K. Liu, Y.H. Sun, P. Liu, X.Y. Lin, S.S. Fan, K.L. Jiang, Cross-stacked superaligned carbon nanotube films for transparent and stretchable conductors. Adv. Funct. Mater. **21**(14), 2721–2728 (2011)

64. D.H. Kim, J.L. Xiao, J.Z. Song, Y.G. Huang, J.A. Rogers, Stretchable, curvilinear electronics based on inorganic materials. Adv. Mater. **22**(19), 2108–2124 (2010)

65. A.R. Madaria, A. Kumar, C.W. Zhou, Large scale, highly conductive and patterned transparent films of silver nanowires on arbitrary substrates and their application in touch screens. Nanotechnology **22**(24) (2011)

66. V. Scardaci, R. Coull, P.E. Lyons, D. Rickard, J.N. Coleman, Spray deposition of highly transparent, low-resistance networks of silver nanowires over large areas. Small **7**(18), 2621–2628 (2011)

67. J.Y. Lee, S.T. Connor, Y. Cui, P. Peumans, Solution-processed metal nanowire mesh transparent electrodes. Nano Lett. **8**(2), 689–692 (2008)

68. S.H. Jeong, K. Hjort, Z.G. Wu, Tape transfer atomization patterning of liquid alloys for microfluidic stretchable wireless power transfer. Sci. Rep. UK **5** (2015)

69. S. De, T.M. Higgins, P.E. Lyons, E.M. Doherty, P.N. Nirmalraj, W.J. Blau et al., Silver nanowire networks as flexible, transparent, conducting films: extremely high DC to optical conductivity ratios. ACS Nano **3**(7), 1767–1774 (2009)

70. J. Lee, P. Lee, H.B. Lee, S. Hong, I. Lee, J. Yeo et al., Room-temperature nanosoldering of a very long metal nanowire network by conducting-polymer-assisted joining for a flexible touch-panel application. Adv. Funct. Mater. **23**(34), 4171–4176 (2013)
71. C. Yang, H.W. Gu, W. Lin, M.M. Yuen, C.P. Wong, M.Y. Xiong et al., Silver nanowires: from scalable synthesis to recyclable foldable electronics. Adv. Mater. **23**(27), 3052–3056 (2011)
72. J. Lee, P. Lee, H. Lee, D. Lee, S.S. Lee, S.H. Ko, Very long Ag nanowire synthesis and its application in a highly transparent, conductive and flexible metal electrode touch panel. Nanoscale **4**(20), 6408–6414 (2012)
73. Y.J. Jung, S. Kar, S. Talapatra, C. Soldano, G. Viswanathan, X.S. Li et al., Aligned carbon nanotube-polymer hybrid architectures for diverse flexible electronic applications. Nano Lett. **6**(3), 413–418 (2006)
74. M. Amjadi, A. Pichitpajongkit, S. Lee, S. Ryu, I. Park, Highly stretchable and sensitive strain sensor based on silver nanowire-elastomer nanocomposite. ACS Nano **8**(5), 5154–5163 (2014)
75. J. Yoon, S.Y. Hong, Y. Lim, S.J. Lee, G. Zi, J.S. Ha, Design and fabrication of novel stretchable device arrays on a deformable polymer substrate with embedded liquid-metal interconnections. Adv. Mater. **26**(38), 6580–6586 (2014)
76. W.J. Ma, L. Song, R. Yang, T.H. Zhang, Y.C. Zhao, L.F. Sun et al., Directly synthesized strong, highly conducting, transparent single-walled carbon nanotube films. Nano Lett. **7**(8), 2307–2311 (2007)
77. L. Wang, J. Liu, Printing low-melting-point alloy ink to directly make a solidified circuit or functional device with a heating pen. Proc. Roy. Soc A-Math. Phys. Eng. Sci. **470**(2172) (2014)
78. S.Z. Guo, X.L. Yang, M.C. Heuzey, D. Therriault, 3D printing of a multifunctional nanocomposite helical liquid sensor. Nanoscale **7**(15), 6451–6456 (2015)
79. D.H. Kim, Y.S. Kim, J. Amsden, B. Panilaitis, D.L. Kaplan, F.G. Omenetto et al., Silicon electronics on silk as a path to bioresorbable, implantable devices. Appl. Phys. Lett. **95**(13) (2009)
80. R. Luginbuehl, U. Rosler, M. Wipf, Biological evaluation according to ISO 10993-1: methods and pitfalls. Eur. Cells Mater. **27** (2014)
81. U. Roesler, R. Luginbuehl, M. Wipf, Biological evaluation according to ISO 10993-1: approach and structure. Eur. Cells Mater. **27** (2014)
82. G.H. Altman, F. Diaz, C. Jakuba, T. Calabro, R.L. Horan, J.S. Chen et al., Silk-based biomaterials. Biomaterials **24**(3), 401–416 (2003)
83. R.L. Horan, K. Antle, A.L. Collette, Y.Z. Huang, J. Huang, J.E. Moreau et al., In vitro degradation of silk fibroin. Biomaterials **26**(17), 3385–3393 (2005)
84. B. Joseph, S.J. Raj, Therapeutic applications and properties of silk proteins from Bombyx mori. Front Life Sci. **6**(3–4), 55–60 (2012)
85. S. Lee, J. Kim, C. Park, J.Y. Hwang, J.S. Hong, K. Lee et al., Self-adhesive and capacitive carbon nanotube-based electrode to record electroencephalograph signals from the hairy scalp. IEEE Trans. Bio-Med. Eng. (2015)
86. C. Pang, G.Y. Lee, T.I. Kim, S.M. Kim, H.N. Kim, S.H. Ahn et al., A flexible and highly sensitive strain-gauge sensor using reversible interlocking of nanofibres. Nat. Mater. **11**(9), 795–801 (2012)
87. K.I. Jang, S.Y. Han, S. Xu, K.E. Mathewson, Y.H. Zhang, J.W. Jeong et al., Rugged and breathable forms of stretchable electronics with adherent composite substrates for transcutaneous monitoring. Nat. Commun. **5** (2014)
88. T. Yamada, Y. Hayamizu, Y. Yamamoto, Y. Yomogida, A. Izadi-Najafabadi, D.N. Futaba et al., A stretchable carbon nanotube strain sensor for human-motion detection. Nat. Nanotechnol. **6**(5), 296–301 (2011)
89. D.J. Lipomi, M. Vosgueritchian, B.C.K. Tee, S.L. Hellstrom, J.A. Lee, C.H. Fox et al., Skin-like pressure and strain sensors based on transparent elastic films of carbon nanotubes. Nat. Nanotechnol. **6**(12), 788–792 (2011)

90. S. Gong, W. Schwalb, Y.W. Wang, Y. Chen, Y. Tang, J. Si et al., A wearable and highly sensitive pressure sensor with ultrathin gold nanowires. Nat. Commun. **5** (2014)
91. G.-W. Hong, S.-H. Kim, J.-H. Kim (eds.), Flexible pressure sensors for burnt skin patient monitoring. SPIE Smart Structures and Materials+Nondestructive Evaluation and Health Monitoring, in *International Society for Optics and Photonics*, 2015
92. X.L. Zhao, Q.L. Hua, R.M. Yu, Y. Zhang, C.F. Pan, Flexible, stretchable and wearable multifunctional sensor array as artificial electronic skin for static and dynamic strain mapping. Adv. Electron. Mater. **1**(7) (2015)
93. Y.P. Zang, F.J. Zhang, C.A. Di, D.B. Zhu, Advances of flexible pressure sensors toward artificial intelligence and health care applications. Mater. Horiz. **2**(2), 140–156 (2015)
94. K. Chu, D. Kim, Y. Sohn, S. Lee, C. Moon, S. Park, Electrical and thermal properties of carbon-nanotube composite for flexible electric heating-unit applications. IEEE Electr. Device Lett. **34**(5), 668–670 (2013)
95. S. Choi, J. Park, W. Hyun, J. Kim, J. Kim, Y.B. Lee et al., Stretchable heater using ligand-exchanged silver nanowire nanocomposite for wearable articular thermotherapy. ACS Nano **9**(6), 6626–6633 (2015)

Part III
Bioinspired/Implantable Electronics

Chapter 13
Mechanically Compliant Neural Interfaces

Ivan R. Minev and Stéphanie P. Lacour

Abstract Neural interfaces are engineered devices that aim at replacing, restoring, and rehabilitating the injured or damaged nervous system. One of the challenges to overcome to deploy therapeutic neural interfaces as clinical treatments lies in the physical mismatch between biological tissues and artificial engineered devices. This chapter details recent development in materials science and technology focused on reducing this physical mismatch thereby opening the path for long-term biointe-grated neural interfaces.

Keywords Thin film devices · Elastomer · Implantable electrodes · Mechanical compliance

13.1 Introduction

A neural interface aims at establishing a communication channel between the nervous system and an engineered device. The information that is exchanged is complex because events in the nervous system can occur on a temporal scale of milliseconds to years; they can emanate from a single synapse, approximately a micrometer in size, to macroscopic bundles of nerve fibers. Within the spatial confines of neural organs (brain, spinal cord and peripheral nerves), cells communicate by electrical and chemical impulses. Capturing and decoding these signals help to understand the physiological basis of perception, memory, and consciousness, and can serve therapeutic purposes for the restoration of functions lost through injury or disease. If we understand the language of nerve tissue, we can use suitably

I.R. Minev · S.P. Lacour (✉)
School of Engineering, Centre for Neuroprosthetics, Bertarelli Chair in Neuroprosthetic Technology, Laboratory for Soft Bioelectronic Interfaces, Ecole Polytechnique Fédérale de Lausanne EPFL, EPFL LSBI Station 17, CH-1015, Lausanne, Switzerland
e-mail: stephanie.lacour@epfl.ch

I.R. Minev
e-mail: ivan.minev@epfl.ch

© Springer International Publishing Switzerland 2016
J.A. Rogers et al. (eds.), *Stretchable Bioelectronics for Medical Devices and Systems*, Microsystems and Nanosystems,
DOI 10.1007/978-3-319-28694-5_13

designed neural interfaces to modulate brain activity to probe the function of specific circuits or to reduce the sensation of chronic pain, overcome clinical depression, or restore movement. Although currently we could not listen (or talk) to every neuron in a mammalian brain at all times (a human brain contains about 100 billion neurons), let alone process the unimaginable amount of data that would be generated each second, the nervous system is accessible both at the level of single cells and of large networks. Neural interfaces are evolving to expand the temporal, spatial, and throughput scales at which information is exchanged. This chapter outlines contemporary challenges in building the physical devices used in neural interfacing and especially those that are placed inside the body. Emphasis is placed on efforts to improve the biocompatibility and functionality of neural implants by considering biomechanics, novel materials, and fabrication methods.

13.2   Establishing Communication with Neural Tissues

Neural interfacing devices need to fulfill two basic purposes. One is to detect (record) and the other is to modulate neural activity (stimulate or inhibit) as illustrated in Fig. 13.1a, b, respectively. The ionic nature of neuronal communication has made the microelectrode array the standard device for neural recordings. Some more recent recording technologies are optical in nature and employ genetically encoded voltage reporter molecules that modify the intensity of their fluorescence as a result of neuron membrane depolarization [7]. Stimulation devices also largely rely on various electrode designs; however several neuromodulation technologies exist. Optical neuron stimulation can be achieved by genetically encoding light sensitive ion channels in the cell membrane, a technique known as optogenetics [4]. Because communication at synapses is chemical, activity can be elicited by the targeted delivery of chemicals such as neurotransmitters released from an implanted device [29, 51]. Delivery of other forms of energy such as heat [8], or ultrasound agitation [61, 74] has also been explored and it may offer less invasive solutions in stimulating deep brain structures. At other times, it may be necessary to inhibit activity (e.g., to investigate the functional significance of a specific circuit or to interrupt an epileptic seizure). This may be achieved with optogenetic techniques where opsin activation causes membrane hyperpolarisation [75], or by local heat extraction (cooling) [33] where both have the effect of reducing the excitability of neural tissues nearby. In an analogy with electrodes, some of the alternative modes of neuromodulation described above can be dubbed "optrodes" (delivery of light), "chemotrodes" (delivery of pharmacology), and "thermotrodes" (delivery or extraction of heat). The delivery (or extraction) of these various forms of energy is what determines the choice of materials employed in realizing the neural interface as well as the microfabrication processes available.

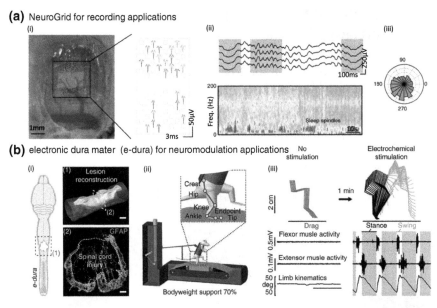

Fig. 13.1 Examples of neural implants for recording and stimulation applications. The NeuroGrid and e-dura implants are examples of the two basic functions of a neural interface; detection and modulation of neural activity. Conceptually, the two modalities can be combined to form a neuroprosthetic system that can bypass a paralyzing spinal cord injury. The system will decode movement intention from brain signals and translate those to motion through functional electrical stimulation. **a** Conformal microelectrodes for recording applications (NeuroGrid). The combination of "neuron sized" ($10 \times 10\ \mu m^2$) recording electrodes and the conformability offered by an ultrathin substrate (4 μm) allows for the simultaneous recording of local field potentials (LFP) and single neuron action potentials (AP) from the surface of the brain. In (*i*) the implant is placed over the somatosensory cortex in rat. The spikes (APs) in the grid on the *right* are obtained by high pass filtering and averaging signals from corresponding electrodes on the array. In (*ii*) the grid is used to record LFP. The *upper panel* shows traces obtained during sleep where typical slow wave activity (*orange box*) and sleep spindles (*grey box*) can be observed. *Below* is a time-frequency spectrogram of LFP recorded during sleep. In (*iii*), a polar plot demonstrates that single units are phase locked to oscillations in the LFP signal (*grey box* slow oscillation in (*ii*)). Adapted from [32]. **b** Conformal microelectrodes for neuromodulation in the spinal cord (electronic dura mater). The use of soft materials has enabled the long-term integration of the implant in the subdural space of the spinal cord. The implant delivers electrical and pharmacological modulations which are tailored to restore locomotion in spinal injured rats. In (*i*), histological examination of the thoracic contusion injury reveals only a small proportion of spared spinal tissue. The implant is placed below the injury. In (*ii*), depicted is a robotic support system for bipedal locomotion and monitoring of hind limb kinematics in injured and implanted rats. In (*iii*), the implant is used as an electrochemical neuroprosthesis. Stick diagram decompositions of hind limb movements are shown together with leg muscle activity and hind limb oscillations. Without input from the prosthesis, the hind limbs of the rat are paralyzed and drag. Upon the synergistic application of electrical and pharmacological stimulation, step-like movement is restored. Adapted from [47]

13.3 Soft Neural Interfaces Maintain Proximity to Interfacing Targets

Functional devices, whether electrical, optical, or chemical, have to coexist (in close proximity) with living neural organs (Fig. 13.2a–k). The neural interface has to form physical contact with neural parenchyma, because the signals generated by neurons are small, bulk tissue filters out fast events, and neurons themselves are closely packed. Proximity to the source of signals therefore improves the selectivity

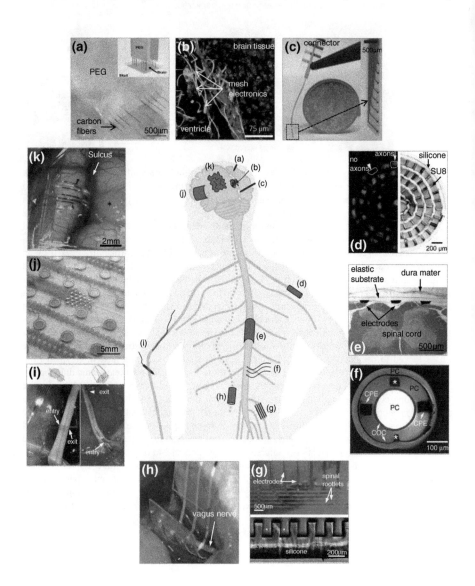

◀ **Fig. 13.2** Mechanically compliant implants are integrated with soft tissues of the nervous system. Since the ultimate goal of neural engineering is the development of new clinical treatments, the envisaged applications of the implants (**a–k**) are presented in the central diagram in the context of the human anatomy. **a** *Ultrathin and flexible intracortical electrode arrays.* Individual electrodes are single carbon fiber electrodes ($d = 8.4$ μm) insulated with parylene and functionalized with a conductive polymer. A temporary water soluble coating of poly(ethylene glycol) holds the electrodes together and facilitates slow insertion without buckling. Such electrodes can be used to record single unit activity chronically from animal brains while minimizing the scarring response. Adapted from [53]. **b** *Syringe injectable electronics.* Electronic components built on flexible polymer meshes are inserted in the lateral ventricle of rats through a hollow needle. Once in place, the mesh unfurls and records local field potentials and single unit brain activity. The image is a brain tissue slice where the filaments of the mesh are visible in *green*. Labeled in *green* are also neurons in the brain parenchyma visible in the *top right corner*. Cell nuclei are in blue. Adapted from [43]. **c** *Flexible neural implants for deep brain interfacing.* The use of polyimide as substrate allows the probe to bend to a 1 mm radius without altering device performance. Electrode tracks are made of platinum and are placed in the neutral plane. The active sites of the electrodes (*inset to the right*) are made from glassy carbon, which apart from electrical functionality could allow chemical sensing of dopamine, as may be required in a deep brain stimulation therapy for Parkinson's disease. Coin diameter is 17 mm. Adapted from [68]. **d** *Microchannel-based regenerative scaffold. Right panel* an optical micrograph of the device is presented. Visible are the spacer walls (SU-8) that delimit regenerative microchannels. The concentric spiral is made from silicone allowing fabrication in 2D and rolling to produce the final 3D structure. *Left panel* histological image of axons (in *red*) that have grown through the channels. Most but not all channels host regenerated axons. The channel dimensions are 150×100 μm^2. Adapted from [64]. **e** *Electronic dura mater.* Through the use of elastomers, this spinal implant mimics the mechanical properties of dura mater, the protective membrane around the brain and spinal cord. Functional components such as stretchable interconnects (gold), electrodes (platinum–silicone composite), and microfluidics are integrated on the elastic substrate. These functionalities have been used to deliver electrochemical spinal neuromodulation that restored locomotion after paralyzing spinal cord injury in rats. The image is a cross-section micrograph of a rat spinal cord that had carried the implant for 8 weeks in the (intrathecal) space under dura mater. Adapted from [47]. **f** *Multi-modal fiber probes.* The fiber supports simultaneously optical, electrical, and pharmacological modes of neuromodulation and recording. Fibers form stable brain–machine interfaces in vivo in mouse brain and spinal cord where they can operate for at least two months. The image is a cross-section of a polycarbonate (PC) fiber. PC and cyclic olefin copolymer (COC) form the waveguide core of the probe. Conductive polyethylene (CPE) forms electrodes and *asterisks mark* microfluidic channels. Adapted from [6]. **g** *Microchannel spinal root interface.* Nerve roots in the spinal canal can be surgically teased apart into smaller "rootlets" and placed in parallel microchannels equipped with electrodes (*upper panel*). This allows for recording nerve impulses with high selectivity as has been demonstrated with monitoring bladder fullness in rats. The mechanical softness of the channels prevents rootlet damage during chronic implantation. The bottom panel is a cross-section of the microchannel array. Asterisks indicate the microchannel lumens to be occupied by nerve rootlets. Following rootlet placement, microchannels are sealed with a silicone lid. Adapted from [10]. **h** *Shape memory nerve cuff.* A shape memory polymer (thiol–ene/acrylate) undergoes a stiffness (600–6 MPa) and 2D to 3D shape transition when temperature is raised form ambient to 37 °C. Equipped with thin film conductors, the device forms a cuff electrode that in the micrograph shown here coils around the vagus nerve of a rat. Adapted from [71]. In more recent work, self-shaping cuffs have been integrated with organic thin film transistors [56]. **i** *Intraneural electrode arrays.* These are often realized from flexible plastics such as polyimide and thin metal films allowing the nerve to remain mechanically decoupled from the implant anchoring point. The micrograph presents the two most common intraneural electrode designs. The *left panel* shows the transverse intrafascicular multichannel electrode (TIME) that can contact axons in several fascicles at a time. The *right panel*

◀ shows a longitudinal intrafascicular electrode (LIFE) which allows several electrode contacts to be placed inside a single fascicle. Both electrode designs have demonstrated high selectivity and low activation thresholds for distal muscles. Adapted from [2]. Intraneural electrodes have recently been used to restore sensory feedback in a bidirectional human hand prosthesis [55]. **j** *ElectroCorticoGraphy (ECoG) array for large brain area monitoring.* This technology relies on laser structured metallic foils embedded in an elastomer substrate which ensures conformability to the global curvature of the brain. Flexibility of the array is further enhanced by machining interconnects in a meandering geometry. The technology can find applications in mapping brain activity or for providing feedback during deep brain stimulation treatments. Visible in the micrograph are large grid (diameter of 2.4 mm) and small cluster (diameter of 870 μm) electrodes which can be used for global and local brain activity recordings, respectively. Image adapted from: CorTec, http://cortec-neuro.com/en/company **k** *Intrasulcal ECoG array.* This technology illustrates the use of thin plastic substrates to achieve conformability to the local topography of the brain. Here, the substrate is a 20 μm thin parylene film which allows the array to be placed in the superior temporal sulcus (STS) of the macaque brain. Electrodes are patterned from thin gold films and are coated with platinum black. Intrasulcal stimulation was found to require lower current thresholds as compared to stimulation on the surface of the brain gyri. Adapted from [45]

of recordings (conversely the specificity of stimulation) and the signal-to-noise ratio of the obtained signals but requires an invasive approach where a foreign body is installed inside or on the surface of neural organs (Fig. 13.3a). For example, recordings made from the surface of the brain (e.g. using ElectroCorticoGraphy electrode arrays) possess spatiotemporal resolution that is superior to those obtained from electrodes attached to the scalp [5].

13.3.1 Interfaces to the Brain

On the surface of the cortex, simultaneous proximity to multiple areas of interest can be achieved by ensuring the electrode array conformability to the macroscopic curvature of the brain (Fig. 13.2j) [59]. If the device can accommodate smaller bending radii then the selectivity of the interface can be further increased as electrodes come into contact with additional cortical surface hidden in sulci (Fig. 13.2k) [35, 45]. If mechanical compliance is combined with miniaturization of the electrode sites, then it is also possible to record single units from surface placed micro-electrodes [32]. From a topographical point of view, the interfacing device is required to conform to a non-developable surface that has two nonzero principle curvatures as illustrated in Fig. 13.3b. Since fabrication technologies are still largely confined to two dimensions, conformability to a curvilinear 3D object requires the interface to tolerate mechanical strain.

Proximity to deep cortical structures can be achieved by introducing a penetrating probe below the surface of the neural organ. Accessing structures deep in the brain, comes at the cost of damaging tissue in the path of the advancing implant as well as activating a sequence of glial cell and local neurodegenerative processes associated with a chronically placed foreign body [54]. Opportunities to record the activity of multiple single neurons [28], and to stimulate deep brain nuclei [13],

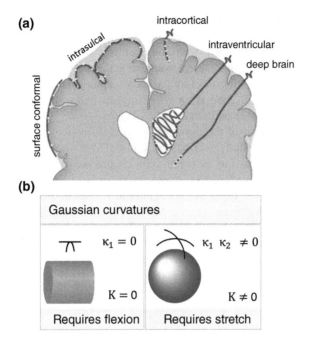

Fig. 13.3 Neural interfaces have to be mechanically compliant. **a** Brain interfaces are designed to keep in close contact with brain tissues. Surface conformal interfaces follow the general curvature of the brain (typical ranges for whole brain bending radii in human 5–7 cm, in rat 6–9 mm). Intrasulcal electrode arrays conform to local features on the cortex surface (bending radii are on the order of several millimeters in human and primate brains). Intracortical implants penetrate grey matter several millimeters below the cortex surface. Intraventricular probes unfurl to conform the inner surface of brain cavities. Deep brain probes access deep brain nuclei. Penetrating brain probes need to be mechanically decoupled from anchoring points (indicated by a *wiggle*). **b** Developable surfaces with zero Gaussian curvatures can be conformed by flexible foils. Non-developable surfaces with nonzero Gaussian curvature require a stretchable foil

however have driven sustained efforts to mitigate the adverse effects of chronic implantation of penetrating probes. Conventional wisdom dictates that the brain dwelling portion of the probe should be mechanically decoupled from (but electrically communicating with) its anchoring point for example, via a flexible electronic cable so as to allow the probe to 'float' with brain tissue (Fig. 13.2a, c) [16, 28, 63, 68]. Brain (micro) motions are largely caused by the pulsatile nature of blood flow [69]. During systole, brain tissue volumetric strains of 3×10^{-4} have been reported [27]. The fact that the brain is a dynamic organ places a demand on the mechanical properties of an intracortical neural interface. To avoid stress concentration at the interface between implant and parenchyma, the implant must have viscoelastic properties similar to those of brain tissue [65]. Since elastic moduli for brain are measured in hundreds of Pascal [9, 11] and the brain tissue is a hydrated material, various natural and synthetic hydrogels have been used as coatings [36] or for transient structural elements of the intracortical probe [18, 62].

Beyond this macroscopic view of mechanical biocompatibility, there is evidence that mechanosensing at the cell level may play a role in how glial cells respond once they come into contact with the surface of a foreign body [48, 49]. Engineering the mechanical microenvironment at the implant surface may therefore offer a route for modulating the foreign body response [46]. This is expected to reduce the neuronal kill zone around the implant hence maintaining a physical proximity between device and target.

A challenging aspect of the development of soft intracortical probes concerns their insertion into similarly soft tissue. Some of the strategies, currently under investigation, involve use of insertion vehicles that are subsequently withdrawn [72], dissolved [53], or undergo a stiffness change driven by a phase transition [70]. An interesting strategy to bring the neural interface closer to deep brain structures is to utilize naturally occurring cavities in the CNS. The ventricular system provides a space where an implant can expand to establish contact with a larger tissue area (Fig. 13.2b) [43]. Because the implant has to be packed in a tight delivery vehicle, such as a syringe, the materials it is built of need to have low flexural stiffness and survive bending radii on the order of 100 μm without electrical failure as illustrated in more detail in Fig. 13.4a.

13.3.2 Interfaces to the Spinal Cord

A stable and reliable neural interface to the spinal cord requires maintaining a fixed relative position between tissue and implant. In freely behaving vertebrates, ventroflexion (bending forward) causes tensile elongation accompanied with Poisson compression, while extension (bending backwards) causes folding (wrinkling) of soft tissues inside the spinal canal. In humans, both the spinal cord and its meningeal protective membranes can experience as much as 10–20 % tensile strain and displacements (relative to the spinal canal) on the order of centimeters during normal postural movements. For a series of comprehensive reviews on the dynamics of the human central nervous system, the reader is referred to Harrison et al. [24, 25]. Spinal cord deformations in animal models such as rodents or nonhuman primates are likely larger.

Our laboratory recently demonstrated that implants with elastic properties similar to the outermost meningeal layer (dura mater), integrate seamlessly on the surface of the spinal cord in rats for extended periods of time (Fig. 13.2e) [47]. The implant is capable of withstanding repeated cycles of physiologically relevant strains without electrical or mechanical failure as illustrated in more detail in Fig. 13.4b). This enabled us to bring a functional interface closer to target neural structures by placing the implant in the intrathecal space (over the spinal cord but bellow dura mater) where we have demonstrated topical delivery of drugs and selective electrical stimulation. Materials with tissue-similar elastic moduli have also enabled implants that target spinal roots within the vertebral canal. The interface takes the form of a series of microchannels, each of which houses a spinal

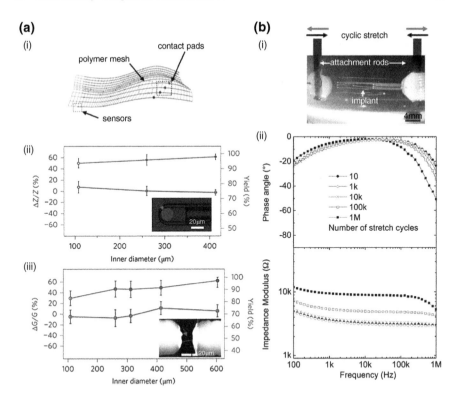

Fig. 13.4 Functional performance of implants subjected to mechanical deformations. **a** Syringe injectable mesh electronics. In order to fit inside the injecting needle, electronic components integrated in this type of device have to withstand bending radii on the order of 100s of micrometers. In the device scheme in (*i*), sensors take the form of exposed platinum disks (electrodes) and single nanowire electrochemical field effect transistors (FET) embedded in a flexible polymer mesh (SU-8). In (*ii*), impedance change (*red*) and yield (*blue*) of electrodes injected through needles of varying sizes are shown. The *red arrow* in the inset points to one such sensing electrode. In (*iii*), the conductance change (*red*) and yield (*blue*) of electrochemical FETs injected through needles of various sizes are shown. The *red arrow* in the *inset* points to a nanowire attached to two contacts thus forming a two terminal electrochemical FET. In both (*ii*) and (*iii*), the small bending radii do not appear to have a detrimental effect on the device functionality. Adapted from [43]. **b** Electronic dura mater implant. In order to match the mechanical properties of natural dura mater, electronic components (e.g. interconnects and electrodes) integrated in this type of device have to withstand tensile elongation on the order of 20 %. Panel (*i*) describes a benchtop setup to characterize the impedance of electrodes as a function of the number of applied stretch cycles (εpeak = 20 %). The ends of the implant are glued to two probes that are clamped to an extensimeter. The implant and (partially) the probes are submerged in physiological saline solution. In (*ii*) shown are impedance spectra recorded from one representative electrode. The spectra are recorded at 0 % applied strain after 10, 1,000, 10,000, 100,000, and 1 million stretch cycles. Despite extensive fatigue cycling, electrodes maintain their electrical functionality. Adapted from [47]

root filament (Fig. 13.2g). Our laboratory has demonstrated that roots housed in the soft channels continue to support healthy axons and vasculature after chronic implantation in rats [10]. The anatomical organization of sensory and motor information transmitted along spinal roots can enable a highly selective neural interface. Neural recordings obtained with electrode equipped microchannels can, for example, form the sensing branch of a bladder neuroprosthesis device [14]. Dynamical compliance with mobile tissues of the spinal cord can also be achieved using single ultra-compliant functional polymer fibers carrying various sensing and stimulation modalities (Fig. 13.2f) [44], or microwires arranged in arrays [30]. These devices offer a route towards establishing neural interfaces inside the spinal cord. In an analogy with intracortical probes, these may allow unprecedented access to specific spinal circuits using optogenetic methods or tracking single unit activity over extended periods of time.

13.3.3 Interfaces to the Peripheral Nervous System

In peripheral nerves, flexible electrode arrays have enabled the implementation of intraneural interfaces with fascicular and sub-fascicular selectivity [2, 3, 55]. Intraneural interfaces are inserted transversally or longitudinally with respect to the long axis of the nerve and rely on the formation of a fibrous capsule that binds nerve and implant together (Fig. 13.2i). Neural cuffs are tubular devices placed around the epineurium of a whole nerve and have electrodes on their inner surface. Since the nerve cannot be transected during the implantation procedure, it is the implant that has to conform to wrap around the target nerve. Shape memory, stiffness transition polymers can be useful in this respect as the implant can be fabricated as a 2D structure (equipped with electrodes) that softens and coils around the nerve once heated to body temperature (Fig. 13.2h) [56, 71]. Similarly, in peripheral nerve implants that rely on nerve regeneration, flexible materials allow rolling of a 2D + microfabricated structure into a 3D implant for guidance and electrical interfacing of regenerating axons (Fig. 13.2d) [39, 64]. In the future, regenerative microchannel arrays realized from soft materials may simultaneously serve as a conduit, functional interface and a mechanically tuned environment permissive to regeneration. The integration of functionality in such devices may extend beyond recording applications and incorporate stimulating electrodes [50] and optical elements [67].

13.4 Materials and Technologies for Soft Interfaces

Materials used in neural interfaces can be classed as functional and substrate. Functional materials support the transfer of energy or information from tissue to external processing devices. The functional material has to integrate with the substrate in such a way that delamination, cracking, leakages (e.g. of current or light), or release of harmful substances and particles into tissue do not occur. The roles of the substrate are to hold the functional elements together, provide means for insertion, and ensure mechanical and biochemical compatibility with host tissue.

13.4.1 Functional Materials

As most interfacing applications rely on the exchange of electrical charges, electrodes are the functional elements most frequently integrated with soft/compliant neural interfaces. Bulk metals have elastic moduli on the order of GPa (6 orders of magnitude higher than neural tissues) and have a narrow (1 % strain) elastic regime (compared with ~ 20 % for tissues); their use in soft neural interfaces is usually in the form of foils (several micrometers in thickness), thin films (several 10s of nanometers in thickness), and microwires (diameters < 50 µm). Metal foils can be processed using laser micromachining [60]; thin films are most often deposited using thermal evaporation or sputtering [22], microwires can be extruded [57].

In recent years, thin metal films have allowed several strategies for fabricating elastic conductors to thrive. One method is based on thin films (currently made of gold) that, when deposited on an elastomeric silicone substrate, spontaneously develop a random microtopology of percolating filaments delimited by micron sized cracks. Upon mechanical stretch, the microcracks open but do not propagate or coalesce to cause macroscopic failure of the conductivity of the film [23, 40]. Traditional rigid electronic materials such as silicon can be integrated on elastic substrates if they are processed into thin membranes (ribbons), typically several micrometers in thickness. Strain engineered fractal inspired meshes [17] and out of plane microscopic structures [73] have exploited the high flexural compliance of the silicon ribbons to enable system level stretchability. Stretchable thin metal films deposited on plastic foils have also been achieved by forcing the formation of buckles out of the plane of stretch by prestretching the substrate [15]. Although metal foils are not intrinsically stretchable, they can be micromachined in a meandering geometry which allows a moderate degree of strain relief at stretch [21]. A limitation of metals in elastic electronic devices is their poor adhesion to substrates or encapsulation layers made from polymers.

Conductive organic polymers, of which poly(3,4-ethylenedioxythiophene) polystyrene-sulfonate (PEDOT:PSS) mixtures and polypyrrole are notable examples, are an alternative to bulk metals because of their lower stiffness and ability to transport ions as well as electrons. Methods for their deposition as films include spin coating and electrodeposition. Conductive polymers are finding applications: as coatings for implanted electrodes [12], where they can reduce impedance and increase charge injection capacity, as the semiconducting layer in ionic filed effect transistors [31], as interpenetrating conductive networks within tissue [58] or hydrogel based coatings [1, 34], and also as a conductive constituent of mechanically soft blends [38]. The electronic conductivity of conductive polymers still remains low (1000 S.cm^{-1} [19]) which currently hinders their integration as interconnects in microelectrodes.

In optogenetics, the delivery of light for neuromodulation can be facilitated by mounting miniaturized light sources such as LEDs directly on a flexible or elastic substrate [37, 42]. Power is delivered via substrate embedded flexible/stretchable interconnects similar to those used as electrodes. Neuromodulation with opsins currently in widespread use requires optical power densities on the order of 1 mW/mm^2 [52], which raises issues with heat dissipation when using surface mounted μ-LEDs as local temperature change in brain tissue should be limited to < 1 °C. Miniaturizing the μ-LED source without compromising on power can be a solution for the problem of heat management. Cell-sized μ-LEDs based on GaN on sapphire have been achieved and integrated on flexible shank type penetrating brain probes [37]. Optogenetic neural interfaces with off-board light sources rely on a waveguide to deliver energy to the target area. Optical fibers for the telecoms industry traditionally prepared by thermal drawing of glass may be too stiff for interfacing mobile tissues. The need for low flexural stiffness fibers has motivated efforts to adapt thermal drawing processes for the realization of all-polymer multifunctional probes [6]. Materials explored for light guiding in thermally drawn fibers include polycarbonate and cyclic olefin copolymer as core and cladding, respectively (Fig. 13.2f) [6], microfabricated short waveguides have also been made from epoxy (SU-8 photoresist) and ITO [42].

13.4.2 Substrate Materials

The substrate material forms the bulk of the implantable device and serves to hold functional elements of the interface together, insulate them electrically (or optically), and facilitate insertion in tissue. They are often dielectric organic polymers and are chosen for their bio inertness, mechanical properties, and processing compatibility. The two most commonly used plastics in the form of foils are parylene and polyimide, both are thermoplasts with elastic moduli of 3.2 and 2.5 GPa, respectively. A comprehensive review of the two materials and their use in neural implants can be found in [26]. Parylene is deposited using vapor phase deposition at room temperature and polyimide can be formed into thin membranes by spin coating. Both materials can be etched physically or structured using laser micromachining. The two materials undergo plastic deformation at several percent tensile strains and are therefore employed in applications where the implant undergoes deformation in the bending regime. In applications where tensile strain exceeding several percent is experienced, commonly used substrates are silicones and polyurethanes. Silicones are compatible with a wide array of microfabrication techniques including soft lithography (moulding on the micro and nanoscale), spin coating, and plasma activated bonding of layers but cannot be chemically etched and are difficult to process with micromachining lasers. Plastics and elastomers tend to be stable over extended periods of time in vivo and are biologically inert; their use in implantable devices is well documented by regulatory bodies.

An ongoing challenge concerns making implantable neural interfaces more tissue-like. Albeit tolerated by the host, state-of-the-art neural interfaces persist as a bulk foreign body for the duration of their operation. They do not permit, revascularisation and complete healing of the blood–brain barrier, invasion of cells, and free diffusion of metabolites. Strategies to incorporate morphogenic cues with the implant substrate can be physical or biochemical. The microtopographical organization of the extracellular matrix can be mimicked using macroporous solids whose interconnected system of voids can host both cells and functional components of the interface [66]. As an example, macroporous structures can be created by lyophilization of hydrogels. Additionally, hydrogels can mimic biochemical signals of the extracellular matrix and be functionalized with adhesion promoting peptide sequences or immobilized growth factors [20, 62]. Porous scaffolds with intrinsic electrical functionality have also been reported, where 3D graphene conductive foams have been interfaced with neuronal stem cells as a potential way to regenerate lost neural tissues [41].

13.5 Conclusion

In summary, functional neural tissue implants that are mechanically compliant allow for establishing and maintaining proximity to target neural structures. This is expected to improve the reliability, longevity, and selectivity of neural interfaces as part of efforts to develop the next generation of neuroprostheses and bioelectronic regenerative therapies. Neural interfaces will continue to be indispensable in neuroscience research and be integrated in clinical practice by expanding on the successes of cochlear implants, deep brain stimulation and opening new frontiers in bioelectronic medicine.

References

1. A.D. Amella, A.J. Patton, P.J. Martens, N.H. Lovell, L.A. Poole-Warren, R.A. Green, Freestanding, soft bioelectronics. In: 2015 7th International IEEE/EMBS Conference on, Neural Engineering (NER), 22–24 April 2015, pp. 607–610. doi:10.1109/ner.2015.7146696
2. J. Badia, T. Boretius, D. Andreu, C. Azevedo-Coste, T. Stieglitz, X. Navarro, Comparative analysis of transverse intrafascicular multichannel, longitudinal intrafascicular and multipolar cuff electrodes for the selective stimulation of nerve fascicles. J. Neural Eng. **8**, 036023 (2011)
3. T. Boretius, J. Badia, A. Pascual-Font, M. Schuettler, X. Navarro, K. Yoshida, T. Stieglitz, A transverse intrafascicular multichannel electrode (TIME) to interface with the peripheral nerve. Biosens. Bioelectron. **26**, 62–69 (2010)
4. E.S. Boyden, A history of optogenetics: the development of tools for controlling brain circuits with light. F1000 Biol. Rep. **3**, 11 (2011). doi:10.3410/b3-11

5. G. Buzsáki, C.A. Anastassiou, C. Koch, The origin of extracellular fields and currents—EEG, ECoG, LFP and spikes. Nat. Rev. Neurosci. **13**, 407–420 (2012). http://www.nature.com/nrn/journal/v13/n6/suppinfo/nrn3241_S1.html
6. A. Canales et al., Multifunctional fibers for simultaneous optical, electrical and chemical interrogation of neural circuits in vivo. Nat. Biotech. **33**, 277–284 (2015). doi:10.1038/nbt.3093, http://www.nature.com/nbt/journal/v33/n3/abs/nbt.3093.html#supplementary-information
7. M. Carandini, D. Shimaoka, L.F. Rossi, T.K. Sato, A. Benucci, T. Knöpfel, Imaging the awake visual cortex with a genetically encoded voltage indicator. J. Neurosci. **35**, 53–63 (2015). doi:10.1523/jneurosci.0594-14.2015
8. R. Chen, G. Romero, M.G. Christiansen, A. Mohr, P. Anikeeva, Wireless magnetothermal deep brain stimulation. Science **347**, 1477–1480 (2015). doi:10.1126/science.1261821
9. S. Cheng, E.C. Clarke, L.E. Bilston, Rheological properties of the tissues of the central nervous system: A review. Med. Eng. Phys. **30**, 1318–1337 (2008). doi:10.1016/j.medengphy.2008.06.003
10. D.J. Chew et al. A microchannel neuroprosthesis for bladder control after spinal cord injury in rat. Sci. Trans. Med. **5**, 210ra155 (2013). doi:10.1126/scitranslmed.3007186
11. A.F. Christ et al., Mechanical difference between white and gray matter in the rat cerebellum measured by scanning force microscopy. J. Biomech. **43**, 2986–2992 (2010). doi:10.1016/j.jbiomech.2010.07.002
12. X. Cui, J. Wiler, M. Dzaman, R.A. Altschuler, D.C. Martin, In vivo studies of polypyrrole/peptide coated neural probes. Biomaterials **24**, 777–787 (2003). doi:10.1016/S0142-9612(02)00415-5
13. C. de Hemptinne, N.C. Swann, J.L. Ostrem, E.S. Ryapolova-Webb, M. San Luciano, N.B. Galifianakis, P.A. Starr, Therapeutic deep brain stimulation reduces cortical phase-amplitude coupling in Parkinson's disease. Nat. Neurosci. **18**, 779–786 (2015) doi:10.1038/nn.3997, http://www.nature.com/neuro/journal/v18/n5/abs/nn.3997.html#supplementary-information
14. E. Delivopoulos, D.J. Chew, I.R. Minev, J.W. Fawcett, S.P. Lacour, Concurrent recordings of bladder afferents from multiple nerves using a microfabricated PDMS microchannel electrode array. Lab. Chip. **12**, 2540–2551 (2012)
15. M. Drack, I. Graz, T. Sekitani, T. Someya, M. Kaltenbrunner, S. Bauer, An Imperceptible Plastic Electronic Wrap. Adv. Mater. **27**, 34–40 (2015). doi:10.1002/adma.201403093
16. A. Ersen, S. Elkabes, D.S. Freedman, M. Sahin, Chronic tissue response to untethered microelectrode implants in the rat brain and spinal cord. J. Neural Eng. **12**, 016019 (2015)
17. J.A. Fan et al., Fractal design concepts for stretchable electronics. Nat. Commun. **5** (2014). doi:10.1038/ncomms4266
18. W. Fan, L.W. Tien, C. Fujun, J.D. Berke, D.L. Kaplan, Y. Euisik, Silk-backed structural optimization of high-density flexible intracortical neural probes. J. Microelectromech. Syst. **24**, 62–69 (2015). doi:10.1109/jmems.2014.2375326
19. Y. Fang, X. Li, Y. Fang, Organic bioelectronics for neural interfaces. J. Mater. Chem. C **3**, 6424–6430 (2015). doi:10.1039/c5tc00569h
20. U. Freudenberg et al., A star-PEG–heparin hydrogel platform to aid cell replacement therapies for neurodegenerative diseases. Biomaterials **30**, 5049–5060 (2009) doi:http://dx.doi.org/10.1016/j.biomaterials.2009.06.002
21. M. Gierthmuehlen et al., Mapping of sheep sensory cortex with a novel microelectrocorticography grid. J. Comp. Neurol. **522**, 3590–3608 (2014). doi:10.1002/cne.23631
22. O. Graudejus, P. Gorrn, S. Wagner, Controlling the morphology of gold films on Poly (dimethylsiloxane). ACS Appl. Mater. Inter. **2**, 1927–1933 (2010). doi:10.1021/am1002537
23. I.M. Graz, D.P.J. Cotton, S.P. Lacour, Extended cyclic uniaxial loading of stretchable gold thin-films on elastomeric substrates. Appl. Phys. Lett. **94**, 071902–071903 (2009)
24. D.E. Harrison, R. Cailliet, D.D. Harrison, S.J. Troyanovich, S.O. Harrison, A review of biomechanics of the central nervous system—Part II: Spinal cord strains from postural loads. J. Manipulative Physiol. Ther. **22**, 322–332 (1999). doi:10.1016/S0161-4754(99)70065-5

25. D.E. Harrison, R. Cailliet, D.D. Harrison, S.J. Troyanovich, S.O. Harrison, A review of biomechanics of the central nervous system—part III: Spinal cord stresses from postural loads and their neurologic effects. J. Manipulative Physiol. Ther. **22**, 399–410 (1999). doi:10.1016/S0161-4754(99)70086-2
26. C. Hassler, T. Boretius, T. Stieglitz, Polymers for neural implants. J. Polym. Sci., Part B: Polym. Phys. **49**, 18–33 (2011). doi:10.1002/polb.22169
27. S. Hirsch, D. Klatt, F. Freimann, M. Scheel, J. Braun, I. Sack, In vivo measurement of volumetric strain in the human brain induced by arterial pulsation and harmonic waves. Magn. Reson. Med. **70**, 671–683 (2013). doi:10.1002/mrm.24499
28. L.R. Hochberg et al., Neuronal ensemble control of prosthetic devices by a human with tetraplegia. Nature **442**, 164–171 (2006) doi:http://www.nature.com/nature/journal/v442/n7099/suppinfo/nature04970_S1.html
29. A. Jonsson, Z. Song, D. Nilsson, B.A. Meyerson, D.T. Simon, B. Linderoth, M. Berggren, Therapy using implanted organic bioelectronics vol 1, 4. (2015) doi:10.1126/sciadv.1500039
30. I. Khaled, S. Elmallah, C. Cheng, W.A. Moussa, V.K. Mushahwar, A.L. Elias, A flexible base electrode array for intraspinal microstimulation. IEEE Trans. Biomed. Eng. **60**, 2904–2913 (2013). doi:10.1109/tbme.2013.2265877
31. D. Khodagholy et al., In vivo recordings of brain activity using organic transistors. Nat. Commun. **4**, 1575 (2013). doi:http://www.nature.com/ncomms/journal/v4/n3/suppinfo/ncomms2573_S1.html
32. D. Khodagholy, J.N. Gelinas, T. Thesen, W. Doyle, O. Devinsky, G.G. Malliaras, G. Buzsaki, NeuroGrid: recording action potentials from the surface of the brain. Nat. Neurosci. **18**, 310–315 (2015). doi:10.1038/nn.3905, http://www.nature.com/neuro/journal/v18/n2/abs/nn.3905.html#supplementary-information
33. H. Kida et al., Focal brain cooling terminates the faster frequency components of epileptic discharges induced by penicillin G in anesthetized rats. Clin. Neurophysiol. **123**, 1708–1713 (2012). doi:10.1016/j.clinph.2012.02.074
34. D.-H. Kim, M. Abidian, D.C. Martin, Conducting polymers grown in hydrogel scaffolds coated on neural prosthetic devices. J. Biomed. Mater. Res. Part A **71A**, 577–585 (2004). doi:10.1002/jbm.a.30124
35. D.-H. Kim et al. Dissolvable films of silk fibroin for ultrathin conformal bio-integrated electronics. Nat. Mater. **9**, 511–517 (2010) http://www.nature.com/nmat/journal/v9/n6/suppinfo/nmat2745_S1.html
36. D.-H. Kim, J.A. Wiler, D.J. Anderson, D.R. Kipke, D.C. Martin, Conducting polymers on hydrogel-coated neural electrode provide sensitive neural recordings in auditory cortex. Acta Biomater. **6**, 57–62 (2010)
37. T.-I. Kim et al., Injectable, cellular-scale optoelectronics with applications for wireless optogenetics. Science **340**, 211–216 (2013). doi:10.1126/science.1232437
38. C.L. Kolarcik et al., Elastomeric and soft conducting microwires for implantable neural interfaces. Soft Matter **11**, 4847–4861 (2015). doi:10.1039/c5sm00174a
39. S.P. Lacour, R. Atta, J.J. FitzGerald, M. Blamire, E. Tarte, J. Fawcett, Polyimide micro-channel arrays for peripheral nerve regenerative implants. Sens. Actuators, A **147**, 456–463 (2008)
40. S.P. Lacour, S. Wagner, Z. Huang, Z. Suo, Stretchable gold conductors on elastomeric substrates. Appl. Phys. Lett. **82**, 2404–2406 (2003)
41. N. Li et al., Three-dimensional graphene foam as a biocompatible and conductive scaffold for neural stem cells. Sci. Rep. **3** (2013). http://www.nature.com/srep/2013/130403/srep01604/abs/srep01604.html#supplementary-information
42. W. Li, K.Y. Kwon, H.-M. Lee, M. Ghovanloo, A. Weber, Design, fabrication, and packaging of an integrated, wirelessly-powered optrode array for optogenetics. Appl. Front. Syst. Neurosci. **9** (2015). doi:10.3389/fnsys.2015.00069
43. J. Liu et al., Syringe-injectable electronics. Nat. Nano. **10**, 629–636 (2015). doi:10.1038/nnano.2015.115, http://www.nature.com/nnano/journal/v10/n7/abs/nnano.2015.115.html#supplementary-information

44. C. Lu et al., Polymer fiber probes enable optical control of spinal cord and muscle function In Vivo. Adv. Funct. Mater. **24**, 6594–6600 (2014). doi:10.1002/adfm.201401266

45. T. Matsuo et al., Intrasulcal Electrocorticography in Macaque Monkeys with Minimally Invasive Neurosurgical Protocols. Front. Syst. Neurosci. **5**, 34 (2011). doi:10.3389/fnsys. 2011.00034

46. I.R. Minev, P. Moshayedi, J.W. Fawcett, S.P. Lacour, Interaction of glia with a compliant, microstructured silicone surface. Acta Biomater. **9**, 6936–6942 (2013). doi:10.1016/j.actbio. 2013.02.048

47. I.R. Minev et al., Electronic dura mater for long-term multimodal neural interfaces. Science **347**, 159–163 (2015). doi:10.1126/science.1260318

48. P. Moshayedi, LdF Costa, A. Christ, S.P. Lacour, J. Fawcett, J. Guck, K. Franze, Mechanosensitivity of astrocytes on optimized polyacrylamide gels analyzed by quantitative morphometry. J. Phys.: Condens. Matter **22**, 194114–194124 (2010)

49. P. Moshayedi et al., The relationship between glial cell mechanosensitivity and foreign body reactions in the central nervous system. Biomaterials **35**, 3919–3925 (2014). doi:10.1016/j. biomaterials.2014.01.038

50. K.M. Musick et al., Scientific Reports—accepted for publication (2015)

51. P. Musienko, R. van den Brand, O. Maerzendorfer, A. Larmagnac, G. Courtine, Combinatory electrical and pharmacological neuroprosthetic interfaces to regain motor function after spinal cord injury. IEEE Trans. Biomed. Eng. **56**, 2707–2711 (2009). doi:10.1109/tbme.2009. 2027226

52. R. Pashaie et al., Optogenetic brain interfaces. IEEE Rev. Biomed. Eng. **7**, 3–30 (2014). doi:10.1109/rbme.2013.2294796

53. P.R. Patel, K. Na, H. Zhang, T.D.Y. Kozai, N.A. Kotov, E. Yoon, C.A. Chestek, Insertion of linear 8.4 μ m diameter 16 channel carbon fiber electrode arrays for single unit recordings. J. Neural Eng. **12**, 046009 (2015)

54. V.S. Polikov, P.A. Tresco, W.M. Reichert, Response of brain tissue to chronically implanted neural electrodes. J. Neurosci. Methods **148**, 1–18 (2005)

55. S. Raspopovic et al., Restoring natural sensory feedback in real-time bidirectional hand prostheses. Sci. Trans. Med. **6**, 222ra219 (2014). doi:10.1126/scitranslmed.3006820

56. J. Reeder et al., Mechanically adaptive organic transistors for implantable electronics. Adv. Mater. **26**, 4967–4973 (2014). doi:10.1002/adma.201400420

57. H.J. Reitboeck, Fiber microelectrodes for electrophysiological recordings. J. Neurosci. Methods **8**, 249–262 (1983). doi:10.1016/0165-0270(83)90038-9

58. S.M. Richardson-Burns, J.L. Hendricks, D.C. Martin, Electrochemical polymerization of conducting polymers in living neural tissue. J. Neural Eng. **4**, L6–L13 (2007)

59. B. Rubehn, C. Bosman, R. Oostenveld, P. Fries, T. Stieglitz, A MEMS-based flexible multichannel ECoG-electrode array. J. Neural Eng. **6**, 036003 (2009)

60. M. Schuettler, S. Stiess, B.V. King, G.J. Suaning, Fabrication of implantable microelectrode arrays by laser cutting of silicone rubber and platinum foil. J. Neural Eng. **2**, S121 (2005)

61. D. Seo, J.M. Carmena, J.M. Rabaey, M.M. Maharbiz, E. Alon, Model validation of untethered, ultrasonic neural dust motes for cortical recording. J. Neurosci. Methods **244**, 114–122 (2015). doi:10.1016/j.jneumeth.2014.07.025

62. W. Shen et al., Extracellular matrix-based intracortical microelectrodes: Toward a microfabricated neural interface based on natural materials. Microsyst. Nanoeng. **1**, (2015). doi:10.1038/micronano.2015.10http://www.nature.com/articles/micronano201510#supplementary-information

63. S. Spieth et al., A floating 3D silicon microprobe array for neural drug delivery compatible with electrical recording. J. Micromech. Microeng. **21**, 125001 (2011)

64. A. Srinivasan et al., Microchannel-based regenerative scaffold for chronic peripheral nerve interfacing in amputees. Biomaterials **41**, 151–165 (2015). doi:10.1016/j.biomaterials.2014. 11.035

65. J. Subbaroyan, D.C. Martin, D.R. Kipke, A finite-element model of the mechanical effects of implantable microelectrodes in the cerebral cortex. J. Neural Eng. **2**, 103 (2005)

66. B. Tian et al., Macroporous nanowire nanoelectronic scaffolds for synthetic tissues. Nat. Mater. **11**, 986–994 (2012). doi:http://www.nature.com/nmat/journal/v11/n11/abs/nmat3404. html#supplementary-information
67. C. Towne, K.L. Montgomery, S.M. Iyer, K. Deisseroth, S.L. Delp, Optogenetic control of targeted peripheral axons in freely moving animals. PLoS ONE **8**, e72691 (2013). doi:10. 1371/journal.pone.0072691
68. J.J. VanDersarl, A. Mercanzini, P. Renaud, Integration of 2D and 3D thin film glassy carbon electrode arrays for electrochemical dopamine sensing in flexible neuroelectronic implants. Adv. Funct. Mater. **25**, 78–84 (2015). doi:10.1002/adfm.201402934
69. M. Wagshul, P. Eide, J. Madsen, The pulsating brain: A review of experimental and clinical studies of intracranial pulsatility. Fluids Barriers CNS **8**, 1–23 (2011). doi:10.1186/2045-8118-8-5
70. T. Ware, D. Simon, D.E. Arreaga-Salas, J. Reeder, R. Rennaker, E.W. Keefer, W. Voit, Fabrication of responsive, softening neural interfaces. Adv. Funct. Mat. **22**, 3470–3479 (2012). doi:10.1002/adfm.201200200
71. T. Ware et al., Three-dimensional flexible electronics enabled by shape memory polymer substrates for responsive neural interfaces. Macromol. Mater. Eng. **297**, 1193–1202 (2012). doi:10.1002/mame.201200241
72. A. Williamson et al., Localized neuron stimulation with organic electrochemical transistors on delaminating depth probes. Adv. Mater. (2015). doi:10.1002/adma.201500218
73. S. Xu et al., Assembly of micro/nanomaterials into complex, three-dimensional architectures by compressive buckling. Sci. **347**, 154–159 (2015). doi:10.1126/science.1260960
74. S.-S. Yoo et al., Focused ultrasound modulates region-specific brain activity. NeuroImage **56**, 1267–1275 (2011). doi:10.1016/j.neuroimage.2011.02.058
75. M. Zhao, R. Alleva, H. Ma, A.G.S. Daniel, T.H. Schwartz, Optogenetic tools for modulating and probing the epileptic network. Epilepsy Res. (2015). doi:10.1016/j.eplepsyres.2015.06.010

Chapter 14
In Vitro Neural Recording by Microelectrode Arrays

Hongki Kang and Yoonkey Nam

Abstract Neural interface plays an important role in monitoring and modulating brain activity. In order to study the neural information processing in vitro, microelectrode array (MEA) platform is used with cell culture or brain slice. To measure neural signals simultaneously from multiple cells for long-term period, extracellular neural recording technique is preferred and subcellular-scale microelectrodes, dense array, and flexible substrates are ideal. In this chapter, we will introduce the state-of-the-art in vitro neural recording technology based on microfabricated electrodes or transistors. MEAs with metal-type microelectrodes are passive types, and MEAs with active electronic components (field-effect transistors or integrated circuits) are active types. The motivation, operation principles, fabrication processes and materials, and current trends are reviewed.

Keywords Microelectrode array (MEA) · Neural interface · Neural recording · Action potentials · Extracellular recording

14.1 Introduction

Electrical signaling is one of the main components in the operation of nervous system. Neurons generate and conduct electrical signals. Because of the signal, we can think, remember, and move our bodies. The electrical signal generated by neurons is called action potential (AP). In our brain, there are over 100 billions of neurons forming over 100 trillions of synapses. In neural circuits, the APs generated at the membrane of neurons spread along the cell membrane, and they are transferred from one neuron to another at the synapse junctions using neurotransmitters. Therefore, technology is required to monitor and modulate the neural signals for

H. Kang · Y. Nam (✉)
Department of Bio and Brain Engineering, Korea Advanced Institute of Science
and Technology (KAIST), Daejeon, Republic of Korea
e-mail: ynam@kaist.ac.kr

© Springer International Publishing Switzerland 2016
J.A. Rogers et al. (eds.), *Stretchable Bioelectronics for Medical Devices and Systems*, Microsystems and Nanosystems,
DOI 10.1007/978-3-319-28694-5_14

disease treatment, mind reading, and so on. In order to understand neural information processing mechanism, various types of neural tissue models are used: cell culture and brain slices. In case of cell culture models, a designated cell type is extracted from embryonic stage brain of rat or mouse and they are cultivated in an incubating condition to obtain an interconnected neural tissue. As for the brain slice, a brain is sliced into a thin piece that contains a circuit of interest, e.g., hippocampal circuits (e.g., DG–CA1, CA1–CA2) or thalamo-cortical circuits. To interrogate neural circuits, one need to measure action potentials from multiple sites or cells, which requires microelectrode arrays for the simultaneous access of neurons in the circuit. Moreover, microscale electrodes ('microelectrodes') are needed to record action potentials from a single cell. As the APs are only measured by directly measuring the membrane voltages, an alternative measurement technique that measures the extracellular field potential is often used. Extracellular recording technique gives an advantage of noninvasiveness to a single cell, and allows to acquire neural signals for long-term period, which makes it possible to investigate the change of neural firing during the development.

MEAs measure extracellular field potentials generated by action potentials and membrane currents and they are used for microstimulation by injecting charges through the same electrodes. MEAs can be classified into two types depending on the included functionality within the arrays: (1) active MEAs include active circuit elements such as amplifiers, analog-to-digital converter (ADC) or multiplexer (MUX) for signal amplification and data processing within the MEA chip itself; (2) passive MEAs are often composed of only passive electrodes on the MEA chip, which can detect and convey the neural signal to the external electronic system for further signal processing. In this chapter, we will review the electrical recording of neural signals using microelectrode arrays for in vitro neural tissues.

14.2 Passive Microelectrode Array Recording

Planar passive MEAs have become a promising experimental platform for electrophysiological studies of neural networks, ranging from dissociated cell cultures to slices of brain. MEA technology can provide very useful spatiotemporal measurement platform with simultaneous signal recording/stimulation from typically several tens of microelectrodes. The measurement can be noninvasive, allowing long-term recording and stimulation even for months. Planar passive MEAs interfaced with micron-scale cells using microfabricated multichannel metal electrode array on a rigid substrate such as glass or silicon wafer are shown in Fig. 14.1. Multielectrode channels are wired out and connected to external modules that are designed for neural signal amplification, filtering, data processing, and analysis.

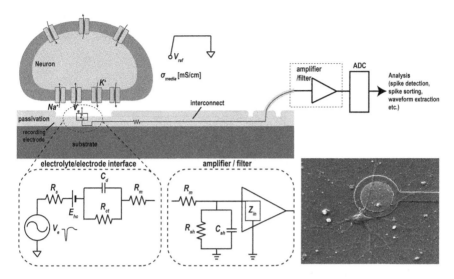

Fig. 14.1 Schematic cross-section of passive MEA (not to scale). Electrical connection of overall system is described. An equivalent electrical circuit at the electrolyte/electrode interface is also described (V_e extracellular potential (30–200 µV), R_s solution resistance, E_{hc} half-cell potential, C_d double-layer capacitance, R_{ct} charge-transfer resistance, R_m metal interconnect resistance, R_{sh} and C_{sh} shunt resistance and capacitance for the amplifier, Z_{in} input impedance of the amplifier, V_{ref} reference voltage of the media bath)

14.2.1   Physics

The electrolyte in extracellular fluid and metal recording electrodes form electrode–electrolyte interface where electrochemical reaction occurs. This interface can be simplified as an RC circuit with one resistor and one capacitor in parallel as shown in Fig. 14.1, which has different electrochemical interpretations. Electrons in the metal electrode can be transferred to the electrolyte by chemical oxidation or reduction of ions in the extracellular solution. This charge-transfer process (or faradaic process) by electrolysis is modeled as a charge-transfer resistor that determines the relationship between faradaic current and applied overpotential. On the other hand, even in the condition where charge-transfer cannot occur, capacitive currents can be generated due to the induced chemical species by the charges at the recording electrode. Depending on the types of the chemical species, characteristics of the adsorption are different. For instance, water molecules and ions form the first inner layer (called Helmholtz layer), which are adsorbed onto the interface. Other species such as solvated ions are attracted by long-range coulombic force forming diffusion layer. Since the amount of charges induced across the double layer vary by voltage, this non-faradaic capacitive current is characterized as double-layer capacitor (C_d). At the interface, electrochemical half-cell potential is formed and it depends on the type of metal electrode, surrounding ion concentration and temperature. As a result, extracellular field potential measurements using metal MEAs suffer from the slow drift of poorly defined metal-electrolyte interface potential. In order to minimize electrochemical reaction at the interface, noble metal, such as platinum or gold, has been preferably used.

14.2.2 Fabrication and Materials

The most common microfabricated MEA structure is a patterned metal layer passivated by an insulating layer with only the measurement electrode areas exposed to the cells of interest. Conventional microfabrication processes are used to deposit and pattern the metal thin film. Since the cells need to stay alive on the passivation layer, the insulating layer (and also the exposed metal electrodes) must be biocompatible. Moreover, the insulating layer needs to have inert surface that does not react with abundant ions in the aqueous solution. Design parameters such as materials used for the electrode and passivation, the size of each electrode, inter-electrode spacing, and coverage area vary depending on the application and the type of cells of interest. After the microfabrication processes, the typical planar MEA is packaged onto a printed circuit board (PCB) such that the MEA can be connected to external circuits for further data processing. For in vitro experiments, a glass or teflon ring chamber that can keep cell culture medium is installed around the MEA.

As electrode materials, various metals such as platinum, gold, gold nanostructures, and aluminum, conductive polymers such as PEDOT (poly(3,4-ethylenedioxythiophene), transparent conductive oxides such as indium tin oxide (ITO), nanomaterials such carbon nanotube (CNT) or nanowire, and 2-D materials such as graphene, have been used. As insulator, silicon oxide, nitride, photoresists, parylene, PDMS, etc., have been used. As substrates, both rigid ones such as glass or wafer and flexible ones such as plastic, PDMS (polydimethylsiloxane) and parylene were used. See Table 14.1 for more details of recent developments.

14.2.3 Operation of Measurement Circuitry

The electrical recording of extracellular action potential signal measures the change of local electrical potential near a cell with respect to the reference electrode potential. In the measurement, the extracellular field potential can be modeled as a time-varying voltage source. In order to perform the voltage measurement, the fully fabricated and packaged passive MEA is typically connected with a multichannel amplifier unit that includes a bandpass filter with an appropriate range of passband frequency (e.g., 300 Hz–5 kHz for spikes, and 10–200 Hz for local field potentials) and a low noise amplifier (LNA) (e.g., gain: 1200 V/V). The processed signal after the amplifier unit is converted to digital signal by analog-to-digital converter

Table 14.1 Recent development of passive and active MEAs

Year	Ref.	Electrode material	No. of electrodes	Size[a] (μm)	$\lvert Z \rvert$ @1 kHz (kΩ)	Electrode thickness (nm)	Insulator thickness (μm)	Substrate	In vitro testing[b]
2009	[14]	Au	4096	21 (sq)	NA	Al	SiO_2	CMOS	(r) Rat neurons (h)
2009	[47]	PtBK	11,011	7 (dia)	NA	Pt (200)	SiO_2–Si_3N_4 (1.6)	CMOS	(r/s) Rat neurons (h) Brain slice
2010	[48]	Au spine	62	NA	NA	Au (45–65)	SiO_2 (0.3)	Glass	(r) Aplysia neuron
2010	[49]	Au nanoflake	60	5–50 (dia)	12–150	Au (200)	Si_3N_4	Glass	(r) Rat neurons (h)
2010	[50]	Au nanoporous	64 (8 × 8)	32 (dia)	30	Au (120)	SU-8 (2)	Glass	(r) Brain slice
2011	[7]	PEDOT:PSS	60	120 (dia)	400–1000	PEDOT:PSS	PDMS	PDMS	(r) Brain slice
2011	[51]	Au nanopillar	NA	15 (dia)	NA	Au (200)	ONO (0.8)	Si wafer	(r) HL-1 cell
2012	[5]	Si nanowire	16 (4 × 4)	0.15 (dia) 3 (H)	NA	Doped Si	Al_2O_3 (0.1)	Si wafer	(r) Rat neurons (h)
2012	[52]	Pt black	64	30 (sq)	100	ITO	NA	Glass	(r) Mouse neurons (c) P19-derived neurons
2012	[53]	Pt nanopollar	16 (4 × 4)	0.15 (dia), 1.5 (H), 3 × 3	6000	Pt (100)	Si_3N_4–SiO_2 (0.35)	Quartz	(r) Cardiac myocytes
2013	[54]	Au nanograin	60	30 (dia)	61	Au (200)	Si_3N_4 (0.5)	Glass	(r/s) Rat neurons (h)
2013	[55]	CNT	64	50 (sq)	NA	ITO	acrylic imide	Glass	(r) Rat neurons (h)
2013	[56]	PEDOT:PSS	64	20 (sq)	100	ITO	–	–	(r) Rat neural stem cells
2013	[57]	PEDOT:PSS	16	20 (sq)	23	Au (100)	Parylene (2)	Glass	(r) Rat brain slice
2014	[58]	IrOx nanotube	64	34 $μm^2$	NA	Pt (80)	Si_3N_4–SiO_2 (0.25/0.05)	Quartz	(r) Cardiac myocytes

(continued)

Table 14.1 (continued)

Year	Ref.	Electrode material	No. of electrodes	Size[a] (μm)	$\lvert Z \rvert$ @1 kHz (kΩ)	Electrode thickness (nm)	Insulator thickness (μm)	Substrate	In vitro testing[b]
2014	[59]	Pt nanocavity	NA	12 (dia)	200	Pt	ONONO (0.8)	Glass	(r) HL-1 cell
2014	[60]	Doped diamond	64	20 (dia)	NA	Doped diamond	SU-8 (1.5)	Si wafer	(r) HL-1 cell
2014	[11]	Pt	26400	9.3 × 5.4	NA	Pt	SiO_2–Si_3N_4	CMOS	(r/s) Rabbit retina
2015	[9]	Au	20 (4 × 5)	30 (dia)	7500	Au	Parylene	Pary. (8 μm)	(r) Rat myocytes
2015	[61]	PEDOT-CNT	59	30 (dia)	19	Au	–	Glass	(r/s) Rat retinal ganglion cells
2015	[8]	MWCNT	16	100 (dia)	55	MWCNT	PDMS	Flex.	(r/s) Embryonic chick retina
2015	[62]	PtBK/pDA	60	50 (dia)	40	Au (200)	Si_3N_4 (0.5)	Glass	(r/s) Rat neurons (h)

PET Polyethylene terephthalate; *Pary.* Parylene; *Flex* various flexible substrates (adhesive tape, Pary., polyimide and PDMS)

PtBK platinum black; *pDA* polydopamine

[a] *dia* diameter; *sq* square

[b] *r* recording, *s* electrical stimulation, *h* primary hippocampal culture, *c* primary cortical culture

(ADC) and then analyzed further for spike detection, spike sorting, and waveform extraction. Although the amplitude of the extracellular potential is small (<1 mV), DC offset of the signal can be much higher and unpredictable due to the electrode–electrolyte interface half-cell potential and the cell/electrode junction potential. Therefore, capacitively coupled input terminal is preferred to filter the DC component of the input signal. For example, a cascaded current mirror operational transconductance amplifier where the input terminals are the gate terminal of metal–oxide–semiconductor field-effect transistors (MOSFETs), which is basically an open circuit, is used [1]. The high input impedance of the amplifier also helps to minimize any signal attenuation that occurs due to the voltage division by high impedance of the electrode–electrolyte interface. Therefore, in case of the voltage recording, this AC coupled LNAs leads to nearly zero DC current. Because the input node potential could be floating, measuring DC voltage signal such as resting potential would be very challenging. In case of electrical stimulation, it can be either high or low impedance node depending on the stimulation conditions (current-controlled or voltage-controlled) [2, 3].

14.2.4   High-Density Passive MEA

While electrodes of cell size (i.e., tens of µm) are commonly used for recording, those electrodes cannot analyze local information of an individual neuron cell (e.g., propagation of AP along the dendrite, activity of ion channels in different locations on a cell and so on). Smaller electrodes, for instance in nano-scale, would allow the transition from large neural network to cellular level in-depth analysis. In addition, ideal goal would be to cover as large area as possible (potentially up to a size of primate brain) while maximizing the spatial resolution by high density and small electrodes.

In order to increase the electrode density, 256 or 512 channel microelectrodes were developed through standard photolithography. 256 channel MEAs are commercially available from multi channel systems (Reutlingen, Germany). The conductor line is ITO, and the electrode material is TiN. The spacing between the electrodes is as small as 30 µm, which is converted to the electrode density of 1,264 electrodes per mm^2 (=0.126 electrodes per 100 $µm^2$). 512 channel MEAs were reported by Mathieson et al. [4]. They used standard photolithography to pattern the ITO on a single plane. The electrode size was 5 µm and the inter-electrode spacing was 60 µm. The electrode density was 281 electrodes per mm^2 (=0.028 electrodes per 100 $µm^2$). There is a maximum density that can be obtained through a conventional single-layer metallization process. The line width and electrode size limit the minimal spacing.

Electrode size could be further reduced to nanometer scale through nanofabrication processes. Recently, nano-sized electrodes were developed using silicon nanowire (SiNW). Park's group integrated a group of silicon nanowires (9 wires in 16 $µm^2$) on a single electrode to improve the cell-electrode contact [5]. Cui's group reported hollow-type iridium oxide nanotube electrodes to improve cell-electrode interface with high density (26 tubes per 100 μm^2) [6]. While these nanowire arrays showed significantly increased electrode density, each nanowire cannot independently operate as individual recording electrodes.

14.2.5 Flexible Passive MEA

Flexible devices for neural recording are also of significant interest. When brain slices are used, especially, it is beneficial if the MEA is soft and flexible enough to have good pliability to form tight physical contact to the nonplanar surface of brain slice for better signal recording. Therefore, various flexible passive MEAs have been demonstrated: PEDOT electrodes on PDMS [7], CNT on various flexible substrates such as medical adhesive tape, Parylene-C, polyimide and PDMS [8], or gold thin film on Parylene-C [9]. Perforated flexible MEAs made of polyimide film was also developed to enhance the brain slice recoding through efficient oxygen supply [10].

14.3 Active Microelectrode Array Recordings

14.3.1 Motivation

There are several expected advantages of the neural signal measurement by active MEAs compared to the passive ones. One of the motivations is to improve signal level by placing amplifiers such as transistors to cells as close as possible. The on-chip amplification and filtering right by the cells or at the recording sites could minimize noise signal generated from parasitics and interferences. Compact designs of the entire measurement platform within the chip area are also possible by the implementation of the circuit blocks on the integrated circuits (IC). These minimized measurement system area would be significantly beneficial, e.g., for implanted in vivo neural recording. Using state-of-the-art semiconductor technologies gives possibilities to add more functionalities on the same neural interface chip such as cyclic voltammetry, pH or temperature sensors, or light emitting diode (LED) array integration for optical stimulation. The active MEA could also enable unprecedentedly high electrode density through multiplexing of a number of electrodes for simpler data processing [11].

There are two types of active MEAs. The first type is to use field-effect transistors (FETs) at the cell-recording electrode interface as front-end amplifiers: neuron-FET coupling approaches have been explored since early 70s [12, 13]. The other is silicon complementary metal–oxide–semiconductor (CMOS) integrated circuits (IC)-based MEAs of which one of the main goals is to increase the spatial resolution of neural interface devices [14, 15].

14.3.2 Fabrication and Materials

When submicron CMOS technologies are used, fabrication of most active MEAs is done in commercial semiconductor foundries. However, since the final device structures and measurement environments are different from conventional CMOS processes (e.g., multiple recording electrodes opening in the center of the chip and their exposure to electrolyte, etc.), additional post-processing steps in research-oriented cleanroom facilities are often necessary. While those additional steps are not too much different from the passive MEA fabrication processes, more complexity and difficulty in the post-processing steps are expected due to the complex structure of the multilayer IC chips.

Silicon has been and will be the most widely used semiconductor materials in CMOS IC. In front-end-of-line (FEOL) of IC fabrication, highly-doped silicon or polysilicon are also used as conductive contacts or as resistors. Since gold creates deep-level traps in silicon, gold is not a CMOS compatible material for MOSFETs and in FEOL of IC process. As gate dielectric materials, high dielectric constant (high-k) insulator materials such as HfO_2 are used in more advanced technologies. In back-end-of-line (BEOL), copper and aluminum (or its alloy) are quite commonly used as interconnections. As described in Fig. 14.2, the top metal layer, which is usually composed of aluminum (or aluminum alloy) in conventional IC chips, is used for recording electrodes. Additional biocompatible metals such as gold, platinum, or platinum black can be either vacuum-deposited or electroplated onto the aluminum electrodes through post-processing [16–19]. Between metal stacks, low-k dielectric insulators are used to minimize parasitic interlayer capacitances. CMOS IC chips are typically passivated with silicon oxide and silicon nitride by plasma-enhanced chemical vapor deposition (PECVD) process, and lastly a polyimide layer. These insulation materials can be used to passivate the MEA but the exposed recording electrodes.

14.3.3 Physics

Fundamental physics governing the neuron-MEA interface in active MEAs varies depending on the types and the structures of the devices. Since CMOS MEAs have metal electrodes directly exposed to the cell, the same metal-electrolyte interface is formed as in passive MEAs, having double-layer capacitance and possible faradaic process [11]. On the other hand, most of FET-based MEAs reported so far have an insulation layer interfacing with the electrolyte instead (insulator-electrolyte interface). Therefore, electrolysis is prevented, and the capacitance of this insulation layer is dominant at the interface. There have been mainly two different structures of FET-based MEAs as shown in Fig. 14.2. One type is conventional MOSFETs without gate electrodes, thus neurons in direct contact with the gate dielectric [12, 13]. In this structure, gate dielectric capacitance coupling of the extracellular

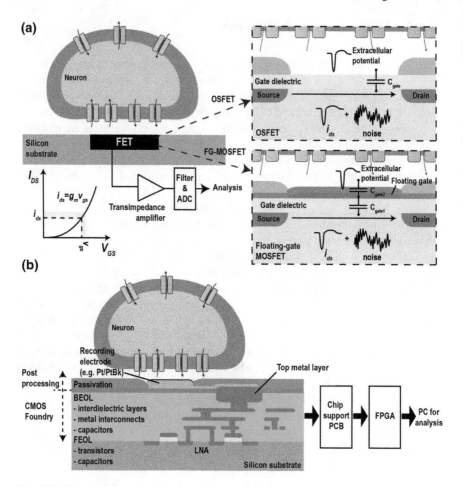

Fig. 14.2 Schematic cross-section of active MEAs. (not to scale) **a** FET-based MEA. Recorded current signal by the FET is further amplified by transimpedance amplifier, and the amplified voltage signal is analyzed. Two common FET-based MEA structures are described: OSFET (MOSFET without gate metal) and FG-MOSFET (floating-gate MOSFET). *Inset graph on the left* describes the transfer characteristic of FETs that visualize how small voltage signal is converted to small drain current. **b** Schematic cross-section of high-density CMOS IC MEA. The top metal layer was used to define recording electrodes. The recorded signal is transferred to active low noise amplifier (LNA) circuits under the electrode area

potential with the transistor channel potential governs the signal recording. The other structure is the floating-gate MOSFETs [18, 20–22]. In this structure, either the actual gate or extended gate electrodes of MOSFETs are passivated with an insulation layer. Neurons are placed on top of the insulation layer. The neuron signal is, therefore, coupled to the transistor channel potential through two capacitors in series.

14.3.4   Operation of FET-Based MEA

The FETs used in active MEAs act as common-source amplifier which is a transconductance amplifier that converts the AC neural voltage signal at the gate terminal with respect to grounded source electrode to an amplified drain current signal that flows between drain and source electrodes as shown in Fig. 14.2a with the following relationship:

$$i_{ds} = g_m v_{gs} + i_{noise}$$

where i_{ds} is drain current, v_{gs} is neuron action potential at the gate, g_m is transconductance defined as $\partial I_{DS}/\partial V_{GS}$, and i_{noise} is mostly a combination of $1/f$ noise and thermal noise. In order to maximize the recorded signal, therefore, it is important to maximize the gain of the transistor (g_m), which is a function of various device parameters such as field-effect mobility, gate dielectric capacitance, geometry of the FET, DC biases, etc. While the transconductance is maximized at relatively high DC gate bias, non-zero DC bias at the gate is particularly not desirable in electrode–electrolyte interface due to the electrochemical corrosion. In depletion-mode MOSFETs that are already 'turned on' at zero gate bias, however, reasonably high transconductance can be achieved at zero gate bias. Cohen et al. suggested this approach to maximize the sensitivity of the neuron-FET interface and maximize the durability of the system [21].

Because of the weak neuron signal, minimization of intrinsic noise in MEAs is crucial. Unlike passive elements such as resistors that show thermal noise, in active electronic devices, $1/f$ noise (also called flicker noise) that shows $1/f$ behavior in its power spectral density is dominant in low frequency range where LFP and spikes are detected. Since the strength of $1/f$ noise is proportional to DC drain current due to more number of carrier electrons that can contributes to the noise, Cohen et al. also operated their FETs in lowest drain current condition to maximize signal-to-noise ratio (SNR) [21]. For silicon MOSFETs, it is known that the channel carriers are trapped/de-trapped by/from the traps within gate dielectric, causing fluctuation of drain current (thus, noise). Therefore, when OSFETs have the gate dielectric layer exposed to ions in electrolyte, the noise characteristics of the OSFETs could be degraded due to enhanced impurity scattering [23].

We can also think about the ideal electrical characteristics of the MEA at different levels. For FET-based active MEAs, in order to maximize SNR, the FETs must have maximum transconductance (e.g., from higher field-effect mobility) while minimizing lowest intrinsic noise signals. For example, transconductance value of 0.18 μm MOSFETs with typical width could be on the order of 100 μA/V which leads to tens of nA for extracellular voltage amplitude of ∼100 μV. So, $1/f$ noise of those FETs in the frequency range from 10 Hz to 5 kHz must be smaller than a few nA. When new transistor technologies are adopted, these electrical characteristics must be carefully characterized. For the HD-MEAs, the on-chip front-end amplifiers must have the input-referred noise on the order of only a few μV or below to detect weak spikes as small as tens of μV.

14.3.5 SiNW-FET or Organic FET MEA

In order to reduce the recording area to nanoscale, nanoelectronic devices such as nanowire FETs (NWFETs) have been applied for neural recording devices [24–27]. With the usage of active nanoscale devices (as small as 10 nm) recording area can be reduced to several orders of magnitude from typical micron-scale MEAs. The small size of the device significantly increases the spatial resolution of MEAs and even allows the access of intracellular spaces for electrical recordings [27]. Nanowire devices could be fabricated on nonplanar flexible substrates. More details of the possibilities of nanoelectronic devices are summarized in [28].

As discussed in previous sections, FETs operate by the modulation of the channel conductance with respect to the change of gate field. It is shown that the change of extracellular field potential near the channel of NWFETs through gate capacitance coupling is converted to the change of channel conductance [29]. NWFETs reported in the work also have shown depletion-mode characteristics as was discussed in a previous section. Non-zero conductance of the channel and its slope allows the signal conversion at near zero gate bias. In other works, modified NWFETs are fabricated such that a small part of the channel area can be directly exposed to intracellular medium using vertical hollow probes [27].

Flexible organic FETs (OFETs) have also been suggested for neural signal recording and stimulation [30]. Unlike the previous FETs that use gate capacitance coupling for the extracellular signal to be converted to drain currents, in these OFETs, neurons are placed far away from the channel of the transistors on the opposite side of the gate electrode. However, relatively poor mobility of the organic semiconductor could be a concern because it could limit SNR and bandwidth of the recording system. On the other hand, unique characteristics of the novel materials could pave a new way of detection and add more functionality in neural interfaces. For example, the new interface between organic materials and biology has been studied and suggested for new applications [31–33]. Further, new fabrication methods such as various printing techniques that are compatible with flexible substrates and the new electronic materials can be used to open up new concepts of neural interface devices at lower fabrication costs.

14.3.6 High-Density CMOS MEA

The circuits designed for HD-MEAs are composed of buffer stage, amplifier, active (or passive) filters, multiplexer, and ADC. Different architectures of MEAs are well classified and summarized by Obien et al. [34]. The highest number of channel electrodes reported to date is 26,400 on 3.85×2.1 mm^2 sensing area from Ballini et al. [11]. With 17.5 μm electrode pitch, the electrode density is as high as near 3,200 electrodes per mm^2 (=0.32 electrode per 100 μm^2). Noise level of their system is 2.4 μV$_{rms}$ noise for AP band and 5.4 μV$_{rms}$ for LFP band. Choosing 10 bit digital resolution and 20 kSamples/s, the IC chip recording 1024 electrodes ($\approx$4 % of entire electrodes; 200 Mbit/s data rate) consumes 75 mW power. While the spatial resolution of electrodes is significantly improved, it is certain that adopting more advanced technology which is more energy efficient (e.g. 0.18 μm process that uses V_{DD} of 1.8 V compared to 3.3 V used in this work [11]) is necessary to operate much more electrodes at the same time under tissue heating constraint [35]. More in-depth summary in MEA IC chips research can be found in this work [17].

Power consumption of the IC chip must be low to prevent thermal damage on neural tissues [36, 37]. Therefore, the range of aforementioned parameters will be fundamentally limited by the cell damage through heat dissipation and radiation. For example, if 25 kHz sampling rate is used with digital resolution of 10 bits, the data rate for a single channel is 250 kbit/s. For 1000 channels, the data rate increases to 250 Mbit/s. If on-chip data compression is implemented, we can reduce the data rate and thus power dissipation. However, the data compression circuit itself will also consume power. While the energy consumption per bit for efficient wired data transfer would be low, the power consumption constraint in, for example, fully implanted untethered in vivo measurement will be stricter. If 10 pJ/bit is consumed, overall power used for the 250 Mbit/s raw data transmission would be 2.5 mW. Harrison et al. reported a 100-electrode system designed for data rate of 330 kb/s and a 433 MHz transmitter consumed 13.5 mW of power [1]. As a point of reference, 12.4 mW power dissipated from a chip for 26 min showed 0.26–0.82 °C temperature increase in human bodies [38].

14.4   New Opportunities and Perspective

Various MEA technologies of both passive and active types are summarized in Table 14.1 for viewing the trend of recent development and comparison. As found in the table, different kinds and shapes of electrode materials have been adopted for better recording. Electrodes are getting smaller so that they can be packed much denser down to subcellular resolution. Nanoscale fabrication techniques are applied for higher sensitivity and spatial selectivity. In addition, not only rigid substrates but also flexible substrates are of interest for providing more in vivo-like environment to interface cells and tissues.

Flexible electronics have been of significant interests in both industries and academia. Flexible electronics is a very wide research field since virtually any technology that can offer mechanical flexibility can be called flexible electronics. In other words, it covers a wide range of applications. Even conventional micro/nanofabrication technologies such as CMOS can be a part of flexible electronics through extreme thinning of the semiconductor wafer because brittle inorganic semiconductor materials become flexible when they are thinned [39–41]. In addition to these conventional semiconductor fabrication processes, recent developments of new solution processable active electronic materials and new patterning methods of those materials have opened up much wider range of applications.

New materials such as nanoparticles, nanowires, organic semiconductor, conducting polymers, and graphene have been developed [42]. These new materials can also offer new functionalities such as chemical sensitivity which are useful for various sensor applications. New fabrication techniques such as inkjet printing have also offered biologically active material patterning for tissue engineering [43, 44]. With the growing interests in bioelectronics, wearable electronics and sensor applications of flexible electronics, there have been major developments in the application of flexible electronics technology to neural interface. To name a few, silk-based conformal microelectrode array, optical stimulator for optogenetic platform, flexible microelectrode array for bladder control, and so on [45, 46]. Flexible electronics technology could be the next key technology toward ultrasmall, ultradense, and ultrasoft in vitro neural interfaces in the future.

References

1. R.R. Harrison, The Design of integrated circuits to observe brain activity. Proc. IEEE **96**, 1203–1216 (2008)
2. P. Fromherz, Electrical interfacing of nerve cells and semiconductor chips. Chem. Phys. Chem. **3**, 276 (2002)
3. P. Livi, F. Heer, U. Frey, D.J. Bakkum, A. Hierlemann, Compact voltage and current stimulation buffer for high-density microelectrode arrays. IEEE Trans. Biomed. Circ. Syst. **4**, 372–378 (2010)
4. K. Mathieson, S. Kachiguine, C. Adams, W. Cunningham, D. Gunning, V. O'Shea et al., Large-area microelectrode arrays for recording of neural signals. IEEE Trans. Nucl. Sci. **51**, 2027–2031 (2004)
5. J.T. Robinson, M. Jorgolli, A.K. Shalek, M.H. Yoon, R.S. Gertner, H. Park, Vertical nanowire electrode arrays as a scalable platform for intracellular interfacing to neuronal circuits. Nat. Nanotechnol. **7**, 180–184 (2012)
6. Z.L.C. Lin, C. Xie, Y. Osakada, Y. Cui, B.X. Cui, Iridium oxide nanotube electrodes for sensitive and prolonged intracellular measurement of action potentials. Nat. Commun. **5**, 3206 (2014)
7. A. Blau, A. Murr, S. Wolff, E. Sernagor, P. Medini, G. Iurilli et al., Flexible, all-polymer microelectrode arrays for the capture of cardiac and neuronal signals. Biomaterials **32**, 1778–1786 (2011)
8. M. David-Pur, L. Bareket-Keren, G. Beit-Yaakov, D. Raz-Prag, Y. Hanein, All-carbon-nanotube flexible multi-electrode array for neuronal recording and stimulation. Biomed. Microdevices. **16**, 43–53 (2014)
9. A. Mondal, B. Baker, I.R. Harvey, A.P. Moreno, PerFlexMEA: a thin microporous microelectrode array for in vitro cardiac electrophysiological studies on hetero-cellular bilayers with controlled gap junction communication. Lab Chip **15**, 2037–2048 (2015)
10. S.A. Boppart, B.C. Wheeler, C.S. Wallace, A flexible perforated microelectrode array for extended neural recordings. IEEE Trans. Biomed. Eng. **39**, 37–42 (1992)
11. M. Ballini, J. Muller, P. Livi, Y. Chen, U. Frey, A. Stettler, et al., A 1024-Channel CMOS microelectrode array with 26,400 electrodes for recording and stimulation of electrogenic cells in vitro. IEEE J. Solid-State Circ. **49**, 2705–2719 (2014)
12. P. Bergveld, J. Wiersma, H. Meertens, Extracellular potential recordings by means of a field effect transistor without gate metal, called OSFET. IEEE Trans. Biomed. Eng. **BME-23**, 136–144 (1976)

13. P. Fromherz, A. Offenhausser, T. Vetter, J. Weis, A neuron-silicon junction: a Retzius cell of the leech on an insulated-gate field-effect transistor. Science **252**, 1290–1293 (1991)
14. L. Berdondini, K. Imfeld, A. Maccione, M. Tedesco, S. Neukom, M. Koudelka-Hep et al., Active pixel sensor array for high spatio-temporal resolution electrophysiological recordings from single cell to large scale neuronal networks. Lab Chip **9**, 2644–2651 (2009)
15. U. Frey, U. Egert, F. Heer, S. Hafizovic, A. Hierlemann, Microelectronic system for high-resolution mapping of extracellular electric fields applied to brain slices. Biosens. Bioelectron. **24**, 2191–2198 (2009)
16. D.J. Bakkum, U. Frey, M. Radivojevic, T.L. Russell, J. Muller, M. Fiscella et al., Tracking axonal action potential propagation on a high-density microelectrode array across hundreds of sites. Nat. Commun. **4**, 2181 (2013)
17. A. Hierlemann, U. Frey, S. Hafizovic, F. Heer, Growing cells atop microelectronic chips: interfacing electrogenic cells in vitro with CMOS-Based microelectrode arrays. Proc. IEEE **99**, 252–284 (2011)
18. B. Eversmann, M. Jenkner, F. Hofmann, C. Paulus, R. Brederlow, B. Holzapfl et al., A 128 × 128 CMOS biosensor array for extracellular recording of neural activity. IEEE J. Solid-State Circ. **38**, 2306–2317 (2003)
19. I.L. Jones, T.L. Russell, K. Farrow, M. Fiscella, F. Franke, J. Müller, et al., A method for electrophysiological characterization of hamster retinal ganglion cells using a high-density CMOS microelectrode array. Front. Neurosci. **9**, (2015)
20. A. Offenhäusser, J. Rühe, W. Knoll, Neuronal cells cultured on modified microelectronic device surfaces. J. Vac. Sci. Technol., A **13**, 2606–2612 (1995)
21. A. Cohen, M.E. Spira, S. Yitshaik, G. Borghs, O. Shwartzglass, J. Shappir, Depletion type floating gate p-channel MOS transistor for recording action potentials generated by cultured neurons. Biosens. Bioelectron. **19**, 1703–1709 (2004)
22. S. Meyburg, M. Goryll, J. Moers, S. Ingebrandt, S. Böcker-Meffert, H. Lüth, et al., N-Channel field-effect transistors with floating gates for extracellular recordings. Biosens. Bioelectron. **21**, 1037–1044 (2006)
23. F.N. Hooge, 1/F noise sources. IEEE Trans. Electron. Devices **41**, 1926–1935 (1994)
24. B. Tian, T. Cohen-Karni, Q. Qing, X. Duan, P. Xie, C.M. Lieber, Three-Dimensional, flexible nanoscale field-effect transistors as localized bioprobes. Science **329**, 830–834 (2010)
25. F. Patolsky, B.P. Timko, G. Yu, Y. Fang, A.B. Greytak, G. Zheng, et al., Detection, stimulation, and inhibition of neuronal signals with high-density nanowire transistor arrays. Science **313**, 1100–1104 (2006)
26. Q. Qing, S.K. Pal, B. Tian, X. Duan, B.P. Timko, T. Cohen-Karni, et al., Nanowire transistor arrays for mapping neural circuits in acute brain slices. Proc. Nat. Acad. Sci. **107**, 1882–1887 (2010)
27. R. Gao, S. Strehle, B. Tian, T. Cohen-Karni, P. Xie, X. Duan, et al., Outside looking in: nanotube transistor intracellular sensors. Nano Lett. **12**, 3329–3333 (2012)
28. P.B. Kruskal, Z. Jiang, T. Gao, C.M. Lieber, Beyond the patch clamp: nanotechnologies for intracellular recording. Neuron **86**, 21–24 (2015)
29. M. De Vittorio, L. Martiradonna, J.A. Assad, *Nanotechnology and Neuroscience: Nano-Electronic, Photonic, and Mechanical Neuronal Interfacing* (Springer, New York, 2014)
30. V. Benfenati, S. Toffanin, S. Bonetti, G. Turatti, A. Pistone, M. Chiappalone, et al., A transparent organic transistor structure for bidirectional stimulation and recording of primary neurons. Nat. Mater. **12**, 672–680 (2013)
31. D. Ghezzi, M.R. Antognazza, R. Maccarone, S. Bellani, E. Lanzarini, N. Martino, et al., A polymer optoelectronic interface restores light sensitivity in blind rat retinas. Nat. Photon. **7**, 400–406 (2013)
32. G. Lanzani, Materials for bioelectronics: Organic electronics meets biology. Nat. Mater. **13**, 775–776 (2014)

33. V. Benfenati, N. Martino, M.R. Antognazza, A. Pistone, S. Toffanin, S. Ferroni et al., Photostimulation of whole-cell conductance in primary rat neocortical astrocytes mediated by organic semiconducting thin films. Adv. Healthc. Mater. **3**, 392–399 (2014)
34. M.E.J. Obien, K. Deligkaris, T. Bullmann, D.J. Bakkum, U. Frey, Revealing neuronal function through microelectrode array recordings. Syst. Biol. **8**, 423 (2015)
35. C.M. Lopez, A. Andrei, S. Mitra, M. Welkenhuysen, W. Eberle, C. Bartic et al., An implantable 455-Active-Electrode 52-Channel CMOS neural probe. IEEE J. Solid-State Circ. **49**, 248–261 (2014)
36. W.M. Reichert, *Indwelling Neural Implants: Strategies for Contending with the In Vivo Environment* (CRC Press, Boca Raton, 2008)
37. A. Denisov, E. Yeatman, Ultrasonic versus Inductive Power Delivery for miniature biomedical implants, in *2010 International Conference on Body Sensor Networks (BSN)*, 2010, pp. 84–89
38. K. Gosalia, J. Weiland, M. Humayun, G. Lazzi, Thermal elevation in the human eye and head due to the operation of a retinal prosthesis. IEEE Trans. Biomed. Eng. **51**, 1469–1477 (2004)
39. J.N. Burghartz, W. Appel, C. Harendt, H. Rempp, H. Richter, M. Zimmermann, Ultra-thin chip technology and applications, a new paradigm in silicon technology. Solid-State Electron. **54**, 818–829 (2010)
40. K. Kashyap, L.-C. Zheng, D.-Y. Lai, M. T. Hou, J.A. Yeh, Rollable Silicon IC wafers achieved by backside nanotexturing. IEEE Electron Device Lett. **36**, 829–831 (2015)
41. A.L.X. Jiang, L.C. Ming, J.C.Y. Gao, T.K. Hwee, silicon wafer backside thinning with mechanical and chemical method for better mechanical property (2006), pp. 1–4
42. W.S. Wong, A. Salleo, *Flexible Electronics: Materials and Applications* (Springer, New York, 2009)
43. P. Ihalainen, A. Määttänen, N. Sandler, Printing technologies for biomolecule and cell-based applications. Int. J. Pharm
44. U. Meyer, *Fundamentals of Tissue Engineering and Regenerative Medicine* (Springer, Berlin, 2009)
45. J.W. Lee, D. Kim, S. Yoo, H. Lee, G.-H. Lee, Y. Nam, Emerging neural stimulation technologies for bladder dysfunctions. Int. Neurourol. J. **19**, 3–11 (2015)
46. D.-H. Kim, J. Viventi, J.J. Amsden, J. Xiao, L. Vigeland, Y.-S. Kim, et al., Dissolvable films of silk fibroin for ultrathin conformal bio-integrated electronics. Nat. Mater. **9**, 511–517 (2010)
47. U. Frey, J. Sedivy, F. Heer, R. Pedron, M. Ballini, J. Mueller et al., Switch-matrix-based high-density microelectrode array in CMOS technology. IEEE J. Solid-State Circ. **45**, 467–482 (2010)
48. A. Hai, J. Shappir, M.E. Spira, In-cell recordings by extracellular microelectrodes. Nat. Methods **7**, 200–202 (2010)
49. J.H. Kim, G. Kang, Y. Nam, Y.K. Choi, Surface-modified microelectrode array with flake nanostructure for neural recording and stimulation. Nanotechnology, **21**, 85303, (2010)
50. E. Seker, Y. Berdichevsky, M.R. Begley, M.L. Reed, K.J. Staley, M.L. Yarmush, The fabrication of low-impedance nanoporous gold multiple-electrode arrays for neural electrophysiology studies. Nanotechnology **21**, 125504 (2010)
51. D. Bruggemann, B. Wolfrum, V. Maybeck, Y. Mourzina, M. Jansen, A. Offenhausser, Nanostructured gold microelectrodes for extracellular recording from electrogenic cells. Nanotechnology **22**, 265104 (2011)
52. Y. Takayama, H. Moriguchi, K. Kotani, T. Suzuki, K. Mabuchi, Y. Jimbo, Network-wide integration of stem cell-derived neurons and mouse cortical neurons using microfabricated co-culture devices. Biosystems **107**, 1–8 (2012)
53. C. Xie, Z. Lin, L. Hanson, Y. Cui, B. Cui, Intracellular recording of action potentials by nanopillar electroporation. Nat. Nanotechnol. **7**, 185–190 (2012)
54. R. Kim, N. Hong, Y. Nam, Gold nanograin microelectrodes for neuroelectronic interfaces. Biotechnol. J. **8**, 206–214 (2013)
55. I. Suzuki, M. Fukuda, K. Shirakawa, H. Jiko, M. Gotoh, Carbon nanotube multi-electrode array chips for noninvasive real-time measurement of dopamine, action potentials, and postsynaptic potentials. Biosens. Bioelectron. **49**, 270–275 (2013)

56. Y. Furukawa, A. Shimada, K. Kato, H. Iwata, K. Torimitsu, Monitoring neural stem cell differentiation using PEDOT–PSS based MEA. Biochimica et Biophysica Acta (BBA) - General Subjects, **1830**, 4329–4333 (2013)

57. M. Sessolo, D. Khodagholy, J. Rivnay, F. Maddalena, M. Gleyzes, E. Steidl et al., Easy-to-fabricate conducting polymer microelectrode arrays. Adv. Mater. **25**, 2135–2139 (2013)

58. Z.C. Lin, C. Xie, Y. Osakada, Y. Cui, B. Cui, Iridium oxide nanotube electrodes for sensitive and prolonged intracellular measurement of action potentials. Nat. Commun. **5**, 3206 (2014)

59. A. Czeschik, A. Offenhäusser, B. Wolfrum, Fabrication of MEA-based nanocavity sensor arrays for extracellular recording of action potentials. Physica Status Solidi (A) (2014)

60. V. Maybeck, R. Edgington, A. Bongrain, J.O. Welch, E. Scorsone, P. Bergonzo et al., Boron-doped nanocrystalline diamond microelectrode arrays monitor cardiac action potentials. Adv. Healthc Mater. **3**, 283–289 (2014)

61. R. Samba, T. Herrmann, G. Zeck, PEDOT–CNT coated electrodes stimulate retinal neurons at low voltage amplitudes and low charge densities. J. Neural Eng. **12**, 016014 (2015)

62. R. Kim, Y. Nam, Electrochemical layer-by-layer approach to fabricate mechanically stable platinum black microelectrodes using a mussel-inspired polydopamine adhesive. J. Neural Eng. **12**, 026010 (2015)

Chapter 15
Materials and Designs for Multimodal Flexible Neural Probes

Sung Hyuk Sunwoo and Tae-il Kim

Abstract The use of electrophysiology (EP) signals is the most relevant way to reflect biological activities in cells and tissues. In neuroscience, EP signals are standard indicators enable to display neural activities as action potentials. The action potentials are typically measured by the change of voltage or current from ion channels in the neurons. Usually, conductive electrodes formed on injectable probes that can be penetrated into deep brain tissue for recording EP signals. Over the last few decades, neural probes have been developed using microfabrication technology. Many researchers have attempted to develop and optimize various materials and designs of electrodes and neural probes to effectively minimize their invasive geometry with biocompatible materials. Compared to the rigid and non-flexible neural probes presented in the late 1980s, the shape of deformable neural probes, reported in the late 1990s, has many advantages. A multimodal function (i.e. electric recording with light or drug delivery) for optogenetics technique has also recently been developed as the next generation flexible neural probe. In this chapter, we deal with several examples of flexible neural probes (FNP) in terms of their geometry, materials, and functions. This study will facilitate a new paradigm for less invasive and more flexible multimodal neural probes that can be utilized in many research fields such as materials science, electrical engineering, and fundamental neuroscience.

Keywords Electrophysiological signal · Flexible neural probe · Stiffener · Multimodal probe · Optogenetics

S.H. Sunwoo · T. Kim
Department of Biomedical Engineering, Sungkyunkwan University (SKKU),
2066, Seobu-ro, Suwon, Korea

T. Kim (✉)
School of Chemical Engineering, Sungkyunkwan University (SKKU),
2066, Seobu-ro, Suwon, Korea
e-mail: taeilkim@skku.edu

S.H. Sunwoo · T. Kim
Center for Neuroscience Imaging Research (CNIR), Institute of Basic Science (IBS),
2066, Seobu-ro, Suwon, Korea

© Springer International Publishing Switzerland 2016
J.A. Rogers et al. (eds.), *Stretchable Bioelectronics for Medical Devices and Systems*, Microsystems and Nanosystems,
DOI 10.1007/978-3-319-28694-5_15

15.1   Introduction: History of the Flexible Neural Probe (FNP)

Measuring an electric current through neural activities has become a primary target among researchers, since L. Galvani discovered that a current runs through neurons in an experiment with frogs' legs [3]. At an early stage in the development, two major works of microelectromechanical systems microelectro mechanical system (MEMS)-based neural probes were presented: Utah electrode array (UEA) and Michigan array (MA) . In 1989, a UEA consisting of a three-dimensional (3D) array of targeted-tip silicon shafts which grew from two-dimensional (2D) substrates, was presented. In 1991, Campbell et al. fabricated a 1.5 mm length silicon needle with platinum electrodes [4]. The recent type of smaller UEA is presented in Fig. 15.1a. Meanwhile, an MA made of a rigid silicon shaft with several electrodes (Fig. 15.1b) to measure the different depths of the brain was developed [21]. However, these two approaches raised two major drawbacks. First, the implanting process of the probe damaged the targeted neuron and tissue in both acute and chronic ways. In actue recording, rigid and brittle probe inserted into the brain and the tissue causes direct neuron damage. Also, the activated inflammatory cells near the implantation site release various cytokines, degrading the recording quality. In chronic implantation, the needle itself is recognized as a "foreign body" by the immune system; various immune cells such as neutrophilic leukocytes or macrophages then gather to form a foreign body giant cell (FBGC), in order to eliminate or exclude the foreign body. The packed cell bodies of the astrocytes may serve as an insulation barrier and cause biofouling [1, 37]. Second, the problem of materials was raised, whereby the conventional probes made of silicon (~ 100 GPa high mechanical modulus) show several limitations when implanted into the soft brain tissue (~ 10 kPa) [2, 7, 13]. This mechanical mismatch between the biological tissue and implantable probe can potentially induce micromotion of the brain in the in vivo condition, causing additional damage [18, 37, 40]. It is known that reducing the mechanical mismatch between the biological tissue and implantable probe can minimize side-effects such as inflammation and brain damage. Here, we classify and elucidate recently developed FNPs to resolve both limitations written above, and predict the future developing directions of FNPs in view of materials, designs, and applications.

15.2   Materials for FNPs

The most common polymeric material for FNP is polyimide (PI). It has superior biocompatibility, dielectric property, and high mechanical property (~ 3 GPa) (Fig. 15.1c), and has been widely used as a base material for flexible prostheses [7, 35]. In 2001, a gold electrode for the chronic use of cortical recording was fabricated on a 20 μm thick PI substrate [35]. In addition, ultrathin (10 μm thickness) implantable needles [49] and a double sided neural probe [38, 44] were also

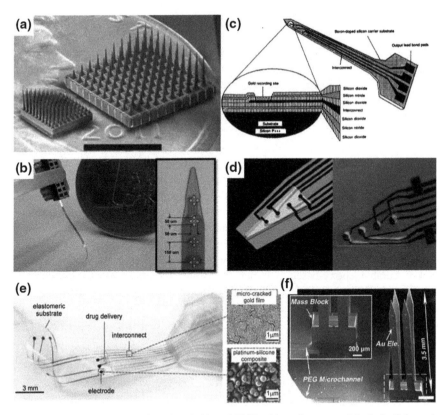

Fig. 15.1 Conventioanl neural probes (a-b) and FNP with various materials (c-f). Schematic design of Michigan Array (MA) and scanning electron microscope (SEM) image of Utah Electrode Array (UEA). **a** SEM image of UEA. The UEA normally has one recording site per shaft, and the shafts are aligned into 2D arrays on the planar substrate. The probe is normally made of silicon shafts fabricated by ion doping and selective etching. The etched shafts are coated with insulating materials, except for the tip part of the shaft. *Scale bar* shows 3 mm [46]. Copyright (2013) from IOP Publishing. **b** Schematic image of MA. The MA has one or more recording sites per shaft. The probe normally contains several stacked layers, including the substrate layer, electrode layer, and encapsulation layer [10]. Copyright (1988) from IEEE. **c** Polyimide-based flexible neural probes (FNPs). The probe has 4 tetrodes in a shaft (*inset*). The electrode itself is made of Ti/Pt/Ti metal sandwiched between two polyimide layers, one is used for the substrate and the other is used for the encapsulation layer. The mechanical flexibility of the probe is thus enforced [32]. Copyright (1988) from Elsevier. **d** Parylene-based FNPs. The parylene-based probe is extremely transparent, with no specific *colors* [25]. Copyright (2013) from Royal Society of Chemistry. **e** Soft and stretchable elastomeric silicone-based FNPs. The silicon-based electronic dura mater is flexible, stretchable, transparent, and soft. Since the mechanical stress versus strain of the silicone is more similar to dura mater or spinal tissue than to the conventional plastic, silicon-based artificial dura causes lower deformation of biological tissue after being implanted [33]. Copyright (2015) from AAAS. **f** Epoxy based uv curable polymer (SU-8) based 3D array of FNPs. The SU-8 base can be folded and fixed into a 3D array structure [7]. Copyright (2011) from Royal Society of Chemistry

demonstrated using a patterned PI. Parylene has also been widely used as an injection carrier of implantable devices as shown in Fig. 15.1d, due to its durability, biocompatibility, corrosion resistance, biostability, transparency, lubricity, flexibility (modulus ~ 3 GPa), and surface consolidation, thus avoiding flaking or dusting [7, 26]. In 2012, Gilgunn et al. presented stretchable serpentine shape electrodes on thin parylene layer [14]. Recently, naturally soft or stretchable materials were used instead of inherently stiff plastic (Fig. 15.1f) [33]. Stretchable, micro-cracked gold electrodes on the surface of soft, stretchable silicone substrate, and polydimethylsiloxane (PDMS) were utilized. They inserted the silicone elastomer probe (defined as e-dura) with similar mechanical properties to those of natural dura material, between the dura and spinal cord of a living mouse. After 6 weeks of implantation, they examined the cross-section of the spine to determine the difference in the deformation of the spine and found no clear deformation of the soft e-dura (120 μm) implanted spine, while serious deformation was observed with thinner (25 μm) PI probes. Moreover, the inflammation caused by foreign body reaction was found to be reduced near the implanted spot of the silicone probe.

UV curable epoxy-based polymer, SU-8 (MicroChem Corp.), is a powerful candidate for FNP, because the others, thin film PI and parylene, need an additional process to enhance their stiffness (Fig. 15.1e) [1, 7]. SU-8 has superior mechanical property and process ability, enabling a 3D array of FNPs.

15.3   Electrode Materials for FNP

Metal is the most common material for electrodes in neural probes because of its high conductivity and workability. In particular, titanium (Ti), gold (Au), tungsten (W), and platinum (Pt) have chemical stability and minimized cytotoxicity. Au has high surficial inertness and minimizes capacitive components on the electrode–electrolyte interface [27]. Pt is much more stable and biocompatible than others and has lower surface impedance. Some alloy metals such as Pt–Ir (iridium) alloy and iridium oxide have also been used [5]. Recently, subcellular dimensions based on single-crystal Au nanowires with a 100 nm diameter have been presented (Fig. 15.2a). The Au nanowires were vertically grown on sapphire by chemical vapor deposition and were mounted on a tungsten tip via conductive glue. The tungsten tip was coated with nail varnish solution as an insulator [20]. A similar geometry was also applied in a carbon nanofiber and conductive polymer. Kozai et al. fabricated a carbon nanofiber probe coated with a conductive polymer, poly (3,4-ethylenedioxythiophene) (PEDOT), as shown in Fig. 15.2b [24]. In 2015, Canales et al. developed multifunctional FNPs containing a microfluidic channel, recording electrode, and light guidance using several polymers with various simultaneous functions using a thermal drawing process (TDP) method (Fig. 15.2c) [5]. These fiber-based probes consist of polyethylene (CPE) as a conductive

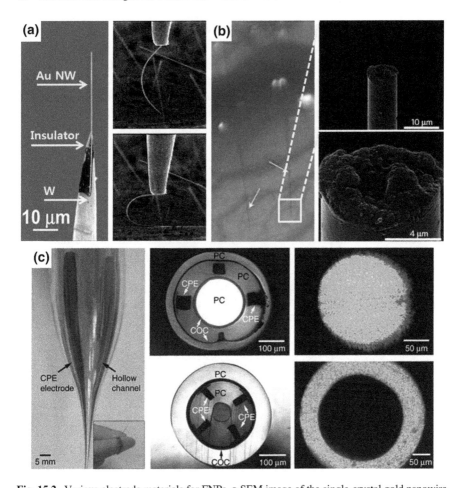

Fig. 15.2 Various electrode materials for FNPs. **a** SEM image of the single-crystal gold nanowire neural probes and their flexibility. The probes are made of a single-crystal gold nanowire attached to a tungsten shaft and covered with an insulator (*left*). Since the thickness of the probe is 100 nm diameter, it has flexibility (*right*) [20]. Copyright (2014) from ACS Publications. **b** Nanoscale carbon fiber coated with conductive polymer, poly(3,4-ethylenedioxythiophene) (PEDOT). The probe was implanted 1.6 mm deep in the cortex (*left*), and SEM images of the probe (*right*) [24]. Copyright (2015) from Nature Publishing Group. **c** Polymer FNP fabricated by TDP method (*left*). The *inset* picture shows that the probe is extremely thin and has super-flexibility. The probe consists of several layers of polymer, all having identical functions (*middle*). The PC layer of the probe can deliver light to the biological tissue (*right*) [5]. Copyright (2015) from Nature Publishing Group

polymer, polycarbonate (PC) as a light waveguide, cyclic olefin copolymer (COC) as the partition wall of each polymer, and microfluidic channels. They were used in neural recording as well as in optogenetics stimulation and drug delivery.

15.4   Implant Techniques for FNPs

We previously mentioned that, in an in vivo experiment, FNP could reduce damage to the tissue and improve the motility of the targeted animal. However, the thin and flexible geometry might cause failure of the implant due to buckling, even in effective moduli of less than 1 mN. Several different techniques have been used for implanting the flexible probes without buckling or failure by the temporary increase of effective moduli of the FNPs.

15.4.1   Implanting with Removable Stiffener

A recent strategy involves inserting an ultra-thinflexible probe with a removable stiffener. In 2009, a FNP attached to a rigid shuttle was first demonstrated [23]. After implanting a stiffener as a shuttle with ultra-thin probe layer together on a rat cerebral cortex as shown in Fig. 15.3a, the shuttle was manually removed using microforceps [23, 47]. Similar approaches with biodegradable materials such as polyethylene glycol (PEG) (Fig. 15.3b) and silk fibroin (Fig. 15.3c) have also been presented [11, 22, 31]. Such softened mechanical properties of FNPs dramatically reduce lesioning, neural loss, gliosis, and immunoreactivity. In Fig. 15.3d, the right image illustrates the thin geometry of the FNP shows substantially less glial activation (2 and 4 weeks) caused by the minimized lesion compared to that with the thick stiffener shown in the left image.

15.4.2   Temporarily-Rigid FNPs

In 2007, Stice et al. found that the optimized thickness and geometry of an implantable prosthesis. It figured out chronic inflammation was more severe with the stiffer FNPs [37]. Today, in order to stiffen the flexible probes, an alternative is to coating or filling the probes with biodegradable materials such as dextrose [9], maltose [49], gelatin [26], carboxy-methylcellulose (CMC) [14], polyglycolic acid [37], poly-(lactic-co-glycolic) [12], PEG [42], tyrosine-based polymers [28], composition of poly(vinyl-alcohol) and cellulose [15], and silk [48] (Fig. 15.4a). In 2014, PI probes with maltose, which can be dissolved in the body were demonstrated by the drawing lithography (Fig. 15.4b). Their thickness was controlled by manipulating the drawing speed and temperature, since the viscosity of maltose varies with the temperature [49]. A parylene-based probe with microfluidic channel presented in 2004, has three main functions (Fig. 15.4c, d): providing a chemical reagent via a channel, recording neural signal through biological saline, and increasing mechanical stiffness during the implantation process [39]. The buckling load of the microfluidic channel filled with water was only about

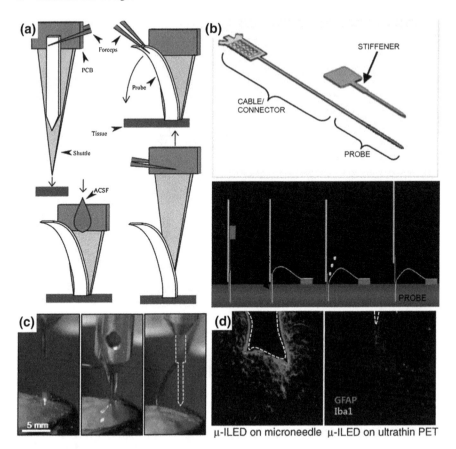

Fig. 15.3 FNPs with removable stiffener. **a** Flexible probe and stiff insertion shuttle is attached and implanted together. The probe is separated from the shuttle by dropping liquid between the probe and the shuttle [23]. Copyright (2012) from Elsevier. **b** The upper image of the flexible probe and its connector with silicon stiffener (*top*). The bottom image shows the implantation process of FNP with silicon stiffener (*bottom*). The probe is implanted with a silicon shuttle then some drops of PBS are added to the PEG in order to promote the dissolving process. After the PEG adhesion dissolves, the shuttle draws back from the tissue [11]. Copyright (2012) from IEEE. **c** Epoxy-based needle inserted with a thick needle into the brain whereby only the needle is removed. The epoxy-based needle is attached to a SU-8 based stiff shuttle with a silk-based adhesive then implanted together. After the silk-based adhesive dissolves, the insertion shuttle is removed from the brain, leaving the integrated device behind. **d** Comparison image of inflammation at the implantation position of Micro-Inorganic Light-Emitting Diodes (μ-ILED) on the microneedle and on the ultrathin PET. The image is taken 4 weeks after implantation. The implantation site of μ-ILED on the microneedle shows a very large scar and high degree of inflammation, while the implantation site of μ-ILED on ultrathin PET does not show notable inflammation [22]. Copyright (2013) from AAAS

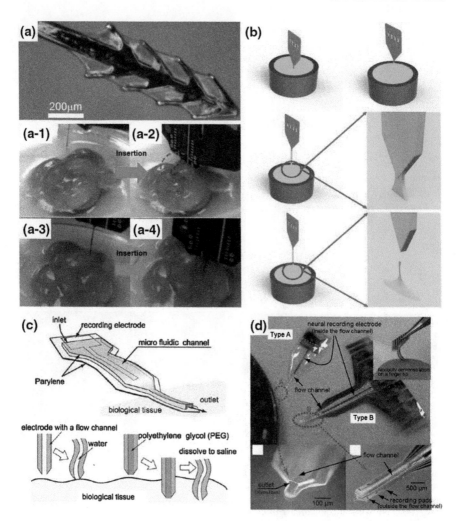

Fig. 15.4 Temporally rigid FNPs. **a** Fish-bone-shaped FNP coated with bio-resolvable silk. The fish-bone-shaped probe without silk coating (a-1) could not penetrate the brain tissue (a-2) while fish-bone-shaped probe with silk coating (a-3) could penetrate the brain tissue (a-4) [48]. Copyright (2011) from IEEE. **b** Schematic process of maltose coating onto the flexible probe. Maltose was coated by dipping the probe then drawn back. The thickness of the coating can be controlled by adjusting the drawing speed and liquid temperature [49]. Copyright (2014) from IOP Science. **c** The microfluidic channel with recording electrode inside is filled with PEG. The probe filled with water could not penetrate the biological tissue, while the probe filled with solid PEG could. After implantation, the PEG dissolved into saline [42]. Copyright (2005) from Royal Society of Science. **d** Single and multichannel probes with microfluidic channel and its close image [39]. Copyright (2004) from IEEE

1 mN, which is insufficient for insertion through the brain tissue. On the other hand, the buckling load of the microfluidic channel filled with solid PEG increased up to ~ 12 mN. Since PEG is dissolved immediately on contact with water, it becomes liquefied as soon as it is implanted in the brain tissue [39, 42].

15.4.3 3D Structured FNPs

Recently, a breakthrough strategy was developed involving the deposit of an electrode in a biological cavity using a syringe (Fig. 15.5a–d). Liu et al. reported thin, net-shaped, tilted mesh polymer/metal electrodes [30] which were dipped in a network structure into a phosphate-bufferedsaline (PBS) and aspirated using a syringe. The syringe needle easily penetrates biological tissue and the electrode mesh is deposited inside the cavity by syringe-injection, and is then spread automatically. The syringe-injected electronics were tested on both the artificial structure and the brain cavity of a mouse. The electrode was well-deposited inside the cavity and showed no notable foreign body reaction. Alternatively, geometry-controllable neural probe by magnetic force was demonstrated in 2004. It is composed of a PI-based flexible 3D multichannel neural probe of 26 μm-thickness and a 6 μm nickel layer (Fig. 15.5e, f). The thick Ni layer substantially increased mechanical property, and it allowed probes to be vertically aligned by magnetic force, so that the probe could be implanted without any additional stiffener [41].

15.5 Multimodal Functions of Implantable FNPs

15.5.1 Recording and Stimulation with FNPs

Multimodal neural probes capable of not only neural recording but also stimulation on specific target nerves were recently demonstrated. Surprisingly, implantable neural probe carrying electrical and chemical stimulation can send artificial signals to a neuron so that it receives a selective reaction to the postsynaptic neurons. They enable to study extremely complicated neural networks by local stimulation in closed-loop circuits. For the other purposes, continuous artificial neural signals can be sent to reinforce or promote the regeneration of nerves or muscles. Recently, some attempts were made to stimulate neurons artificially for esthetic purposes, although implantable FNPs have not yet been commercialized for this purpose. Several trials have also been carried out for non-brain stimulation such as stimulation of the nerve cord of a moth for flying control [45], the bulbus oculi for a prosthesis [29], and sub-dura mater for manipulated walking motion [33].

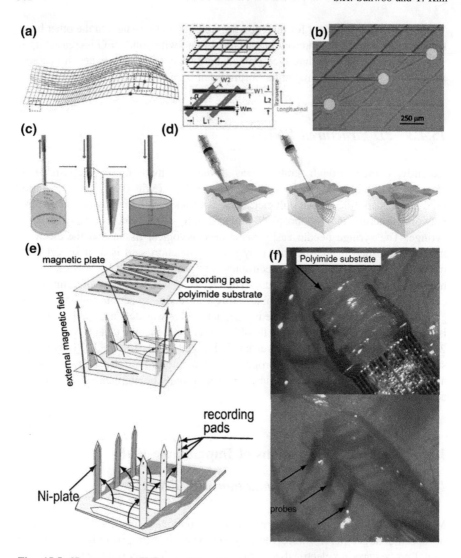

Fig. 15.5 3D structured FNPs. **a** Schematic image of syringe-injectable electronics. The electronics have a mesh-structure so they have natural stress. The stress depends strongly on the angle of the mesh. **b** Optical microscope image of mesh structure and electrode of syringe-injectable electronics. **c**, **d** Schematic process showing the mechanism of syringe-injectable electronics. The syringe absorbs the mesh electronics in the liquid which is then injected into the targeted tissue to be discharged from the syringe through the needle. The emitted mesh spreads automatically owing to its own stress [30]. Copyright (2015) from Nature Publishing Group. **e** Schematic image of 3D structured FNP. Thick nickel (6 µm)-coated probe aligned 3D structure by magnetic field. The nickel coating provides enough stiffness to be implanted into the biological tissue without additional supporting devices. **f** Polyimide-based 3D structural FNP implanted in biological tissue. The probe did not need any additional supporting layers, and was not damaged during the implantation process and is thus reusable [41]. Copyright (2004) from IOP Science

Optical stimulation (optogenetics) with multichannel tetrode is a good example. With recent rise of attention in optogenetics, an artificial stimulation technique was developed using the specific wavelength of light and light sensitive protein. When blue light (470 nm) is sent to the Channelrhodopsin (ChR2)-expressed neuron, the neuron activates. With a similar mechanism, when yellow light (593 nm) is sent to the halorhodopsin from the Natronomonas (NpHR)-expressed neuron, it deactivates. Since these two proteins are able to get stimulated individually, optogenetics may be more effective than conventional electric stimulation, which is invasive and is unable to control single neuron for experimental and therapeutic approaches [6, 36]. To deliver light in the body, PEG [8], silk [34], agarose gel [17], and bacteria [50] have been used for a light-guiding structure. In 2009, the direct-write assembly of pure silk fibroin was used in order to make linear and wavy structures of silk waveguides [34]. In 2013, an in vivo light-guiding implantable hydrogel was reported [8] (Fig. 15.6a, b). Unlike conventional optical fiber, the hydrogel linked with optical fiber was able to deliver light signals to the overall structures. The skin of a living rodent was incised 1 cm in width, and a cell containing hydrogel was implanted in a subcutaneous pocket of the rodent. Interestingly, notable immune cell infiltration was not found.

Alternatively, μ-ILED on an implantable FNP based on an epoxy substrate were demonstrated (Fig. 15.6c–e) (Kim 2013). In this work, ultra-small GaN blue light μ-ILEDs of a 25 μm × 25 μm size and 6.45 μm thickness, two orders smaller than conventional ILEDs (1 × 1 mm^2-size, 100 μm-thickness) were utilized. It demonstrated a multifunctional, injectable probe: a microelectrode to record neurons, a μ-IPD to evaluate the performances of the μ-ILEDs, four μ-ILEDs, and a temperature sensor. Each sensors and devices formed on pattern polyethylene terephthalate (PET) assembled on a releasable SU-8 stiffener. The ultra-thin and small μ-ILEDs and sensors offer the advantage of minimized inflammation or other foreign body reactions during implantation or long-term recordings. Especially, a wireless system based on radio frequency (RF) powering provides in vivo probe utilized in behavior control of a freely-moving mouse.

15.5.2 Drug Delivery and Chemical Stimulation with FNPs

Several cases have been recorded where drugs have been delivered to implanted regions by using FNPs. Two main strategies have been developed for drug delivery. These include (i) coating the probe itself with the mixture of drug and biodegradable materials then delivering the drug automatically while the coating dissolves (Wu 2005) and (ii) drug delivery through a microfluid channel formed on the probe. First, Mercanzini et al. developed adjustable drug delivery coating methods with a common 42 μm thick polyimide probe using a MEMS fabrication technique [32]. They made a poly(p-phenylene sulfide) (PPS) nanoparticle with 120 nm and mixed it with a dexamethasone solution. A fabricated polyimide probe was then dip-coated into the PPS nanoparticle: dexamethasone solution three times. In an in vivo implantation test, the drug is released for 5–7 days, which is sufficient

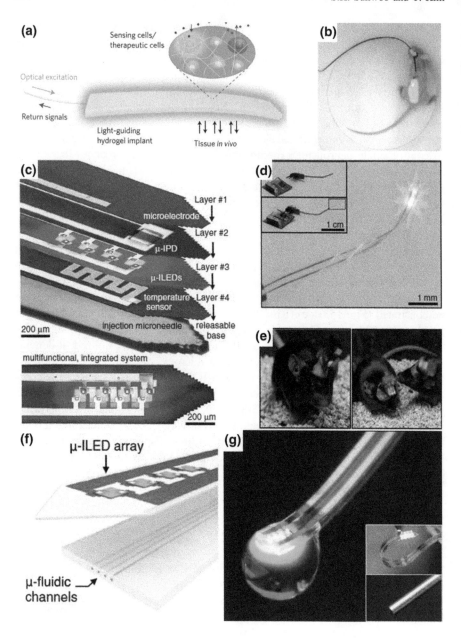

(a)
Sensing cells/therapeutic cells
Optical excitation
Return signals
Light-guiding hydrogel implant
Tissue *in vivo*

(c)
Layer #1
microelectrode
Layer #2
μ-IPD
Layer #3
μ-ILEDs
temperature sensor   Layer #4
injection microneedle   releasable base
200 μm
multifunctional, integrated system
200 μm

(d)
1 cm
1 mm

(f)
μ-ILED array
μ-fluidic channels

◄ **Fig. 15.6** Multimodal, implantable FNPs. **a** Schemes of flexible hydrogel light-guiding system. **b** Light-delivery hydrogel implanted into a living mouse. *Blue light* delivered through the optical fiber cemented on a mouse head [8]. Copyright from Nature Publishing Group. **c** Schemes of multifunctional μ-ILED (inorganic light emitting diode) implant structure. The integrated device consists of several layers, each of which has its own function: microelectrode for neural signal recording, micro inorganic photodiode (μ-IPD) for evaluating ILED working, micro-ILEDs for light delivery, temperature sensor for monitoring heat from ILED, and injection needle for supporting implanting stiffness. **d** The ultra-flexible probe with the *blue light*-on μ-ILED. The *inset* image shows the entire integrated device containing a wireless radio frequency (RF) scavenger. **e** Integrated wireless device is applied to a freely moving mouse with the RF scavengers [22]. Copyright (2013) from AAAS. **f** Schematic image of micro-ILED array integrated with microfluidic channels made on polydimethylsiloxane (PDMS). With the PDMS-based microfluidic channel, the probe can deliver not only light but also drugs or other liquids. **g** Image of microfluidic channel integrated with micro-ILEDs delivering light and drug simultaneously. The *inlet image* shows a comparison of the flexible device (*top*) and conventional metal cannula (*bottom*) [19]. Copyright (2015) from Elsevier

time to resist post-implantation inflammation. Similar work has been performed with silk, a typical biocompatible, biodegradable material. Tien et al. succeeded in delivering drugs loaded in a silk coating. Chondroitinase ABC (ChABC) was loaded in a silk film that has to be released slowly in a PBS buffer solution over one week. ChABC enzyme is known to promote the damage recovery of the spine. Tien et al. mixed ChABC solution in silk solution then evaporated the solution to make drug delivering injectable probes containing ChABC enzyme [43]. Second, drug delivery through a microfluidic channel formed on the probe has been reported. Recently, an in vivo wireless optofluidic system that can be used in pharmacology and optogenetics was presented (Fig. 15.6f, g) [19]. Jeong et al. built a minimized, multifunctional, and wireless-controlled system embedded in μ-ILEDs and mirco-fluidic channels. The system includes power supplies, manipulating electronics, wireless systems, active fluidic controlling systems, and effective light sources that are linked to head-mountable devices and thin, flexible microelectrodes. They fabricated a 50 μm thickness microfluidic system containing four microfluidic channels with a 100 μm² cross-sectional area. The microfluidic channel was ultra-transparent with 95 % transparency in the visual ray area and mechanically flexible (under 1 MPa modulus, 13–18 N/m of bending stiffness) for optical applications. It was suggested that the integrated system dramatically reduced trauma or any other damages in the brain by minimizing microlesions.

15.6 Conclusion

The potential of neural signals is increasing in neuroscience. Moreover, systematically well-organized and significantly improved tools for reliable measurement are critical. Here, we summarized the recently presented FNPs with optimized materials and designs. By using the flexible materials, problems from the mismatch of probes and tissue would be minimized. However, a dilemma occurs where it is very difficult or impossible to implant the probes due to their flexibility. Therefore, the design is optimized as follows. First, some probes are made flexible yet sufficiently stiff to be implanted in the tissue. Second, some flexible probes are implanted with a stiff carrier or shuttle, which is then removed after implantation. Third, some probes are made stiffer using a biodegradable material. Recently, the rapid development of optogenetics has allowed the following progress of light delivery through injectable probes containing optic fibers, light guidance, or μ-ILEDs. Also, the injectable probes can contain a microfluidic channel or biodegradable coating to deliver drugs into the body. However, we still need to challenge works to reduce the expected side-effects during long-term implantation of clinical application and promote non-invasive implantation techniques to reduce the burden of the patients. Also, effort to reduce the fabrication cost for business applications by developing simpler fabrication techniques is needed.

References

1. A. Altuna, G. Gabriel et al., SU-8-based microneedles for in vivo neural applications. J. Micromech. Microeng. **20**, 064014 (2010)
2. M. Asplund, E. Thaning et al., Toxicity evaluation of PEDOT/biomolecular composites intended for neural communication electrode. Biomed. Mater. **4**, 045009 (2009)
3. M. Bresadola, Medicine and science in the life of Luigi Galvani (1737–1798). Brain Res. Bull. **46**, 367–380 (1998)
4. P.K. Campbell, K.E. Jones et al., A silicon-based, three-dimensional neural interface: manufacturing processes for an intracortical electrode array. IEEE Trans. Biomed. Eng. **38**, 758–768 (1991)
5. A. Canales, X. Jia et al., Multifunctional fibers for simultaneous optical, electrical and chemical interrogation of neural circuits in vivo. Nat. Biotech. **33**, 277–284 (2015)
6. J.A. Cardin, M. Carlén et al., Targeted optogenetic stimulation and recording of neurons in vivo using cell-type-specific expression of Channelrhodopsin-2. Nat. Protoc. **5**, 247–254 (2010)
7. C.H. Chen, S.C. Chuang et al., A three-dimensional flexible microprobe array for neural recording assembled through electrostatic actuation. Lab Chip **11**, 1647–1655 (2011)
8. M. Choi, J.W. Choi et al., Light-guiding hydrogels for cell-based sensing and optogenetic synthesis in vivo. Nat. Photon. **7**, 987–994 (2013)
9. S.L. Chorover, and A.M. Deluca, A sweet new multiple electrode for chronic single unit recording in moving animals. Physiol & Behav **9**, 671–674 (1972)
10. K.L. Drake, K.D. Wise et al., Performance of planar multisite microprobe in recording extracellular single-unit intracortical activity. IEEE Trans. Biomed. Eng. **35**, 719–732 (1988)
11. S.H. Felix, K.G. Shah et al., Removable silicon insertion stiffeners for neural probes using polyethylene glycol as a biodissolvable adhesive. Paper presented in international conference of the IEEE engineering in medicine and biology society, San Diego, 28 Aug–1 Sept 2012
12. C.P. Foley, N. Nishimura et al., Flexible microfluidic devices supported by biodegradable insertion scaffolds for convection-enhanced neural drug delivery. Biomed. Microdevices **11**, 915–924 (2009)
13. P.M. George, A.W. Lyckman et al., Fabrication and biocompatibility of polypyrrole implants suitable for neural prosthetics. Biomaterials **26**, 3511–3519 (2005)
14. P.J. Gilgunn, R. Khilwani et al., An ultra-complaint, scalable neural probe with molded biodissolvable delivery vehicle. Paper presented at the MEMS 2012, Paris, 29 Jan–2 Feb 2012

15. A.E. Hess, J.R. Capadona et al., Development of a stimuli-responsive polymer nanocomposite toward biologically optimized, MEMS-based neural probes. J. Micromech. Microeng. **21**, 054009 (2011)
16. A. Jain, A.H.J. Yang et al., Gel-based optical waveguides with live cell encapsulation and integrated microfluidics. Opt. Lett. **37**, 1472–1474 (2012)
17. W. Jensen, K. Yoshida et al., In-vivo implant mechanics of flexible, silicon-based ACREO microelectrode arrays in rat cerebral cortex. IEEE Trans. Biomed. Eng. **53**, 934–940 (2006)
18. J.W. Jeong, J.G. McCall et al., Wireless optofluidic systems for programmable in vivo pharmacology and optogenetics. Cell **162**, 662–674 (2015)
19. M. Kang, S. Jung et al., Subcellular neural probes from single-crystal gold nanowires. ACS Nano **8**, 8182–8189 (2014)
20. B.J. Kim, J.T.W. Kuo et al., 3D Parylene sheath neural probe for chronic recordings. J. Neural Eng. **10**, 045002 (2013)
21. T.I. Kim, J.G. McCall et al., Injectable, cellular-scale optoelectronics with applications for wireless optogenetics. Science **340**, 211–216 (2013)
22. T.D.Y. Kozai, D.R. Kipke, Insertion shuttle with carboxyl terminated self-assembled monolayer coatings for implanting flexible polymer neural probes in the brain. J. Neurosci. Methods **184**, 199–205 (2009)
23. T.D.Y. Kozai, N.B. Langhals et al., Ultrasmall implantable composite microelectrodes with bioactive surfaces for chronic neural interfaces. Nat. Mater. **11**, 1065–1073 (2012)
24. J.T.W. Kuo, B.J. Kim et al., Novel flexible Parylene neural probe with 3D sheath structure for enhancing tissue integration. Lab Chip **13**, 554–561 (2013)
25. G. Lind, C.E. Linsmeier et al., Gelatine-embedded electrode - a novel biocompatible vehicle allowing implantation of highly flexible microelectrodes. J. Neural. Eng. 7, 046005 (2010)
26. S. Kuppusami, R.H. Oskouei, Parylene coatings in medical devices and implants: a review. Univ. J. Biomed. Eng. **3**, 9–14 (2015)
27. K.K. Lee, J. He et al., Polyimide-based intracortical neural implant with improved structural stiffness. J. Micromech. Microeng. **14**, 32–37 (2004)
28. D. Lewitus, K.L. Smith et al., Ultrafast resorbing polymers for use as carriers for cortical neural probes. Acta Biomater. **7**, 2483–2491 (2011)
29. W. Li, D.C. Rodger et al., Parylene-based integrated wireless single-channel neurostimulator. Sens. Actuator A **166**, 193–200 (2011)
30. J. Liu, T.M. Fu et al., Syringe-injectable electronics. Nat. Nanotechnol. **10**, 629–636 (2015)
31. J.G. McCall, T. Kim et al., Fabrication and application of flexible, multimodal light-emitting devices for wireless optogenetics. Nat. Protoc. **8**, 2413–2428 (2013)
32. A. Mercanzini, K. Cheung et al., Demonstration of cortical recording using novel flexible polymer neural probes. Sens. Actuators A **143**, 90–96 (2008)
33. I.R. Minev, P. Musienko et al., Electronic dura mater for long-term multimodal neural interfaces. Science **347**, 159–163 (2015)
34. S.T. Parker, P. Domachuk et al., Biocompatible silk printed optical waveguides. Adv. Mater. **21**, 2411–2415 (2009)
35. P.J. Rousche, D.S. Pellinen et al., Flexible polyimide-based intracortical electrode arrays with bioactive capability. IEEE Trans. Biomed. Eng. **48**, 361–371 (2001)
36. B. Rubehn, S.B.E. Wolff et al., A polymer-based microimplant for optogenetic applications: design and first in vivo study. Lab Chip **13**, 579–588 (2013)
37. P. Stice, A. Gilletti et al., Thin microelectrodes reduce GFAP expression in the implant site in rodent somatosensory cortex. J. Neural Eng. **4**, 42–53 (2007)
38. T. Stieglitz, Flexible biomedical microdevices with double-sided electrode arrangements for neural applications. Sens. Actuators A **90**, 203–211 (2001)
39. T. Suzuki, D. Ziegler et al., Flexible neural probes with micro-fluidic channels for stable interface with the nervous system. Paper presented at the proceedings of the 26th annual international conference of the IEEE EMBS, San Francisco, 1–5 Sept 2004
40. D.H. Szarowski, M.D. Andersen et al., Brain responses to micro-machined silicon devices. Brain Res. **983**, 23–35 (2003)

41. S. Takeuchi, T. Suzuki et al., 3D Flexible multichannel neural probe array. J. Micromech. Microeng. **14**, 104–107 (2004)
42. S. Takeuchi, D. Ziegler et al., Parylene flexible neural probes integrated with microfluidic channels. Lab Chip **5**, 519–523 (2005)
43. L.W. Tien, F. Wu et al., Silk as a multifunctional biomaterial substrate for reduced glial scarring around brain-penetrating electrodes. Adv. Funct. Mater. **23**, 3185–3193 (2013)
44. A. Tooker, V. Tolosa et al., Polymer neural interface with dual-sided electrodes for neural stimulation and recording. Paper presented at the 34th annual international conference of the IEEE EMBS, San Diego, 28 Aug–1 Sept 2012
45. W.M. Tsang, A.L. Stone et al., Flexible split-ring electrode for insect flight biasing using multisite neural stimulation. IEEE Trans. Biomed. Eng. **57**, 1757–1764 (2010)
46. H.A.C. Wark, R. Sharma et al., A new high-density (25 electrodes/mm^2) penetrating microelectrode array for recording and stimulating sub-millimeter neuroanatomical structures. J. Neural Eng. **10**, 045003 (2013)
47. A. Williamson, M. Ferro et al., Localized neuron stimulation with organic electrochemical transistors on delaminating depth probes. Adv. Mater. **27**, 4405–4410 (2015)
48. F. Wu, M. Im et al., A flexible fish-bone-shaped neural probe strengthened by biodegradable silk coating for enhanced biocompatibility. Paper presented in IEEE Transducers'11, Beijing, 5–9 June 2011
49. Z. Xiang, S.C. Yen et al., Ultra-thin flexible polyimide neural probe embedded in a dissolvable maltose-coated microneedle. J. Micromech. Microeng. **24**, 065015 (2014)
50. H. Xin, Y. Li et al., Escherichia coli-based biophotonic waveguides. Nano Lett. **13**, 3408–3413 (2013)

Index

© Springer International Publishing Switzerland 2016
J.A. Rogers et al. (eds.), *Stretchable Bioelectronics for Medical Devices
and Systems*, Microsystems and Nanosystems,
DOI 10.1007/978-3-319-28694-5